MICRODOMAINS IN POLYMER SOLUTIONS

POLYMER SCIENCE AND TECHNOLOGY

Recent volumes in the series:

Volume 20 POLYMER ALLOYS III: Blends, Blocks, Grafts, and Interpenetrating Networks
Edited by Daniel Klempner and Kurt C. Frisch

Volume 21 MODIFICATION OF POLYMERS
Edited by Charles E. Carraher, Jr., and James A. Moore

Volume 22 STRUCTURE PROPERTY RELATIONSHIPS OF POLYMERIC SOLIDS
Edited by Anne Hiltner

Volume 23 POLYMERS IN MEDICINE: Biomedical and Pharmacological Applications
Edited by Emo Chiellini and Paolo Giusti

Volume 24 CROWN ETHERS AND PHASE TRANSFER CATALYSIS IN POLYMER SCIENCE
Edited by Lon J. Mathias and Charles E. Carraher, Jr.

Volume 25 NEW MONOMERS AND POLYMERS
Edited by Bill M. Culbertson and Charles U. Pittman, Jr.

Volume 26 POLYMER ADDITIVES
Edited by Jiri E. Kresta

Volume 27 MOLECULAR CHARACTERIZATION OF COMPOSITE INTERFACES
Edited by Hatsuo Ishida and Ganesh Kumar

Volume 28 POLYMERIC LIQUID CRYSTALS
Edited by Alexandre Blumstein

Volume 29 ADHESIVE CHEMISTRY
Edited by Lieng-Huang Lee

Volume 30 MICRODOMAINS IN POLYMER SOLUTIONS
Edited by Paul Dubin

Volume 31 ADVANCES IN POLYMER SYNTHESIS
Edited by Bill M. Culbertson and James E. McGrath

Volume 32 POLYMERIC MATERIALS IN MEDICATION
Edited by Charles G. Gebelein and Charles E. Carraher, Jr.

A Continuation Order Plan is available for this series. A continuation order will bring delivery of each new volume immediately upon publication. Volumes are billed only upon actual shipment. For further information please contact the publisher.

MICRODOMAINS IN POLYMER SOLUTIONS

Edited by
Paul Dubin

Indiana University–Purdue University at Indianapolis
Indianapolis, Indiana

PLENUM PRESS • NEW YORK AND LONDON

Library of Congress Cataloging in Publication Data

Symposium on Microdomains in Polymer Solutions (1982: Las Vegas, Nev.)
  Microdomains in polymer solution.

  (Polymer science and technology; v. 30)
  "Based on the proceedings of an American Chemical Society Polymer Division Sym-
posium on Microdomains in Polymer Solutions held March 29–April 1, 1982, in Las
Vegas, Nevada"—T.p. verso.
  Includes bibliographies and index.
  1. Polymers   and   polymerization—Congresses.   2.   Solution
(Chemistry)—Congresses. 3. Micelles—Congresses. I. Dubin, Paul. II. American
Chemical Society. Division of Polymer Chemistry. III. Title. IV. Series.
QD380.S927 1982   547.7   85-24146
ISBN-13: 978-1-4612-9255-5   e-ISBN-13: 978-1-4613-2123-1
DOI: 10.1007/978-1-4613-2123-1

Based on the Proceedings of an American Chemical Society Polymer
Division Symposium on Microdomains in Polymer Solutions,
held March 29–April 1, 1982, in Las Vegas, Nevada

© 1985 Plenum Press, New York
Softcover reprint of the hardcover 1st edition 1985
A Division of Plenum Publishing Corporation
233 Spring Street, New York, N.Y. 10013

PREFACE

In the first half of this century, great strides were made in understanding the behavior of polymers in dilute solutions or in the solid state. Concentrated solutions, on the other hand, were commonly regarded as mainly of interest to practitioners, being too complex for the rigorous application of statistical theory. Given the preoccupation with the isolated polymer molecule and the attendant focus on the state of infinite dilution, it is not surprising that aggregation, and inter-polymer association in general, was the bugaboo of experimentalists.

These attitudes have changed remarkably over the last few decades. The application of scaling theory to polymer solutions has stimulated investigation of the semi-dilute state, and the region between infinite dilution and swollen gel is no longer perceived as terra incognita. New techniques, such as dynamic light scattering, have proven to be of much value in such investigations. At the same time, it has become clear that consideration of strong inter- and intra-polymer forces, superimposed on the familiar description of the statistical chain, is prerequisite to the application of polymer science to numerous systems of interest. Paramount among these, of course, are biopolymers, their complexes and assemblies. The isolated random coil must be viewed as a rarity in nature.

Any attempt to apply the principles of macromolecular chemistry to natural systems must confront the phenomena of aggregation. Formerly, many polymer chemists believed that aggregating systems were not amerable to rigorous study, because they were either inherently irreversible or excessively heterogeneous. In actuality, well-defined equilibrium aggregates exist for a number of synthetic polymers and are certainly critically important in biopolymer solutions. The interest in the latter systems has outpaced progress with the former class of compounds, so that explorations of biopolymer complexes may lack the benefit of understanding developed with synthetic polymers.

Intra-molecular ordering in synthetic polymers also merits consideration vis-a-vis biopolymers. In the natural macromolecules, the superposition of hydrogen bonding, ion-pair formation and hydrophobic interactions is well-known, and the native structures are viewed as accomodations to these and other solvation and coulombic effects. With synthetic polymers, one may focus on a few of these phenomena selectively, and much may be gained from the judicious use of the synthetic macromolecules as analogs to the natural ones.

The growing interest in synthetic polymers that display intra- or inter-molecular ordering was the impetus for the symposium "Micro-domains in Polymer Solutions" and, subsequently, this book. In its preparation, several areas emerge as established fields of investigation. One is the study of amphiphilic polymers, in which solvophobic and solvophilic groups are united; these typically display compacted sequences of chain segments, commonly described as intra-molecular micelles. Block copolymers that exhibit such behavior in organic solvents have indeed been commercialized, but the water-soluble polyamphiphiles that have been the most studied are polyelectrolytes containing hydrophobic repeat units. Such copolymers are discussed in Part I in the chapters by Beaumais, Strauss, Sugai, Vert and Yamagishi.

Micelles, with the macromolecular domain also arise from the assoc-iation of surfactants with polymers. Interest in this subject developed first from investigations of protein denaturation by detergents. Then, more recently, from problems arising in enhanced oil recovery with polymer/micelle flooding. The ability of relatively simple molecules to develop a hierarchy of organization accounts in large part for interest in this area. Detergent complexes are formed with nonionic polymers, polyelectrolytes and natural macromolecules. Studies in systems with natural components are described in the reports of Shirahama, Ter-Minassian-Saraga, Tirrell and Yang in Part V (A). The manuscripts by Gilanyi, Goddard, Nagarajan and Zana in Part V (B) deal with purely synthetic polymer-amphiphile complexes.

Inter-polymer ordering is generally a consequence of long-range forces or steric effects. Whether the solid crystalline or gel states provide good molecular models for solution structures is debatable, but ordering in solution presumably must precede the formation of the solid or gel phases and, for some polymers, is significantly present even at high dilution. The lectures by Mandelkern and Miller in Part II both represent rigorous analyses of the consequences of solution conformation with regard to the generation of macroscopic ordered phases. Experimental studies of aggregation in dilute solution are also presented in this section by Cotts.

It is obvious that both intra-molecular and inter-polymer phenomena in polyelectrolyte solutions are dominated by coulomb forces. Repulsive interactions would be expected to diminish aggregation. At the same time, extended chain dimensions and the long-range character of electrostatic effects can promote forms of ordering unique to polyelectrolytes. Hydrodynamic, spectroscopic and thermodynamic methods have all been brought to bear on the coupled problems of conformation, counterion distribution and inter-polymer ordering in polyion solutions. These approaches are well represented in Part III by the works on Paoletti, Berry and Jamieson.

The properties of ionic polymers in nonaqueous media have only recently become the subject of systematic studies. In solvents of low dielectric constant, salt groups resist dissociation and are poorly solvated. Thus, ionic moieties promote intra- and inter-polymer association in organic solvents. The tendency of ionic groups to aggregate or cluster resembles the coalescence of such groups in reversed micelles. Similar considerations underly the formation of ionic "cross-links" that modify the behavior of ionomers in the solid state. Solutions of polyions in nonaqueous media thus provide systems in which a powerful array of experimental techniques can be used to probe phenomena that are important to the bulk properties of a commercially important group of materials. The article by Teyssie and Varoqui in Part IV describe significant explorations in this novel field.

CONTENTS

PART I.   INTRAMOLECULAR MICELLES

1. Microdomains in Hydrophobic Polyacids 1
   U. P. Strauss

2. Hydrophobic Domains of Maleic Acid Copolymers 13
   S. Sugai, K. Nitta, and N. Ohno

3. L-Phenylalanine Oligopeptides Grafted
   Poly(acrylic acid).   Evidence of Specific
   Interactions of Ethidium Bromide and
   Acridine Orange with Hydrophobic Microdomains 33
   J. Beaumais and Th. Ackermann

4. (Acid-Base)-Dependent Globular Structures of
   Partially N-Alkylated Poly(Tertiary Amines) 51
   J. Huguet and M. Vert

5. Hydrophobic Region of Poly(styrenesulfonic acid)
   as Studied by Electric Dichroism Measurements 67
   A. Yamagishi

PART II.   ASSOCIATION, AGGREGATION AND GELATION

6. Association and Complex Formation in
   Stereoregular PMMA Solutions 87
   G. Rehage and D. Wagner

7. Dilute Solutions of Poly(vinylbutyral): Characterization
   of Aggregated and Non-Aggregated Solutions 101
   P. M. Cotts and A. C. Ouano

8. Gelation Accompanying Crystallization from
   Dilute Solutions: Some Guiding Principles 121
   L. Mandelkern, C. O. Edwards, R. C. Domszy,
   and M. W. Davidson

9.   Aggregation, Phase Behavior and the Nature of
     Networks Formed by Some Rod–Like Polymers 143
     W. G. Miller, S. Chakrabarti, and K. M. Seibel

PART III.   ORDERING IN POLYELECTROLYTE SOLUTIONS

10. Polyelectrolytic Aspects of Conformational
     Transitions and Interchain Interactions in Ionic
     Polysaccharide Solutions: Comparison of Theory
     and Microcalorimetric Data 159
     S. Paoletti, A. Cesàro, F. Delben,
     V. Crescenzi, and R. Rizzo

11. Studies on Dilute Solutions of Rodlike
     Macroions III. Integrated Intensity and Photon
     Correlation Light Scattering Investigation of
     Association 191
     Y. Einaga and G. C. Berry

12. Studies of PSM in Aqueous Solution Near
     the Overlap Concentration 211
     R. L. Shogren, A. M. Jamieson, and J. Blackwell

PART IV.   MICRODOMAINS IN NONAQUEOUS MEDIA

13. Ion Distribution and Polyion Conformation
     Displayed by Amphiphilic Polyacids in
     Aqueous and Organic Media 225
     R. Varoqui and E. Pefferkorn

14. Association of the Ion Pair End–Groups of Halato
     Telechelic Polymers in Nonpolar Solvents 243
     R. Jerome, G. Broze, and Ph. Teyssie

15. On the Microenvironment of Soluble
     and Cross–Linked Polymers 265
     F. Mikeš, J. Labský, P. Štrop, and J. Králíček

PART V.   ORDERED POLYMER–LIGAND COMPLEXES

A.   SYSTEMS WITH BIOLOGICAL COMPONENTS

16. Interaction Between DNA and
     Dodecylpyridinium Cation 299
     K. Shirahama, T. Masaki, and K. Takashima

17. Ordered Conformation of Poly(L-Lysine) and
its Homologs in Anionic Surfactant Solutions................. 311
J. T. Yang and S. Kubota

18. Structural Complexes of Cationic
Polysoap-Phospholipid 333
L. Ter-Minassian-Saraga

19. High-Sensitivity Differential Scanning
Calorimetry of Polymer-Phospholipid Mixtures 343
D. A. Tirrell, A. B. Turek, K. Seki, and
D. Y. Takigawa

B. SYNTHETIC SYSTEMS

20. Fluorescence Probe Studies of the
Aggregation State of Surfactants in
Aqueous Polymer Solutions....................................... 357
R. Zana, J. Lang, and P. Lianos

21. Viscometric Investigation of Complexes
Between Polyethyleneoxide and
Surfactant Micelles 369
R. Nagarajan and B. Kalpakci

22. Complex Formation Between Ionic Surfactants
and Polymers in Aqueous Solution 383
T. Gilányi and E. Wolfram

23. Complexes of Cationic Polymers and
Anionic Surfactants 407
E. D. Goddard and P. S. Leung

24. Cooperative Interaction of Anionic Dyes
with Imidazole-Containing Polymers 417
J. S. Tan and T. M. Handel

25. Electron Transfer Process in the Domain Formed
by Intermacromolecular Complexes 429
E. Tsuchida and H. Ohno

CONTRIBUTORS 443

INDEX... 451

PART I

INTRAMOLECULAR MICELLES

MICRODOMAINS IN HYDROPHOBIC POLYACIDS

Ulrich P. Strauss
Department of Chemistry
Rutgers - The State University of New Jersey
New Brunswick, NJ 08903

ABSTRACT

The alternating copolymers of maleic acid and n-alkyl vinyl ethers constitute a versatile group of polyacids which are ideally suited for studying the opposing effects of hydrophobic and ionic interactions in macromolecular behavior. The members with intermediate alkyl group size (n = 4-8) have been of special interest because they undergo conformational transitions from compact to random coil structures upon neutralization by base. The compact conformation is stabilized by the hydrophobic forces between the alkyl groups which form micelle-like microdomains inside the macromolecules. A number of selected studies which have been carried out in our laboratory to characterize the hydrophobic microdomains and the conformational transitions are reviewed here. These studies include fluorescence spectroscopy of dansylated copolymers, viscosity and phase separation investigations, and potentiometric titrations. Two recently developed methods for extracting novel information from the latter are included. One involves the resolution of appropriately chosen subunits into species differing in their states of deprotonation, which for the butyl copolymer leads to bimodal population distributions in the pH range where this copolymer undergoes its conformational transition. The other involves the determination of the micelle size for the case where the polyacid is large compared to the micelle and, therefore, may contain many micelles.

INTRODUCTION

Synthetic polyelectrolytes containing hydrophobic microdomains provide a challenging subject for investigation not only for their own intrinsic

1

interest, but also for their similarities to proteins, in many of whose special functions such microdomains play an essential role. The family of alternating copolymers of maleic anhydride and n-alkyl vinyl ethers has been found a convenient and versatile means for studying such hydrophobic microdomains. On dissolution in water these copolymers hydrolyze to form poly-diacids for which the intrinsic first and second pK-values of the diacid groups differ by about three units. Of special interest to us have been the copolymers with intermediate n-alkyl group size (n = 4-8) in which the hydrophobic domains may be reversibly destroyed or created by changes in the pH of the medium. The hydrophobic domains are associated with the hypercoiled conformation which these polyacids assume at low pH; the domains disappear at high pH where the polyacids are random coils. The members with small (n = 1-3) and large (n>10) alkyl groups behave as random coils and hypercoils, respectively, over the entire pH range and serve as convenient reference compounds for the behavior of these forms in the analysis of the conformational transitions exhibited by the copolymers with intermediate alkyl group size. What follows will be a brief review of the results obtained in studies carried out in our laboratory with these macromolecules. While the first evidence for the existence of the conformational transitions was obtained by potentiometric titration and viscosity studies,[1,2] we shall discuss here first the later results obtained by fluorescence from chemically attached optical probes because these results illustrate most directly the existence of hydrophobic microdomains.

FLUORESCENCE OF DANSYLATED COPOLYMERS

An optical probe whose fluorescence is very sensitive to the polarity of its solvent environment is the 1-dimethylamino-naphthalene-5-sulfonyl (dansyl, DNS) group. This group was chemically attached to our macromolecules by the reaction of the primary amino group of dansyl(β-aminoethyl)amide with the anhydride group of the copolymers.[3,4] The extent of dansylation was kept small enough, (2-7% of the anhydride groups), to minimize possible alterations of the polymer properties by the probe. The probe is characterized by a large fluorescence peaking near 500 nm in nonpolar media. The fluorescence decreases sharply and the peak shifts to higher wave-lengths with increasing polarity of the medium.

The fluorescence, F, at 520 nm of the butyl copolymer is shown as a function of α, the degree of deprotonation,[5] in Figure 1.[6] It is seen that the fluorescence starts out high at low values of α and decreases sharply as α increases. The high values of F are typical of a nonpolar environment and indicate that the dansyl groups are imbedded in hydrophobic microdomains which limit or prevent their exposure to water. With increasing ionization the butyl copolymer undergoes a transition from a compact to a randomly coiled structure in which the hydrophobic domains are broken up. The effect of ionic strength is also shown in Figure 1. The addition of sodium chloride shifts the transition to higher values of α,

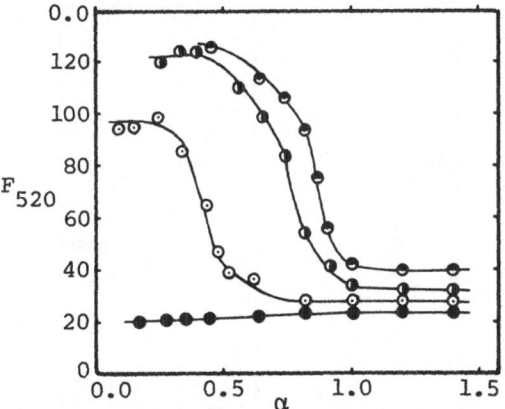

Figure 1. Dependence of F_{520}, the fluorescence emitted at 520 nm by dansylated copolymers, on α: ($\bigcirc$, $\ominus$, $\ominus$) butyl copolymer in water, 0.2 M NaCl, and 0.5 M NaCl, respectively; ($\bullet$) methyl copolymer in water. From Reference 6.

indicating a stabilizing effect on the hypercoiled form by the electrolyte, presumably due to its reducing the electrostatic repulsions between the ionic groups of the macromolecule. Changes in the polymer concentration were shown to have no significant effect on the fluorescence emitted per unit amount of polymer, indicating that the microdomains, as well as the transition, are intra- rather than intermolecular. In contrast to the butyl copolymer, the dansylated methyl copolymer whose fluorescence is also shown in Figure 1 is seen to undergo no transition. Its fluorescence is nearly constant over the whole range of α. Its low magnitude is very close to that of the free probe in water and is characteristic of the aqueous environment which the probe experiences when the macroion has the form of a random coil.

The fluorescence method can also demonstrate the effects of alkyl group size and of specific counterion interactions on the conformational transition as is summarized in Figure 2.[4] The quantity r_F is defined as the ratio of the emission at 480 nm to that at 580 nm. It serves as a convenient measure of the extent of the hydrophobicity in the environment of the probe. The curves in Figure 2 illustrate the following features: (1) The ethyl copolymer shows an almost constant low value of r_F characteristic of the probe being in an aqueous environment as expected from the many observations indicating that this copolymer behaves as a normal non-hydrophobic polyacid. The overlapping of the tetramethylammonium (TMA) and lithium (Li) data indicates that these ions have no direct effects on the fluorescence. (2) The curves for the butyl copolymer decrease with increasing α from initially large values of r_F until they reach the ethyl copolymer curve. The coincidence of these curves beyond the completion of the conformational transition of the

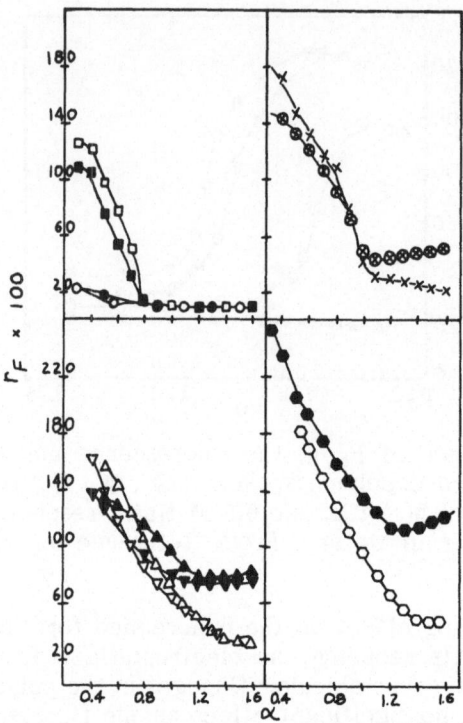

Figure 2. Fluorescence ratio, r_F, as a function of α for copolymers in universal buffer solutions containing 0.1 M TMACl (first symbol in parentheses) or 0.1 M LiCl (second symbol in parentheses): ($\bigcirc$, $\bullet$) ethyl; ($\square$, $\blacksquare$) butyl; (X, $\boxtimes$) pentyl; ($\triangle$, $\blacktriangle$) hexyl, high M.W.; ($\bigtriangledown$, $\blacktriangledown$) hexyl, low M.W.; ($\hexagon$, $\hexagon$) octyl copolymer. From Reference 4.

butyl copolymer show that the butyl side chains do not affect the fluorescence once the random coil conformation has been attained. The observation that the TMA ion produces a somewhat larger value of r_F than does the Li ion in the region where the butyl copolymer is hypercoiled suggests that the TMA ion may contribute to the hydrophobicity by being preferentially attracted into the hydrophobic domains. (3) The curves for the pentyl, hexyl and octyl copolymers start at higher values of r_F than that of the butyl copolymer indicating that the compact conformation becomes increasingly stabilized with increasing alkyl group size. This stablization would be expected to result in an increased shielding of the dansyl groups from water as well as in an increased electrical charge needed to bring about the conformational transition. For the TMA curves one may indeed notice that the completion of the conformational transition, as indicated by the flattening of the curves, is shifted to higher values of α with increasing alkyl group size. Moreover, the terminal values of these curves do not reach those of the ethyl and butyl copolymers, indicating a slight residual water-shielding effect of the alkyl on the dansyl groups. As might be expected, this effect increases with increasing alkyl group size. (4) The diagrams for the pentyl, hexyl and octyl copolymers also show an increasing deviation between the Li and TMA curves at higer values of α. It is known from potentiometric titrations and dilatometry[7] that in the secondary ionization region (corresponding to $1 < α < 2$) the lithium ion is specifically chelated by a dicarboxylate group and the adjacent ether oxygen while the TMA ion is not. As can be seen from the curves of the ethyl and butyl copolymers this chelation has no direct effect on the dansyl fluorescence. However, for the pentyl, hexyl and octyl copolymer for which the conformational transiton is not completed at α = 1, the replacement of bound protons by bound Li ions seems to result in a reversal of the transition as indicated by the slightly ascending r_F curves in the secondary ionization region. Since the lithium ion is bound more weakly than the proton it replaces, the lithium ion must have a greater contracting effect on the molecular dimensions. This conclusion is in accord with results from viscosity and phase separation studies.[8]

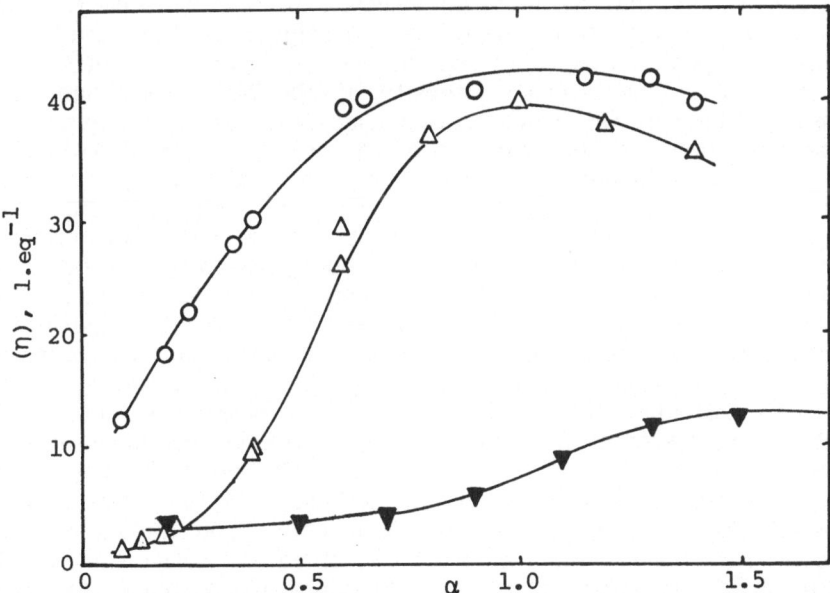

Figure 3. Intrinsic viscosity as a function of α in 0.04 M NaCl for ethyl (○), butyl (△) and hexyl (▼) copolymers. Adapted from Reference 9.

VISCOSITY AND PHASE SEPARATION

The intrinsic viscosity, expressed in liters per equivalents of carboxylate, of samples of the ethyl, butyl and hexyl copolymers dissolved in 0.04 M NaCl at 30°C are presented as a function of α in Figure 3.[2,9] The ethyl copolymer shows an expansion in molecular dimensions as α increases from 0 to 1 typical of normal weak polyacids. In contrast, the butyl copolymer whose degree of polymerization is close to that of the ethyl copolymer (intrinsic viscosities of the anhydrides of the ethyl and butyl copolymers in tetrahydrofuran were 11.8 and 9.9 ℓ eq^{-1}, respectively) shows an abnormally low intrinsic viscosity at low α, followed by a steep rise as α increases until its value becomes close to that of the ethyl copolymer at $\alpha = 1$. These findings indicate an exceptional compactness of the butyl copolymer at low degrees of ionization which can only be ascribed to intramolecular micelle formation induced by the hydrophobic forces between butyl groups. With increasing ionization the electrostatic repulsion between the ionic groups eventually overcomes the hydrophobic forces, resulting in the conformational transition to normal random coil behavior, in conformity with the fluorescence results.

The behavior of the hexyl copolymer is qualitatively similar to that of the butyl copolymer, in that the intrinsic viscosity is very small at low α and rises as α becomes larger. Considering that the intrinsic viscosity of the anhydride of this copolymer in tetrahedrofuran is 20.4 ℓ eq^{-1}, i.e. about twice as large as those of the ethyl and butyl copolymers, it appears remarkable that its intrinsic viscosity in aqueous 0.04 M NaCl remains so far below those of the other two polyacids. This result may be ascribed to specific binding of sodium ion. In Figure 4 the effects of various cations on the reduced viscosity of the hexyl copolymer are compared.[8] The polymer concentration used in these experiments was low enough to allow the interpretation of the data in terms of molecular dimensions of the polymer molecules. It is seen that in the presence of tetramethylammonium ion the polymer shows the large expansion expected from a transition from compact to random coil conformation. However, this expansion is largely suppressed by the alkali metal ions, somewhat more strongly by the sodium than by the lithium ion. The explanation for this effect in terms of the decreased solvent compatibility of dicarboxylate-alkali metal ion complexes has already been mentioned in connection with the fluorescence results. The specificity of these effects is even more pronounced with liquid-liquid phase separation effects which occur at high values of α in the presence of sufficient sodium or lithium ion concentration.[8] For instance, at $\alpha = 1.88$ and 30°C the hexyl copolymer becomes insoluble when the sodium chloride concentration becomes 0.20 M, but to precipitate the same polymer at the same α and temperature requires a lithium chloride concentration higher than 1.1 M. Since the binding affinities of the dicarboxylate groups for lithium and sodium ions are about equal, these results provide further evidence that the bound sodium depresses the solvent affinity more strongly than does the bound lithium.

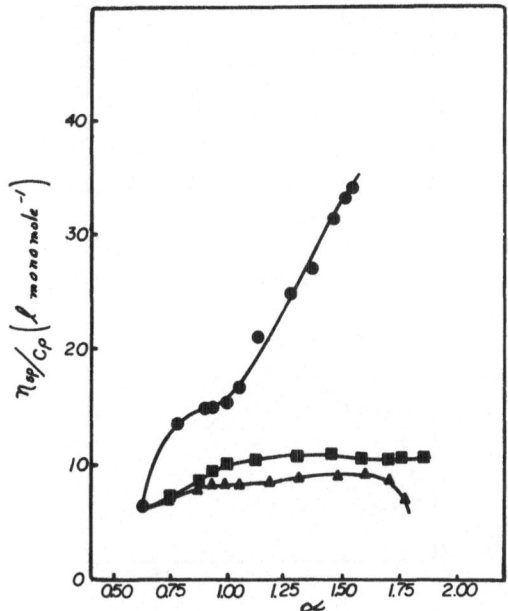

Figure 4. Reduced viscosity of 3.2×10^{-3} monomolar hexyl copolymer
solutions as a function of α in 0.2 M electrolyte solutions. (●)
TMACl; (■) LiCl; (▲) NaCl.   From Reference 8.

POTENTIOMETRIC TITRATIONS

        Potentiometric titrations provide a versatile means for obtaining
information concerning conformational transitions brought about by
changes in the ionization of the polyacids. In Figure 5 such titration
results are shown for the methyl, ethyl, buty, and hexyl copolymers.[2]
Between $\alpha = 0$ and $\alpha = 1$ the curves for the methyl and ethyl copolymers
coincide. Further analysis of these curves show that they are typical of
the behavior of normal polyacids. The curves for the butyl and hexyl
copolymers are seen to start out more steeply, then flatten out and
eventually join the curves of the other two polyacids. This behavior is
consistent with what would be expected from an initially compact
conformation which, with increasing ionization, expands to a random coil.
The transtion for the hexyl copolymer occurs at higher values of α than
that of the butyl copolymer, in conformity with the fluorescence and
viscosity results. It can be shown that the free energy change, ΔG_t^{0},
accompanying the hypothetical transition from the uncharged compact to
the uncharged random coil conformation is proportional to the area
between the curve representing the polyacid undergoing the conformational transition and the curve representing the same polyacid if it
behaved as a random coil over the whole ionization range. The latter
curve was taken as that of the ethyl copolymer. The resulting values of
ΔG_t^{0} at 25°C were 310 and 1110 cal per monomole for the butyl and hexyl

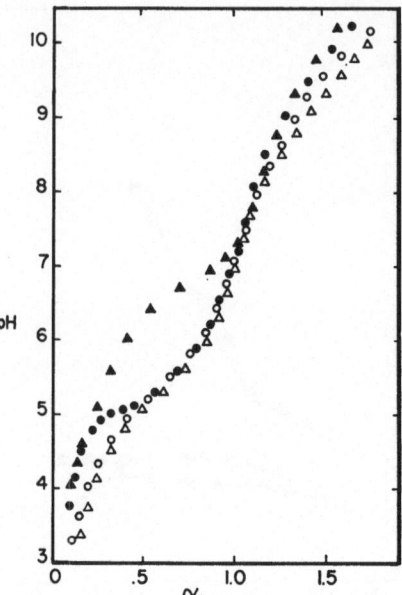

Figure 5. Potentiometric titrations in pure water for methyl ($\triangle$), ethyl
($\bigcirc$), butyl ($\bullet$) and hexyl ($\blacktriangle$) copolymers. From Reference
2.

copolymers, respectively.[2] One half this difference, i.e. 400 cal, should
give the contribution of one methylene group to $\Delta G_t{}^0$. This value is
considerably smaller than the value of 900 cal calculated for the transfer
of a methylene groups from a pure hydrocarbon environment into aqueous
solution.[10] The difference may arise from restraints due to the polymer
chain which may interfere with efficient packing of the hydrocarbon
groups in the compact conformation as well as with complete separation
of the hydrocarbon groups in the random coil conformation.

     With certain assumptions one may calculate θ, the fraction of
residues in the random coil form, from potentiometric titration data by
the equation:[2,11]

$$\theta = (\alpha - \alpha_m)/(\alpha_r - \alpha_m) \qquad (1)$$

where α_r and α_m are the values of α of the hypothetical completely
random coil and completely micellar forms, respectively, at the pH
corresponding to the experimental value of α. The values of α_r and α_m
are obtained from the titration curves of suitable copolymers of the
homologous series whenever possible, otherwise by appropriate extrapolation. The assumptions are that all acid groups can be considered
to belong to either the micellar or the random coil states, and that the
titration behavior for each type of group in the actual polymer is

identical with that in the completely micellar or random coil state. The course of the transition determined in this way resembles quite closely that exhibited by the fluorescence and viscosity results.[2,6]

A novel method for treating potentiometric titration data of polyacids has been developed which can throw light on the cooperativity of conformational transitions.[12,13] The data are treated as if they originated from an oligo-acid having $\underline{N}$ acidic groups and $\underline{N}$ overall dissociation constants, $\beta_1, \beta_2, \ldots \beta_i, \ldots \beta_N$. This procedure results in the empirical equation:

$$\alpha = \frac{\alpha_{max}}{N} \frac{\sum_{i=1}^{N} i\beta_i h^i}{1 + \sum_{i=1}^{N} \beta_i h^i} \tag{2}$$

relating α to h, the antilogarithm of the pH, i.e. the reciprocal of the hydrogen ion activity. The degree of deprotonation, α, is defined so that its value at complete deprotonation, α_{max}, equals the number of acidic groups per repeat unit of the polyacid. Thus, for the maleic acid copolymers considered here, $\alpha_{max} = 2$. The coefficients β_i may be determined from the set of simultaneous linear equations, one for each data point, obtained by multiplying both sides of eq. 2 by the denominator on the right hand side. The value of $\underline{N}$ must be chosen large enough so that eq. 2 fits the experimental data within their precision limits.

The oligo-acid to which eq. 2 applies may be given the physical interpretation that it represents any chain segment containing $\underline{N}$ acidic groups. Its ionization constants, β_i, arise from interactions with groups located both within and outside this chain segment. Because the chain segment samples the behavior of the whole polyacid, it has been denoted by the term Representative Sample Subunit or RSSU.[13]

If one selects at random a large number of RSSU's, the mole fraction of these RSSU's having $\underline{i}$ dissociated protons may be shown to be expressed by the relation

$$x_i = \beta_i h^i / (1 + \sum_{i=1}^{N} \beta_i h^i) \tag{3}$$

With the definition, $\beta_0 = 1$, this relation also applies to x_0, the mole fraction of undissociated RSSU's.

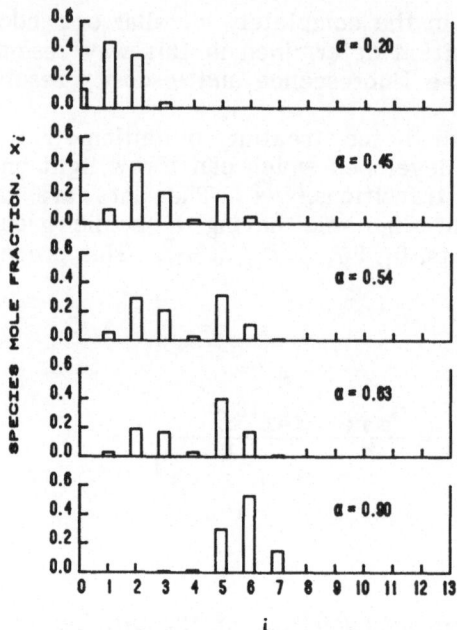

Figure 6.   Progress of population distribution of <u>RSSU</u>'s with i dissociated
protons through conformational transition of butyl copolymer
in 0.2 M LiCl for N = 13.   From Reference 12.

    The results of the application of this method to potentiometric
titration data obtained for the butyl copolymer in 0.2 M LiCl is illustrated
in Figure 6, where the progress of the species population distribution is
followed as the polyacid passes through the conformational transition.[12]
Over the range of α in which the transition is known to occur, the
distribution functions are bimodal, with a deep minimum at i = 4. One
may assign the species with i < 4 to the compact conformation and those
with i > 4 to the random coil conformation.   In contrast, only single-
peaked distribution functions were obtained for the methyl copolymer,
which is known not to undergo a conformational transition.   The method
thus clearly establishes the two-state nature of the conformational
transition of the butyl copolymer.

    It can be shown that at least two conditions have to be fulfilled for
obtaining bimodal population distribution functions by this method.   One
of these is that the two conformations must have a significant gap in
their ionization states, so that the transition occurs with a discontinuous
jump across the state characterized by the population minimum.   The
second condition is that the <u>RSSU</u> must be small compared to the
cooperative unit.   The latter is easily demonstrated by the following
considerations: The macromolecule, which in the transition range generally consists of alternating stretches of the two conformations, may be

visualized as being scanned by the RSSU which reports the ionization behavior of N consecutive ionizable groups. If in such a scan the RSSU were to overlap predominantly two or more stretches of these conformations, the different ionization properties of the conformations would average out, and only single peak distribution functions would be obtained. In order to observe bimodal distribution functions, the scanning RSSU must for the most part cover one or the other conformation. This will be the case if the RSSU is small compared to the cooperative unit.

The size of the cooperative unit may be determined from the pH dependence of the transition parameter, θ. For the special case that the whole polyacid molecule encompasses the cooperative unit, i.e. that the whole polyacid molecule must be in one conformation or the other, the number of repeat units, n', in a cooperative unit is given by the relation:[14]

$$n' = \frac{1}{\alpha_r - \alpha_m} \frac{d \ln[\theta/(1 - \theta)]}{d \ln h} \qquad (4)$$

For the butyl copolymer sample in 0.2 M LiCl, this equation leads to values of n' ranging from 4 to 15 over the range of the conformational transition. Since the weight-average degree of polymerization of the sample was 1800, the result for n' is not consistent with the basis of eq. 4. If one assumes instead that the micelle size is constant and much smaller than the degree of polymerization, and, furthermore, that the random coil sequences separating the micelles may be of any size, one obtains the following equation for n, the number of repeat units per micelle:[15]

$$n = 1 + (n' - 1)/\theta \qquad (5)$$

where n' is given by eq. 4. The application of eq. 5 leads to a value of n which is reasonably constant from $\alpha = 0.35$ to $\alpha = 0.70$, consistent with the theoretical basis of eq. 5. Over the indicated range the average value of n was found to be 18, with a standard deviation of 2.

The difference between eqs. 4 and 5 has been explained as follows. In the derivation of eq. 4 it is implicity assumed that the sizes of the micelles and the random coil units are equal, whereas in the derivation of eq. 5 the restriction for the random coil size has been removed. Starting with the compact conformation, the number of statistical configurations arising from the conversion of the first micelles to random coils is greatly underestimated by eq. 4 relative to eq. 5. This statistical effect facilitates the break-up of micelles, and as a consequence the conformational transition is spread out over a larger pH-range. Equation 4 neglects this statistical effect and therefore leads to unrealistically low values for the micelle size.[15]

It is noteworthy that the RSSU carrying 13 ionizable groups used for the histograms of Figure 6 is indeed small compared to the cooperative unit having 36 ionizable groups, in accord with one of the conditions necessary for observing a bimodal population distribution. It has also been observed that as N, i.e. the size of the RSSU, is increased, the minimum in the distribution becomes shallower, and eventually disappears.

CONCLUSION

The hydrophobic microdomains in the macromolecules discusssed here occur also in soaps and proteins. Indeed, one may consider our hydrophobic polyacids as a link between these two substances. Because these synthetic macromolecules allow the investigator great latitude in the design of their chemical structure, they promise to continue to enhance our insight into hydrophobic phenomena, as well as to permit their exploitation for a variety of attractive applications.

ACKNOWLEDGMENT

The support of this research by a grant from the United States Public Health Service is gratefully acknowledged.

REFERENCES

1. Dubin, P.; Strauss, U. P. J. Phys. Chem. 71, 2757 (1967).
2. Dubin, P. L.; Strauss, U. P. J. Phys. Chem. 74, 2842 (1970).
3. Strauss, U. P.; Vesnaver, G. J. Phys. Chem. 79, 1558 (1975).
4. Strauss, U. P.; Schlesinger, M. S. J. Phys. Chem. 82, 1627 (1978).
5. Because the repeat unit has two carboxylic acid groups, α is defined to be equal to 2 at complete deprotonation.
6. Strauss, U. P.; Vesnaver, G. J. Phys. Chem. 79, 2426 (1975).
7. Begala, A. J.; Strauss, U. P. J. Phys. Chem. 76, 254 (1972).
8. Lane, P.; Strauss, U. P. Adv. Chem. Ser. 142, 31 (1975).
9. Dubin, P. L.; Strauss, U. P. in "Polyelectrolytes and their Applications," Rembaum, A.; Sélegny, E., eds., Reidel, Dordrecht, 1975, pp. 3-13.
10. Nemethy, G.; Scheraga, H. A. J. Chem. Phys. 36, 3401 (1962). Calculated for the series methane-butane from collected vapor pressure and solubility data.
11. Nagasawa, M.; Holtzer, A. J. Amer. Chem. Soc. 86, 538 (1964).
12. Strauss, U. P.; Barbieri, B. W.; Wong, G. J. Phys. Chem. 83, 2840 (1979).
13. Strauss, U. P. Macromolecules, 15, 1567 (1982).
14. Ptitsyn, O. B.; Birshtein, T. M. Biopolymer 7, 435 (1969).
15. Strauss, U. P.; Barbieri, B. Macromolecules 15, 1347 (1982).

HYDROPHOBIC DOMAINS OF MALEIC ACID COPOLYMERS

Shintaro Sugai, Katsutoshi Nitta, and Nobumichi Ohno*
Department of Polymer Science, Faculty of Science
Hokkaido University
Sapporo, Hokkaido 060, Japan

and

*Department of Industrial Chemistry
Akita Technical College,
Akita, Akita 011, Japan

ABSTRACT

As an introduction, our previous studies on the conformations of maleic acid copolymers with aromatic vinyl monomers are summarized. To characterize the compact form and the pH-induced conformational transition of the maleic acid copolymer with styrene in aqueous NaCl, 400 MH_z ^1H-NMR spectra were measured. The spectral form depended on the molecular conformation. Because each of proton resonance peaks could not be separated, the spin-lattice relaxation time T_1 was estimated by using the inversion recovery technique (π-τ-$\pi/2$). The T_1's for both side chain and backbone protons reflected the transition, and the protons were considered to be in a more restricted motional state in the compact form than in the coil form. Also, from temperature dependence of each T_1, motion of the copolymer in the coil form was described in terms of the local segmental jump (D) combined with the isotropic rotational motion (O), when a ratio between both the correlation times τ_D and τ_O was about 0.07. For the compact form, the ratio was found to be about 10. By referring to theoretical diagram of T_1 vs. τ_D for the methylene protons on the backbone, value of τ_D for the compact form was compared with that for the coil form at 35°C.

INTRODUCTION

The hydrophobic polyelecytrolyte is known to be in the compact conformation with the hydrophobic domains in interior of the molecule at low degrees of ionization, and the domains are considered to be stabilized by short range interaction between the hydrophobic side chains. Long range electrostatic interaction becomes predominant at high degrees of ionization, and the compact conformation is converted into the extended coil form. Such a pH-induced conformational transition seems to be a good model for denaturation of globular proteins. Poly(methacrylic acid) is a typical hydrophobic polyelectrolyte, and its compact form at low degrees of ionization of carboxyl groups in the polymer α and conformational transition to the extended coil at high degrees were extensively studied with various measuring techniques.[1-5] Recently, poly(ethacrylic acid) has also been investigated in aqueous media,[6,7] and a similar transition of conformation induced by change in pH has been studied. The standard free energy change of transition of poly(ethacrylic acid) at zero charge was compared with that for poly(methacrylic acid).

On the other hand, copolymers of maleic acid with various vinyl monomers had been investigated in studying the influence of local charge density on polyelectrolyte properties,[8,9] and some alternating copolymers of maleic acid with n-alkyl vinyl ethers were found to undergo the conformational transitions from the compact form to the extended coil upon ionization of primary carboxyl groups in maleic acid.[10-15] The hydrophobic interaction seems to be responsible for stabilization of the compact form. Viscometric, potentiometric and calorimetric measurements have been used to study the transition mechanisms.

The alternating copolymers of maleic acid with aromatic vinyl monomers such as styrene had also been investigated by several researchers. The dependence of intrinsic viscosity and apparent dissociation constant of primary carboxyl groups in maleic acid pK_1 on the degree of ionization of the carboxyl groups α_1 were not as simple[16] as was indicated by Ferry et al.[17] The pH-induced conformational transition of the maleic acid–styrene copolymer $(MA-St)_n$ had been observed in the dissociation range of the primary carboxyl groups by Sakurada and his collaborators,[18] and the transition was shown to be similar to that of poly(methacrylic acid). We have studied quantitatively the transition being similar to those of the copolymers of maleic acid with n-alkyl vinyl ethers. We have, at first, used viscometric, potentiometric, dilatometric and calorimetric titrations to study the transition.[19] However, as a specific method, we have been able to use the difference spectrum method, which had frequently been applied to studies on denaturation or folding processes of globular proteins, to follow the transition, because this copolymer has an optical probe, phenyl group, in the repeating unit reflecting the conformational transition.[20] With the previous results from other measurements of the fractionated samples in aqueous media, the optical absorption method has given important data of the compact form at low α_1 and the conformational transition.[21]

Another aromatic copolymer of maleic acid with α-methyl styrene $(MA-MSt)_n$ was also confirmed to undergo the conformational transition from the compact to coil form by means of the various techniques including the optical one.[22,23]

The optical probe is appropriate for measurements of the kinetic rates of the transitions, as shown in the kinetic studies on denaturation and folding processes of globular proteins. We applied the optical probe in $(MA-St)_n$ to measurements of the transition rate in aqueous NaCl, and the process was shown to be slower than the helix to coil transition of synthetic polypeptides in solutions.[24] However, NMR measurements seem to be more appropriate to investigate the dynamic characters of the transitions.

In this report, our previous equilibrium studies on the conformational transitions of the maleic acid copolymers $(MA-St)_n$ and $(MA-MSt)_n$ are summarized, and characterizations of the compact conformation and conformational transition of $(MA-St)_n$ in aqueous NaCl from 400 MHz ^{1}H-NMR data are described.

Conformational Transitions of $(MA-St)_n$ and $(MA-MSt)_n$ in Aqueous Salts

The modified titration curves of the fractionated $(MA-St)_n$ in aqueous NaCl at various ionic strengths from 0.009 to 0.27 resembled the potentiometric behaviors of the maleic acid copolymer with n-butyl vinyl ether and poly(methacrylic acid) in aqueous salts.[19] There was no precipitation or turbidity in the deionized polymer solution, except in aqueous NaCl of 0.27 M. The conformational transition from the compact to extended coil form was seen in diagrams of the intrinsic viscosity vs. α_1, and the reduced viscosity in the range $0.1 < \alpha_1 < 0.4$ was independent of polymer concentration.[19] Both the potentiometric and viscometric titration data showed that the pH-induced conformational transition of $(MA-St)_n$ in aqueous NaCl is due to intramolecular interaction. The unperturbed molecular dimensions were calculated from the viscosity data and the acid compact form of this copolymer was found to be not so compact as the conformations of globular proteins.[19] Dilatometry did also indicate existence of the transition,[19] and the transition was observed in aqueous solutions of various salts: chlorides, bromides, perchlorates and thiocyanates of Na^+, K^+, Rb^+, Cs^+ and Li^+.[22]

Temperature dependence of the conformational transition of $(MA-St)_n$ in aqueous NaCl was studied at 15°C to 40°C by the potentiometric method.[21] The titration curve scarcely changed with temperature. The calorimetric measurement expressed an anormaly due to the conformational transition, and the standard enthalpy change of transition ΔH_t^o at 25°C estimated from the calorimetric anormaly was about 360 cal/monomole in aqueous NaCl of 0.03 M.[21] The apparently scarce dependence of the pH-titration curve on temperature in a range between 15°C and 40°C was explained with this value of ΔH_t^o. The enthalpy change was also similar to the transfer heat of benzene into aqueous medium,[25] and therefore the compact form of the copolymer was concluded to be stabilized by the hydrophobic interaction between the phenyl groups in interior of the molecule.

The pH-induced difference spectra of $(MA-St)_n$ in aqueous NaCl at 269 nm indicated the transition, where the copolymer solution in the coil form was used as the reference.[21] By correcting perturbations of the phenyl groups produced by change in charge on nearby carboxyl groups in the compact or coil form, fraction of the exposed phenyl groups was calcualted from the optical titration data at 269 nm. The fraction means that of the repeating units in the coil form under an assumption of two-state transition mechanism. The temperature dependence of the difference extinction coefficient at 269 nm of the copolymer in aqueous NaCl of 0.03 M was investigated between 5°C and 45°C. The scarce dependence of transition on temperature was also confirmed from the optical method. The normalized transition curve, which expresses dependence of the fraction f_c on α_1, calculated from the optical data at 269 nm coicided well with that obtained from the pH-titration curve using the two-state assumption. The midpoint of transition at higher ionic strength corresponds to higher α_1, as shown in Figure 1. Various thermodynamic parameters of transition were calculated from the normalized transition curves. A parameter for intramolecular micelle formation, which does also refer to cooperativity of transition, was estimated from the curves. It is about 10^{-2} at ionic strength I = 0.03 0.09 and 25°C, which indicates the transition of the copolymer is far less cooperative than that of the polypeptide.[26]

A fact that the almost temperature-independent normalized transition curve calculated from the optical data coincided well with that from the pH-titration curve irrespective of ionic strength may mean the two-state mechanism of transition of $(MA-St)_n$ in aqueous salts. Recently Strauss et al.[27] showed the two-state character of transition of the maleic acid copolymer with n-butyl vinyl ether in aqueous solution from the potentiometric data. We have attempted to study the derivative spectra of optical absorption of the copolymer in aqueous NaCl between 240 and 280 nm, because the charge effect on nearby carboxyl groups may be eliminated in the derivative spectra in the wavelength region of 240-280 nm,[28] and change in environment near the optical probe may be discussed from the derivative spectra. In the difference spectrum between the first derivatives, which corresponds to the second derivatives, of the copolymer in aqueous NaCl, we found existence of many iso-derivative points.[29] Existence of such points means the two-state character of transition. Also, environment near the optical probe in the copolymer in the compact form was suggested to have similar hydro-phobicity to aqueous ethanol of 40-50 vol%.

On the other hand, studies on interaction of a hydrophobic dye, acridine orange, with $(MA-St)_n$ in aqueous NaCl gave information about the hydrophobicity in the hydrophobic microdomains.[30] Dilution of the dye along the polymer chain was remarkable for the compact form, but scarce

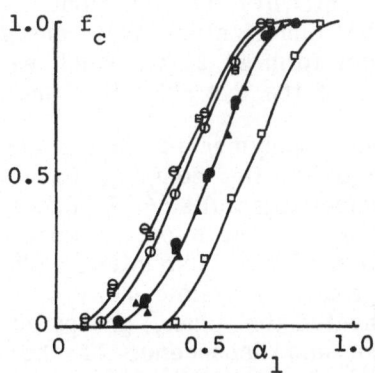

Figure 1.   Fraction f_c of the phenyl groups of (MA-St$_n$) in the coil form in aqueous NaCl at 25°C vs. α_1. Square, circle and triangle show the data from pH-titration, optical absorption and derivative spectrum, respectively. Ionic strength: 0.27 (□), 0.09, 0.03 and 0.009 (⊟, θ).

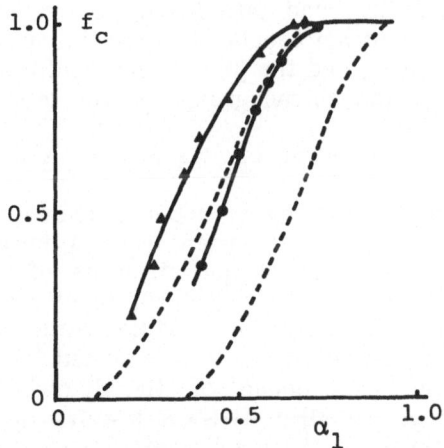

Figure 2.   The normalized transition curves of (MA-MSt)$_n$ in aqueous NaCl at 25°C. Ionic strength: 0.27 (●) and 0.009 (▲). The dotted lines show the curves for (MA-St)$_n$ at ionic strengths 0.27 and 0.009.

for the coil form. The remarkable dilution effect in the domains could be observed as an appearance of bound monomeric dye band at 502.5 nm. Analysis of the peak intensity as functions of polymer and dye concentrations gave the dimerization free energy of the dye -2.1 Kcal/mole in the compact form at 25°C, which was compared with the dimerization free energy of the dye -10.2 Kcal/mole in water at 25°C.

$(MA-MSt)_n$, another copolymer of maleic acid studied previously, was also confirmed to be in the compact form at low α_1 from the potentiometric and viscometric results.[22,23] Exact determination of the standard free energy change ΔG_t^0 at unionized state was not done, because of difficulty of estimation of pK_1^0, the intrinsic dissociation constant of the primary carboxyl groups in maleic acid of the copolymer. The difference spectrum indicated the phenyl groups buried in interior of the compact form molecule, and dependence of the difference in molar extinction coefficient at 261 nm on α_1 is in a sigmoidal fashion. The dilatometric data were consistent with the conformational transition, and the dilution of acridine orange along the chain in the compact form was also found. Therefore, $(MA-MSt)_n$ was concluded to undergo the transition from the compact to extended coil form upon ionization of the primary carboxyl groups in maleic acid. However, the compact form in $(MA-MSt)_n$ was found to be more unstable than in $(MA-St)_n$ from comparison of the transition curve between both the copolymers, as shown in Figure 2. Such an unstability of the compact form in $(MA-MSt)_n$ may refer to motional freedom of the backbone chain.

In summary, $(MA-St)_n$ and $(MA-MSt)_n$ are in the compact forms being stabilized by the intramolecular hydrophobic interactions between the phenyl groups at low α_1, and the pH-induced conformational transitions accompany exposure of the phenyl groups on the molecular surfaces.

^{1}H-NMR Studies on Conformation of $(MA-St)_n$ in Aqueous NaCl

^{1}H-NMR spectra of the hydrophobic polyelectrolytes have been measured by some researchers in order to gain informations on the nature of conformational transition and structure of the compact form. Schriever and Leyte showed 60 MHz ^{1}H-NMR spectra of poly(methacrylic acid) in D_2O and remarkable dependence of linewidth of methyl resonance in the side chain on ionization degree was analyzed.[31] Kay et al. did also obtain the remarkable dependence of the linewidth of side chain resonance on studies of 100 MHz ^{1}H-NMR spectra of the polymer, and they explained it in terms of the conformational transition.[32] Also, the latter authors showed that the linewidths of proton resonances in side chain of the maleic acid copolymer with n-butyl vinyl ether decrease sharply at the transition region determined from the potentiometric and viscometric methods, although such changes do not occur in the maleic acid copolymer with methyl vinyl ether. The NMR results of both the groups mean that the compact form in the hydrophobic polyelectrolyte at low degrees of ionization is associated with restriction to free rotation of

side chain protons. The 100 MHz ^{1}H-NMR spectra of $(MA-St)_n$ in D_2O were previously shown by Takakura and Ishikawa,[33] who obtained a sharp peak of phenyl resonance at about 7 ppm from an internal reference, sodium 4,4-dimethyl-4-silapentanesulfonate (DSS), and some broad peaks of backbone proton resonances at 1 3 ppm for the coiled form at high temperatures 70-80°C.   However, informations about the motional freedoms of side chain protons and backbone protons in the compact form copolymer at room temperature could hardly be obtained from their spectra. Even the peak of phenyl resonance should be composed of some contributions from para, meta and ortho protons. To gain exact data about the motional freedoms of protons in both the side and main chains in $(MA-St)_n$, a higher resolution than in their spectra has to be required.

We have used a 400 HMz ^{1}H-NMR apparatus (JEOL FX-400). $(MA-St)_n$ at various degrees of neutralization of the primary carboxyl groups in maleic acid α_1' in 0.05 M $NaCl-D_2O$ was prepared by the ordinary method. $(MA-St)_n$ used here was fractionated, and its average molecular weight is $2.05 \cdot 10^5$.   As an internal reference, DSS was used. Polymer concentration used for the NMR measurements was $3.7 \cdot 10^{-2}$ monomole/l, which is far more dilute than in the previous NMR studies of the hydrophobic polyelectrolytes.[31-33] In Figure 3, the spectrum of the coiled copolymer at $\alpha_1' = 1.0$ and 35°C is shown. In it, peaks of phenyl proton resonances are found at 7.16(I), 7.08(II), 6.83(III), 6.55(IV) ppm and so on. The multiple peaks at 6~8 ppm is clearly shown on an expanded scale in Figure 3(b). The peaks I and II may correspond to resonances of para and metha protons in the phenyl groups, respectively, and the peak III to ortho protons. Peaks at 4.7~4.8 ppm should be assigned to HDO. Peaks of the backbone proton resonances are found at 1~3 ppm. The 1.72 ppm peak should be assigned to the methylene protons in styrene. In Figure 4, two spectra at the transition range are shown.   The spectrum of the completely compact form at $\alpha_1' = 0.25$ and 35°C is shown in Figure 5. The spectral form of $(MA-St)_n$ in aqueous NaCl depends remarkably on the molecular conformation. The spectrum of the compact form has a broad peak at 6~8 ppm, and the multiple peaks of phenyl resonances shown in Figure 3(b) can not be seen.   Linewidths of the backbone proton resonances increase, as α_1' decreases, and each of the backbone proton resonances can not be separated in Figure 5.   However, even in the compact form at $\alpha_1' = 0.25$ and 35°C, the methylene proton resonance at about 1.7 ppm can be recognized. From Figures 3-5, the side chain and backbone protons are concluded to be in a more restricted motional state in the compact form than in the coil form.

Chemical shift of the main peak I at about 7.1 ppm of the phenyl proton resonances seems to reflect the conformational transition of $(MA-St)_n$, as shown in Figure 6. However, to characterize the compact form and the conformational transition, the motional freedom of each proton, especially of each of backbone protons, has to be analyzed. The freedom can be described in terms of the linewidth of resonance. Unfortunately, each of linewidths of both the side chain and backbone proton resonances

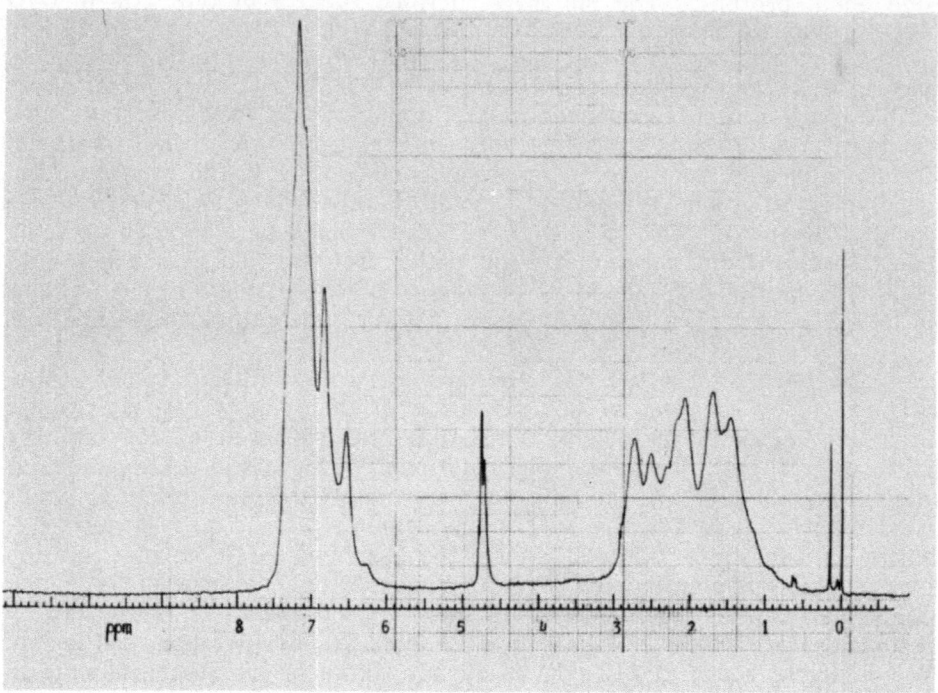

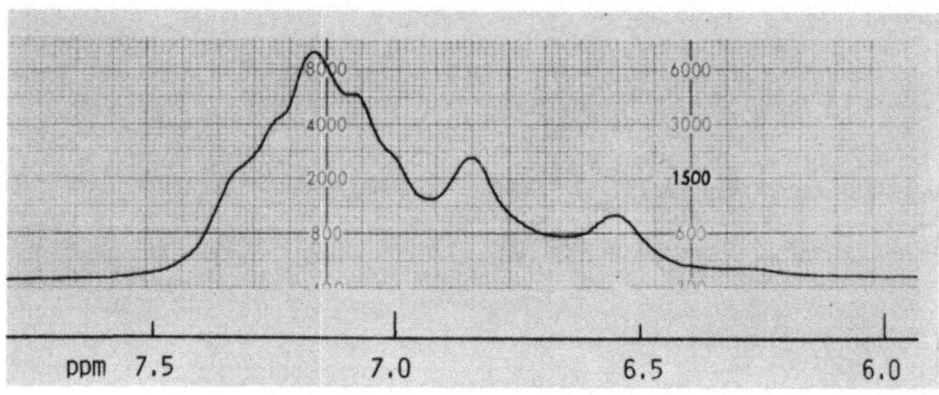

Figure 3(a). ^{1}H-NMR spectrum of (MA-St)$_n$ (MW = 2.05·10^5) in the coil
form at $\alpha_1' = 1.0$ and 35°C.   Polymer concentration and
ionic strength (NaCl) are 3.7·10^{-2} monomole/1 and 0.05,
respectively.

Figure 3(b). The expanded spectrum of (MA-St)$_n$ at $\alpha_1' = 1.0$ and 35°C.

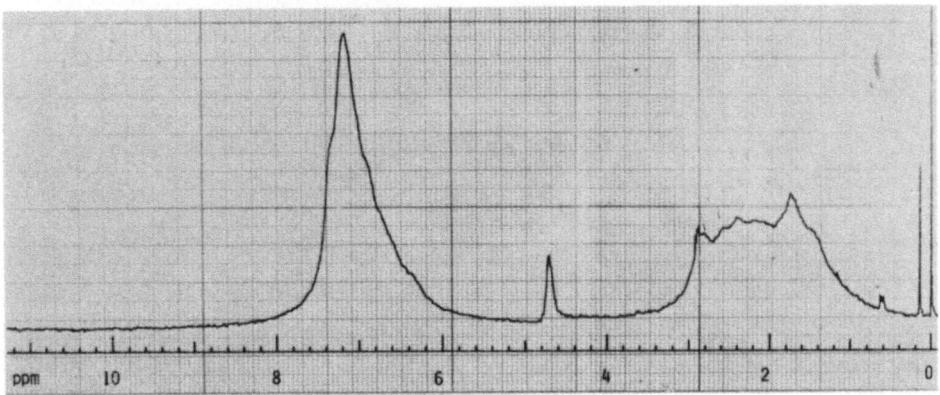

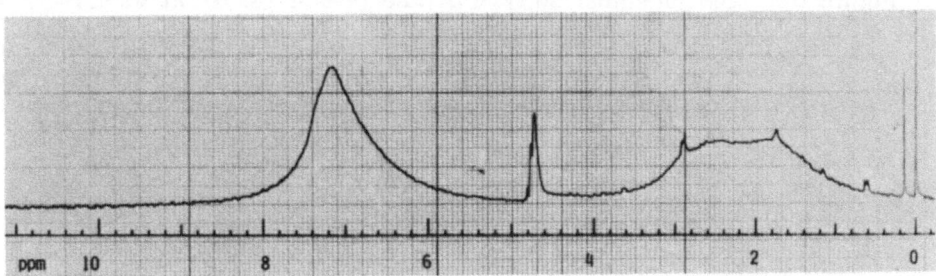

Figure 4. ^{1}H-NMR spectra of (MA-St)$_n$ at $\alpha_1' = 0.63$ (a) and 0.38 (b), and at 35°C. Molecular weight, polymer concentration and ionic strength (NaCl) are same as in Figure 3(a).

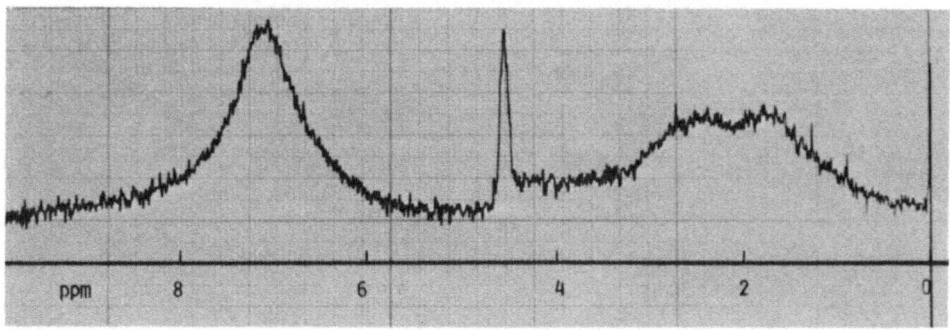

Figure 5. ^{1}H-NMR spectrum of (MA-St)$_n$ in the compact form at $\alpha_1' = 0.25$ and 35°C. Molecular weight, polymer concentration and ionic strength (NaCl) are same as in Figure 3(a).

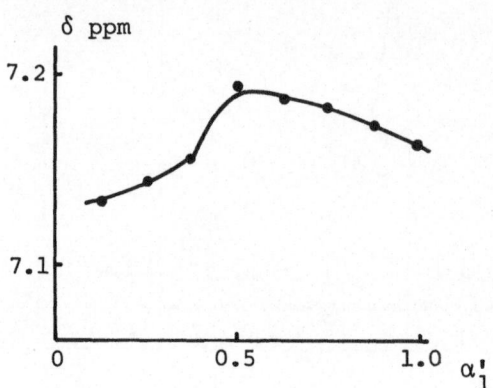

Figure 6.   The chemical shift δ of the peak I vs. α_1' at 35°C.

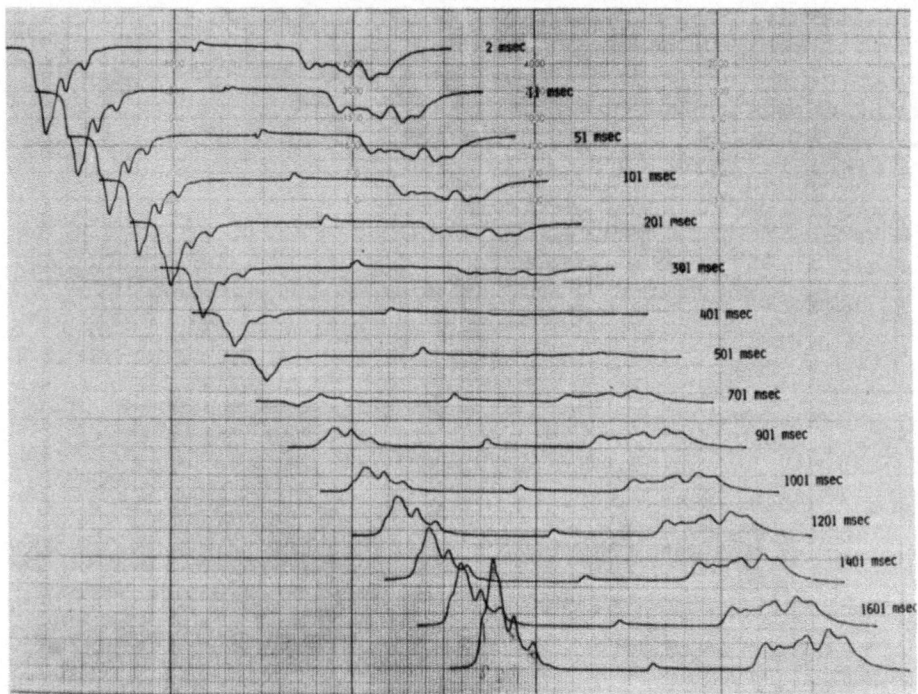

Figure 7.   Inversion recovery curves of the spectrum of $(MA-St)_n$ in the coil form at $\alpha_1' = 0.87$ and 35°C. Ionic strength (NaCl) = 0.05.

can not exactly be determined even in the coil form, as shown in Figure 3.

We measured the spin–lattice relaxation time T_1 by using the standard π–τ–$\pi/2$ sequences. As an example of inversion recovery kinetics, results for $(MA-St)_n$ in the coil form at $\alpha_1' = 0.87$ and 35°C is shown in Figure 7, where the delay time between 180° and 90° pulses is shown. At a glance, it is understood that both the T_1-values of the side chain and backbone protons do not coincide with each other. The T_1 can simply be calculated by the null-method.[34] However, it is prefered to calculate it by use of the following general formulae of inversion recovery kinetics under an assumption of single exponential decay function:

$$M_z = M_o \left(1 - 2 \exp(-\tau/ T_1) \right), \qquad (1)$$

where M_z is the longitudinal magnetization at τ after inversion and M_o = M_z at $\tau = \infty$. The T_1 was calculated by a least square fit of Equation (1) in a range of τ between 0 and 2~3 sec, when the deviation from equilibrium falls to about 10~20% of its initial value. In Figure 8(a), dependence of T_1 on α_1' is shown at 35°C for both the side chain (peak I) and backbone (methylene) protons. The null-method calculates the T_1 from τ_o, when M_z disappears. A following relation is obtained between T_1 and τ_o from Equation (1):

$$T_1 = (\ln 2)^{-1} \tau_o . \qquad (2)$$

The dependence of T_1 calculated from Equation (2) is shown in Figure 8(b). The T_1's of both the side chain and backbone protons reflect the conformational transition of the copolymer. However, the T_1 in Figure 8(a) does not completely coincide with the corresponding one in Figure 8(b). Therefore we checked the inversion kinetics in details.

Figure 9(a) shows the kinetics for the peak I and the methylene proton peak of $(MA-St)_n$ in the coil form at $\alpha_1'=0.87$ and 35°C. Under the assumption of single exponential decay function expressed by Equation (1), $\ln(1 - M_z/M_o)$ has to be proportional to τ. For the phenyl protons, the proportional relationship is apparently well satisfied in a range of τ less than 2 sec, and $\ln(1 - M_z/M_o)$ at $\tau = 0$ is very close to the expected value (= $\ln 2$), but the kinetics for the methylene protons in styrene is not expressed by Equation (1) in the range of τ. Figure 9(b) shows the kinetics for the compact form copolymer at $\alpha_1' = 0.25$. For the peak I, the proportional relationship between $\ln(1 - M_z/M_o)$ and τ is clearly seen in a range of τ less than 3 sec, and for the backbone protons the proportionality is apparently found, but not strictly. As described later, the recovery curve with a correlation time is in principle nonexponential because of coupled relaxation, and therefore we made use as before[35] of concept of an effective relaxation time T_1 from the initial perturbation to the time when the deviation from equilibrium falls to about 25~30% of its initial value. A dotted line in Figure 9(a) shows an example of the

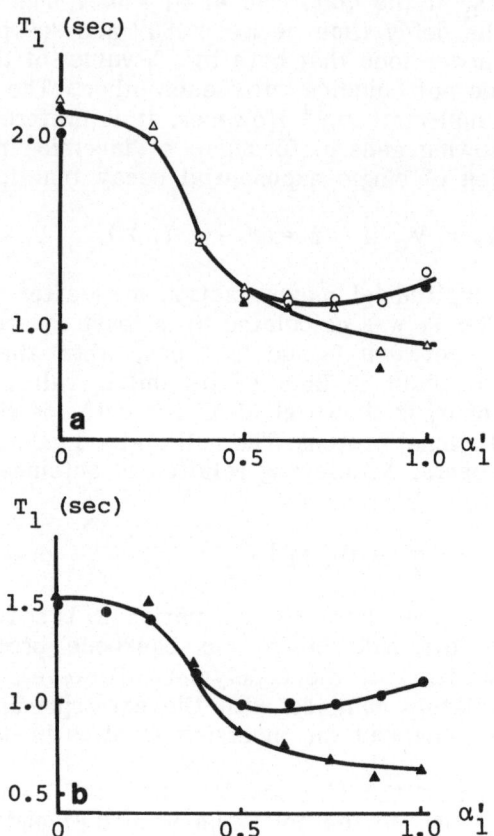

Figure 8(a).   Dependence of T_1 calculated by Equation (1) on α_1' at ionic
strength 0.05 and 35°C. Open circle represents data for
the peak I, open triangle for the methylene proton peak,
and filled symbols for the data of the effective relaxation
coefficient.

Figure 8(b).   Dependence of T_1 calculated by Equation (2) on α_1'. The
symbols express the same resonances as in Figure 8(a).

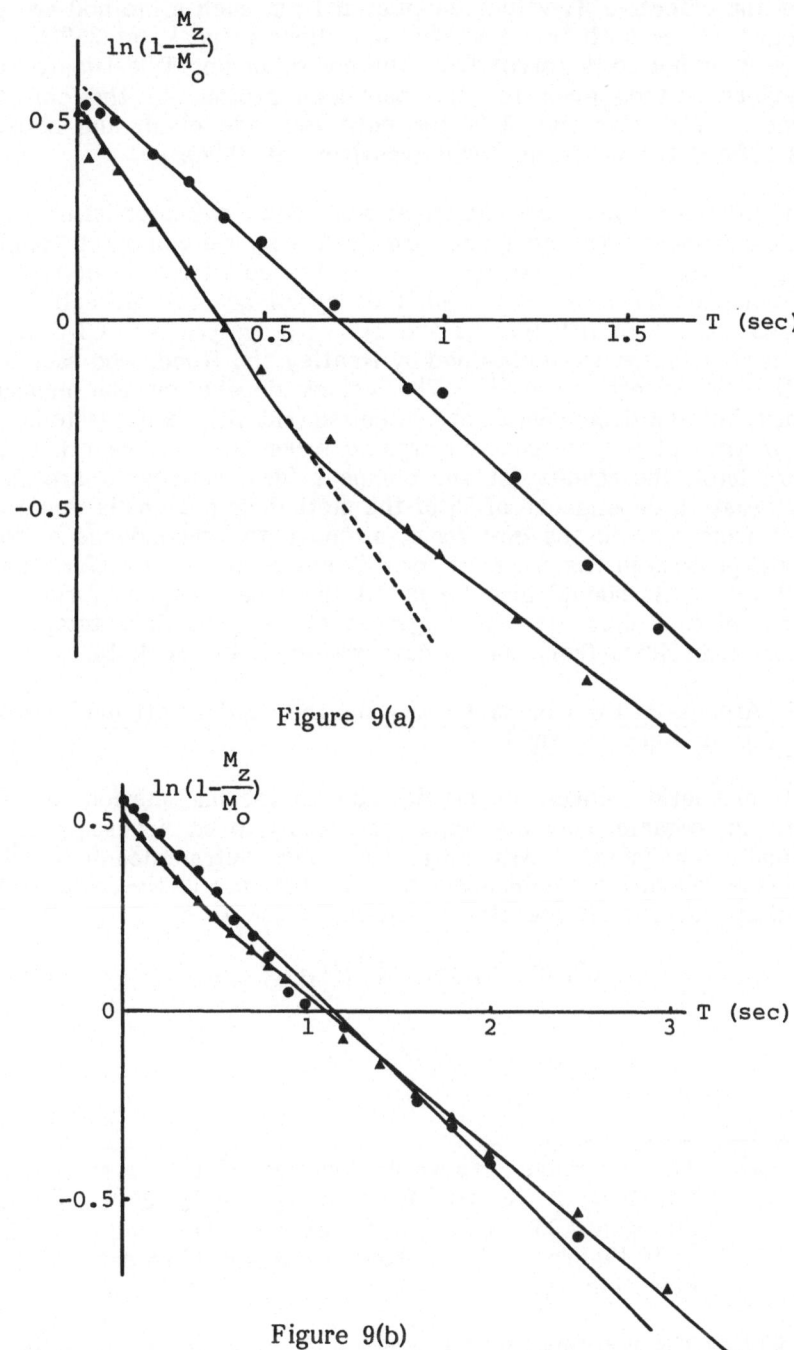

Figure 9(a)

Figure 9(b)

initial recovery kinetics to determine the effective T_1's. In Figure 8(a), some of the effective T_1-values calculated from such a method are plotted against α_1' for both the aromatic and aliphatic peaks at 35°C. The effective T_1 is not so different from the corresponding T_1 estimated from the previous method even for the backbone protons in the coil form copolymer. The effective T_1's for both the side chain and backbone protons reflect the conformational transition at 35°C.

In order to gain more informations about characteristics of the motional freedom of the compact form $(MA-St)_n$, we compared temperature dependence of both the T_1-values for the coil form copolymer with for the compact form. In Figure 10, the dependence for the coil form is shown and compared with that of the T_1's of polystyrene in $CDCl_3$. The results of polystyrene were obtained by Heatley and Wood, who used a 360 MHz ^1H-NMR spectrometer.[35] The values depend on the measuring frequency, but the dependence for the coiled $(MA-St)_n$ is understood to be similar to that of polystyrene in organic solvents. On the other hand, in Figure 10(b), the results for the compact form copolymer are shown. The temperature dependence of T_1 of the methylene protons is remarkably different from that in the coil form, although the dependence of T_1 for the phenyl protons in the compact form is not remarkably different from the coil form. It should also be noted that the compact form in the copolymer at $\alpha_1' = 0.25$ may be suggested to be stable in a temperature range less than 40°C from the T_1-data shown in Figure 10(b).

Dynamic Aspects of the Compact Form and the Conformational Transition of $(MA-St)_n$ in Aqueous NaCl

^1H magnetic relaxation studies on molecular motion of vinyl polymers in organic solvents have previously been investigated and theoretically analyzed. Assuming that only intramolecular $^1H-^1H$ dipole·dipole relaxation mechanisms operate, the spin–lattice relaxation is governed by coupled differential equations of the form:

$$dS_i/dt = \sum_j (S_j - S_j^0)/T_{ij} , \qquad (2)$$

Figure 9(a). The inversion recovery kinetics of the resonances of $(MA-St)_n$ in the coil form at $\alpha_1' = 0.87$, 35°C and ionic strength 0.05. The symbols express the same resonances as in Figure 8. The dotted line was used to determine the effective T_1.

Figure 9(b). The kinetics for the compact form at $\alpha_1' = 0.25$, 35°C and ionic strength 0.05. The symbols express the same resonances as in Figure 8.

where S_i is the longitudinal magnetization of an i-th spin and S^0 is its equilibrium value. The diagonal and cross relaxation coefficients can be expressed in terms of some parameters: number of nuclei of type j with which a nucleus of type i interacts, an effective distance for the interaction between nuclei i and j, and resonance frequency of each nucleous. The evaluation of the distance depends on the nuclei involved in the molecule, and for polystyrene the distances R_{XX}, R_{XP}, R_{MM}, R_{OP}, R_{MP}, and so on were estimated under appropriate assumptions, where X, M, P, and O represent methylene, meta, para and ortho protons, respectively. The spectral density of the normalized auto-correlation function of the dipole–dipole interaction between i and j at the frequency ω, $J_{ij}(\omega)$, has to be given to calculate the diagonal and cross relaxation coefficients, from which the effective relaxation coefficients T^X, T^O, T_1^{M+P}, and so on can be calculated. For the simple case of isotropic rotational motion with a correlation time, $J(\omega)$ is simply given. In many cases, however, polymer motion is better described by a distribution of correlation times. Two such distributions have previously been used: Cole-Cole and log -χ^2 distributions. The two distributions are simply a formal way of representing a non-exponential auto-correlation function. Recently, an analytical form of $J(\omega)$ from a model of vinyl polymer motion consisting of a combination of locally conformational jumps and overall molecular tumbling has been used.[36,37] The jump model yields $J(\omega)$ for backbone protons as follow:

$$
J(\omega) = \frac{\tau_o\,\tau_D\,(\tau_D - \tau_o)}{(\tau_o - \tau_D)^2 + \omega^2\,\tau_o^2\,\tau_D^2}\left\{\left(\frac{\tau_o}{2\tau_D}\right)^{1/2}\quad x\right.
$$

$$
\left[\frac{(1 + \omega^2\,\tau_D^2)^{1/2} + 1}{1 + \omega^2\,\tau_D^2}\right]^{1/2} + \left(\frac{\tau_o}{2\tau_D}\right)^{1/2}\frac{\tau_o\,\tau_D}{(\tau_o - \tau_D)}\quad x
$$

$$
\left.\left[\frac{(1 + \omega^2\tau_o^2)^{1/2} - 1}{1 + \omega^2\,\tau_o^2}\right]^{1/2} -1\right\}\qquad\qquad (4)
$$

where τ_D is the local jump correlation time and τ_o is the overall tumbling time. Equation (4) was found to be applicable to polystyrene and poly(vinyl acetate) in organic solvents. The spectral density functions for interactions between side chain protons and between side chain and backbone protons are described in terms of three correlation times: τ_D, τ_o and an average time between side chain jumps τ_J. Heatley and Wood assumed the relation $\tau_J = \tau_D$ for polystyrene in organic solvents, and they obtained $\tau_D/\tau_o = 0.07\sim0.1$ from temperature dependence of T_1^X and T_1^{M+P} of polystyrene in $CDCl_3$. T_1^{M+P} in their paper[35] corresponds to the T_1 of the peak I in this study. Moreover, because of similarity in shape between

T_1^X vs. temperature plots for all solvents used by them, they attempted to make a master curve of T_1^X on the reduced temperature scale. Choosing $CDCl_3$ as a standard, the reduced temperature T* was defined for another solvent by:

$$\frac{1000}{T*} = p\,\frac{1000}{T} + q \tag{5}$$

The master curve for T_1^{M+P} was also made on the same reduced temperature scale as for T_1^X. Therefore, both T_1^X and T_1^{M+P} suggested that molecular motion of polystyrene in the solvents can be interpreted in terms of a rotational diffusion process combined with the local conformational jumps, of which the correlation time τ_D is about an order of magnitude shorter than the diffusion correlation time τ_0.

Our data of T_1^X in the coil form in Figure 10 can nearly be put over the master curve for T_1^X of polystyrene in the organic solvents, as shown in Figure 11, with the values of the scaling parameters p and q: p = 1.305 and q = 0.045(K^{-1}), by which the data of T_1^{M+P} in the coil form (in Figure 10) can also be put over the master curve for T_1^{M+P} of polystyrene in the organic solvents. Therefore, the molecular motion of $(MA-St)_n$ in the coil form may be described in terms of the conformational jump model combined with isotropic rotational diffusion, when the ratio τ_D/τ_0 seems to be nearly 0.07.

On the other hand, plot of T_1^{M+P} in the compact form on the reduced temperature scale did not fit to the theoretical curve with $\tau_D/\tau_0 = 0.07$, but is nearly put over the theoretical one with $\tau_D/\tau_0 = 10$, although the theoretical curves by Heatley and Wood were calculated for $\omega_i = 360$ MHz. Shape of the temperature dependence curve of T_1^{M+P} seems to be scarcely affected by a difference of ω_i between 360 MHz and 400 MHz. Unfortunately, we could not compare our data of T^X in the compact form with the theoretical results by them, because they did not show the theoretical calculation of T^X of polystyrene. However, a larger τ_D/τ_0 for the compact form may be reasonable than that for the coil form.

To estimate τ_D for the compact form, we calculated dependence of T^X on τ_D for a case of $\tau_D/\tau_0 = 10$ and $\omega = 400$ MHz, when we took the interactions only between the backbone protons into consideration, as a first approximation. The calculated result was very similar to the dependence for poly(vinyl acetate) given by Heatley and Cox,[37] who also took only the backbone proton interactions into consideration. The value of τ_D for the compact form at 35°C estimated from such a theoretical diagram and the observed T^X was about $1 \cdot 10^{-7}$ sec. (The observed T_1^X corresponds to two τ_D's on the theoretical diagram. To determine a true value of τ_D, we attempted to compare the T^X at 400 MHz with that at 100 MHz.[34] We measured T^X for the compact form at 100 MHz (JEOL 100-FX), but we could not obtain any reliable value. We measured T_1^{M+P} for the compact form at 100 MHz, and it was found to be far less than

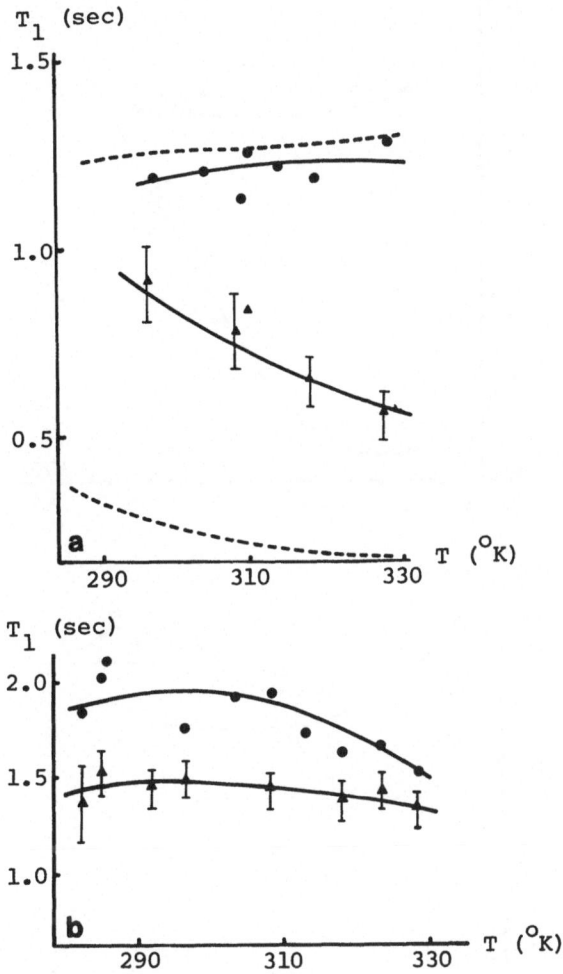

Figure 10(a). Temperature dependence of the resonances of $(MA-St)_n$ in the coil form at $\alpha_1' = 0.87$ and ionic strength 0.05. The symbols express the same resonances as in Figure 8. Dotted lines represent the results for polystyrene in $CDCl_3$ observed by a 360 MHz ^{1}H-NMR spectrometer.

Figure 10(b). The dependence for the compact form at $\alpha_1' = 0.25$ and ionic strength 0.05. The symbols express the same resonances as in Figure 8.

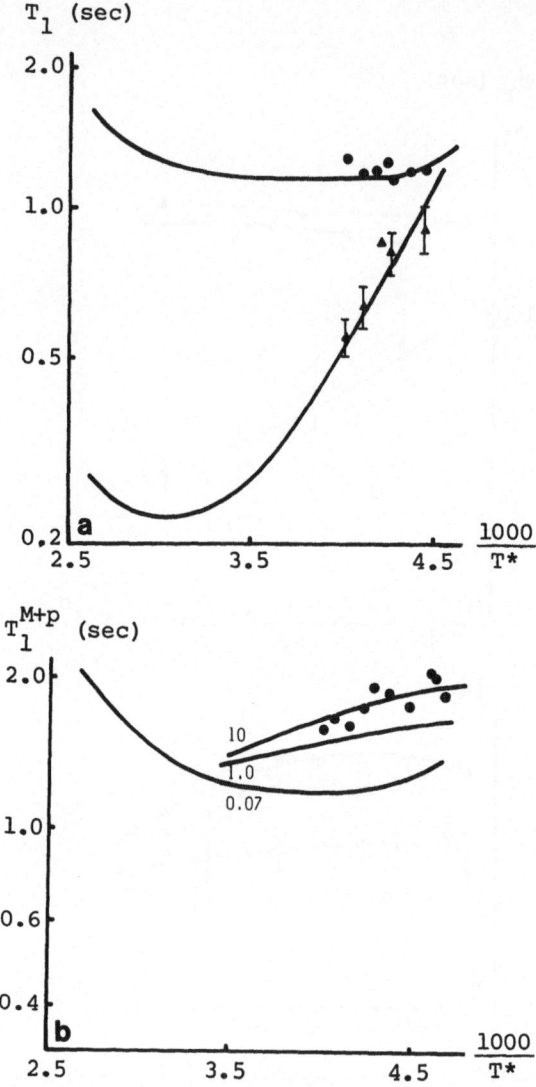

Figure 11(a). Dependence of the effective T_1^X and T_1^{M+P} shown in Figure 10(a) on the reduced temperature T^*. Solid curves show the master curves of T_1^X and T_1^{M+P} for polystyrene in organic solvents.

Figure 11(b). The dependence of the effective T_1^{M+P} shown in Figure 10(b). Solid curves show the theoretical ones calculated by use of the jump model. The value of τ_D/τ_o is inserted in the figure.

T^{M+P} at 400 MHz: $T^{M+P} = 0.30$ sec at $\alpha' = 0.125, \omega_i = 100$ MHz and $35^{\circ}C$. Then we expected a smaller T^X at 100 MHz than at 400 MHz for the compact form.) From a comparison with τ_D in the coil form at $35^{\circ}C$, which was estimated from the dependence of T^X on τ_D for a case of $\tau_D/\tau_0 = 0.07$ and $\omega = 400$ MHz to be about $4.8 \cdot 10^{-9}$ sec, the local jump is concluded to be considerably restricted in the compact form.

Further studies on $^1H-$ and $^{13}C-$NMR seem to be important to analyze the compact conformation and conformational transition of the hydrophobic polyelectrolyte.

ACKNOWLEDGEMENTS

The authors wish to thank the Ministry of Education of Japan and Yamada Kagaku Shinko Zaidan for supports by a Grant in Aid. Also, they thank Professors K. Hikichi and A. Tsutsumi for their valuable suggestions.

REFERENCES

1. J. L. Leyte and M. Mandel, J. Poly. Sci., Phys. Ed., 2, 1879 (1964).
2. M. Mandel, J. C. Leyte and M. G. Stadhouder, J. Phys. Chem., 71, 603 (1967).
3. V. Crescenzi, Adv. Polym. Sci., 5, 538 (1968).
4. C. Braud, G. Müller, J. C. Feno and E. Selegney, J. Poly. Sci., Chem. Ed., 12, 2767 (1974).
5. E. Okamoto and Y. Wada, J. Poly. Sci., Phys. Ed., 12, 2767 (1974).
6. F. Fichter and H. Schonert, Colloid & Poly. Sci., 255, 230 (1977).
7. K. Nitta, N. Ohno, H. Nakano and S. Sugai, Rept. Progr. Polym. Phys. Jpn. 25, 93 (1982); Colloid & Polym. Sci. 261, 159 (1983).
8. M. Nagasawa and S. A. Rice, J. Amer. Chem. Soc., 82, 5010 (1960).
9. E. Binanchi, A. Ciferri, R. Parodi, R. Rapone and A. Teardi, J. Phys. Chem., 74, 1050 (1970).
10. P. L. Dubin and U. P. Strauss, J. Phys. Chem., 71, 2757 (1967).
11. P. L. Dubin and U. P. Strauss, J. Phys. Chem., 74, 2482 (1970).
12. U. P. Strauss and R. Varoqui, J. Phys. Chem., 72, 2657 (1968).
13. P. L. Dubin and U. P. Strauss, Polyelectrolytes and Their Application" A. Rembaum and E. Selegney, Eds., Reidel, Dordrecht, pp 3-13 (1975).
14. U. P. Strauss and G. Vesnaver, J. Phys. Chem., 79, 1558 (1975).
15. U. P. Strauss and M. S. Schlesinger, J. Phys. Chem., 82 571 (1978).
16. N. Ohno, K. Nitta and S. Sugai, Kobunshi Kagaku, 28, 671 (1971).
17. J. D. Ferry, D. C. Udy, F. C. Wu, G. C. Heckel and D. B. Fordyce, J. Colloid Sci., 6, 429 (1951).
18. I. Sakurada, Y. Sakaguchi and H. Uehara, Kobunshi Kagaku, 27, 82 (1970).

19. N. Ohno, K. Nitta, S. Makino and S. Sugai, J. Poly. Sci., Phys. Ed., 11, 413 (1973).
20. S. Sugai, N. Ohno and K. Nitta, Macromolecules, 7, 96 (1974).
21. T. Okuda, N. Ohno, K. Nitta and S. Sugai, J. Polym. Sci., Phys. Ed., 15, 749 (1977).
22. S. Sugai and N. Ohno, Biophys. Chem., 11, 387 (1980).
23. N. Ohno, Polymer J., 13, 719 (1981).
24. N. Ohno, T. Okuda, K. Nitta and S. Sugai, J. Polym. Sci., Phys. Ed., 16, 513 (1978).
25. R. L. Bohon and W. F. Claussen, J. Amer. Chem. Soc., 73, 1571 (1951).
26. P. Doty and J. T. Yang, J. Amer. Chem. Soc., 78, 498 (1956).
27. U. P. Strauss, B. W. Barbleri and G. Wong, J. Phys. Chem., 83, 2840 (1979).
28. T. Ichikawa and H. Terada, Biochim. Biophys. Acta, 580, 120 (1979).
29. S. Sugai, K. Nitta and N. Ohno, Polymer, 23, 238 (1982).
30. N. Ohno, J. Chem. Soc. Japan, 1137 (1978).
31. J. Schriever and J. C. Leyte, Polymer, 18 1185 (1977).
32. P. J. Kay, D. P. Kelly, G. I. Milgata and F. E. Treloar, Makromol. Chem., 177, 885 (1976).
33. K. Takakura and M. Ishikawa, Rept. Progr. Polym. Phys. Jpn., 21, 449 (1978).
34. K. Wüthrich, "NMR in Biological Research", North-Holland Pub. Co., Chapter 4 (1976).
35. F. Heatley and B. Wood, Polymer, 19, 1405 (1978).
36. B. Valeur, J. P. Jarry, F. Geny and L. Monnerie, J. Poly. Sci., Phys. Ed., 13, 2251 (1975).
37. F. Heatley and M. K. Cox, Polymer, 18, 225 (1977).

L-PHENYLALANINE OLIGOPEPTIDES GRAFTED POLY(ACRYLIC ACID). EVIDENCE OF SPECIFIC INTERACTIONS OF ETHIDIUM BROMIDE AND ACRIDINE ORANGE WITH HYDROPHOBIC MICRODOMAINS.

J. Beaumais
Laboratoire de Chimie Macromoléculaire
ERA 471 Université
de Rouen 76130 Mont Saint Aignan France
Th. Ackermann
Institut für Physikalische Chemie II der Universität
Freiburg, Albertstr. 23a, D 7800 Freiburg

ABSTRACT

Hydrophobic moieties were introduced along the macromolecular chains of a hydrophilic polymer, i.e. poly(acrylic acid) (PAA), through the partial grafting ($\tau \simeq 10\%$) of L-phenylalanine (Phe) and some of its oligopeptides (Phe_n) $(1 \leqslant n \leqslant 4)$. The overall conformational state of these new water-soluble polyelectrolytes was evaluated from potentiometric and viscometric data. Hydrophobic domains, mainly due to Phe_n aromatic moieties, lead, at low ionization, to the stabilization of compact states of the polymeric chains.
At constant grafting value ($\tau \simeq 10\%$) the $\Delta G_t/N$ values are:

	PAA.Phe$_1$	PAA.Phe$_2$	PAA.Phe$_3$	PAA.Phe$_4$
$\Delta G_t/N$	$\simeq$ 100 cal.	230 cal.	700 cal.	1100 cal.

The binding of the cationic dye Ethidium Bromide to PAA.Phe$_n$ in aqueous medium was investigated by equilibrium dialysis experiments. The influence of hydrophobic microdomains was evidenced from such measurements. An induced optical activity of bound Acridine Orange was also exemplified.

INTRODUCTION

In the field of water-soluble polymers, hydrophobic interactions are of major importance in two ways: First, they can play a determining role for the stabilization of more or less compact conformations of natural and synthetic polyelectrolytes. Secondly, they can provide the driving forces for non-covalent interactions between small molecules (or ions) and polymers, these interactions depending in their turn on the conformational state and/or on the formation of microdomains.

Regarding this, the interactions of dyes and related compounds, such as antibiotics[1] and of aromatic amino-acids or peptides[2] with polynucleotides have been extensively investigated. The mechanisms of binding seem quite complex, but two interactions differing in kind are generally considered:
- a "strong binding" process involves monomeric dyes which intercalate into the DNA double-helix, by inserting their flat aromatic moieties between adjacent base pairs. These complexes are little sensitive to ionic strength effects.
- a "weak binding" process, when dyes are externally bound to ionic groups of the polynucleotide chains; as it depends mainly on electrostatic interactions, this binding is very sensitive to ionic strength effects. Furthermore, the electrostatically fixed dyes molecules can be stabilized by dye-dye stacking (cooperative binding).[3]

As to synthetic polyelectrolytes, the aims of previous studies may be divided into two major classes, for the sake of convenience:
- those which are related to the effects of more or less "hydrophobic" moieties covalently bound to water-soluble polyelectrolytes.
- those which concern the non-covalent binding of "hydrophobic" species such as dyes, surfactants (ionic and non-ionic), or N-tetraalkylammonium ions.

The Influence of Hydrophobic Interactions on Conformational States of Synthetic Polyelectrolytes

One of the most investigated examples concerns poly-(methacrylic acid) (PMA): a conformational transition from a compact state to an extended one is observed within a rather small range of ionization and is attributed to the presence of the methyl side groups.[4-5] The compact structure was found to be stabilized through the incorporation of benzyl methacrylate within the PMA chains.[6]

Examination of the behavior of 1-1 copolymers of maleic acid and alkylvinyl ethers studied by Dubin and Strauss,[7-9] and Martin and Strauss[10-11] is also significant in this regard. With such copolymers, if the alkyl side groups are small, the conformational properties of the polyelectrolytes can be considered as normal, whereas in the case of increasing size of these groups (up to n-octyl copolymers) compact states stabilized by hydrophobic forces are evidenced. Similar conclusions were drawn by other workers for series of copolymers of maleic acid and alkylvinyl ethers[12] and for amide acid copolymers prepared by partial

hydrolysis of poly N-(sec-butyl)-N-methyl-acrylamide (PAAm).[13]   The
case of a polycation - partially neutralized poly(2-vinylpyridine) (PVP) -
was described by Muller:[14] depending on ionization and dilution of this
polyelectrolyte, "hydrophobic clusters" may exist, as deduced from
hydrodynamic and dielectric behavior of PVP.

Dye-Polyelectrolyte Interactions

In connection with studies concerning the conformational state of
hydrophobic polyelectrolytes, the mode of interaction of dyes (and
related compounds) with such polymers has been investigated.   This
matter appears to be quite complex but some aspects are to be pointed
out. First, the extent of dye binding is generally related to the actual
conformation of polymers chains: as compared to more extended
conformational states, tightly coiled polyanions can bind much more
dye.[15-19]   This is not only observed for charged dyes-polyelectrolytes
interactions, but also when electrostatic effects are lowered or suppressed. Secondly, the spectroscopic properties of the dyes, and particularly
the intensity of fluorescence, are affected by the presence of polyions,
and strongly dependent on their conformational state.[15,18-21]

At this point, we should like to briefly quote some results
previously obtained in our laboratory, concerning a polycondensate
between L-lysine and 1,3-benzenedisulfonylchloride (PLL).

$$\left[SO_2 - \bigcirc - SO_2 - NH - CH - (CH_2)_4 - NH \right]_n \quad PLL$$
$$\underset{COOH}{|}$$

Increasing the degree of ionization of PLL, results in a conformational
change between a very compact structure and an extended coil.   N-
tetraalkylammonium counterions are found to stabilize the compact
conformation, and the same holds for a cationic dye, Acridine Orange
(AO), while these effects are suppressed in aqueous-organic mixtures or
in urea solutions.[22-24] The interactions of this dye with PLL was also
investigated by visible and fluorescence spectroscopy and the results
compared with those dealing with PAA and PMA.[21]   To sum up, the
aggregation (stacking) of bound AO appears to be decreased when the
polyions are in compact coils. But this tendency was found to be less
important in the case of PLL, this difference being ascribed to the
presence of aromatic rings in the main chain of the polyelectrolyte.
Similar findings were reported for polystyrene sulfonic acid-AO systems.[30-31]

It appeared worthwhile to us to obtain further information on the
contribution of the aromatic ring, first as stabilizing group for compact
conformations, secondly as a site interacting in a peculiar way with dyes.
In this approach, we considered that a partial grafting of L-phenylalanine
(Phe) and some of its oligopeptides (Phe_n) along the chain of poly(acrylic

acid) could be suitable to introduce such hydrophobic moieties. More-over, studies of synthetic water-soluble polymers carrying potent biologically active species such as peptides are of current interest.

EXPERIMENTAL SECTION

A series of copolymers with the formula $PAA.Phe_n$ (with $1 \leqslant n \leqslant 4$) was synthesized:

$$PAA - Phe_n$$

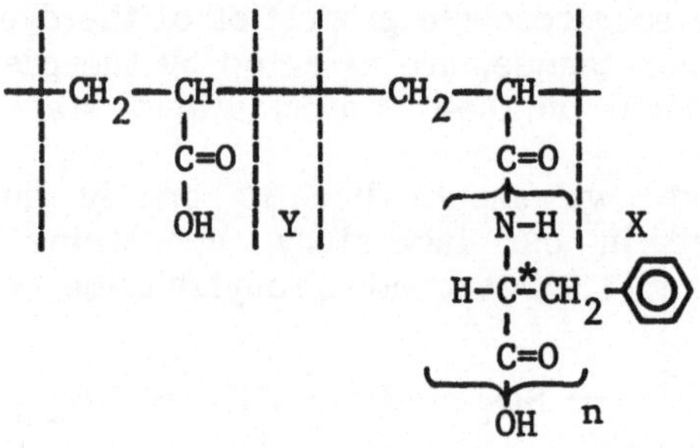

$0 \leq n \leq 4$

X = Grafting Degree ≈ 0.10

C^* = Asymmetric Carbon

A sample of PAA was first prepared as described in a previous paper.[17]   As our aim was to obtain $PAA.Phe_n$ with a relatively low grafting degree ($X \simeq 0.10$) an appropriate way of synthesis was established, as summed up in Table I (experimental details will be given in an other paper).[27]

Table I

SYNTHESIS OF PAA.Phe$_n$

1) Partial esterification of carboxylic groups of PAA:

$$\text{P-C-OH} + \text{HO-N} \underset{\text{O}}{\overset{\text{O}}{\diagdown}} \xrightarrow[\substack{0\text{-}5^\circ \\ \text{(THF or DMF)}}]{\text{DCC}} \text{P-C-O-Su} + \text{DCU}$$

(PAA)   (SuOH)

2) Coupling of the active ester with Phe$_n$:

$$\text{P-C-O-Su} + \text{Phe}_n \xrightarrow[0\text{-}5^\circ]{\text{Et}_3\text{N}} \text{PAA.Phe}_n$$

$$(\text{DMF-H}_2\text{O})$$

3) Recovering of PAA.Phe$_n$:

- Neutralization of the solution with KOH

  (→ PAA.Phe$_n$, K$^+$)

- Dialysis of PAA.Phe$_n$, K$^+$ (8 days)

- Percolation through a cationic exchange resin

  (→ PAA.Phe$_n$, H$^+$)

- Freeze-drying of PAA.Phe$_n$, H$^+$

RESULTS AND DISCUSSION

Potentiometric Titrations

Potentiometric titrations may provide a good insight into the conformational state of a polyelectrolyte. In particular, a conformational transitition can be characterized by taking such results into account. The initial titration can be normalized into plots pK_a = pH + log $(1-\alpha)/\alpha$ versus α. These plots are represented in Figure 1 for PAA.Phe$_n$ and also for ungrafted PAA. (titrations were carried out in the presence of salt, C_{KCl} = 2.10^{-3}M, for further advisable comparisons with dialysis experiments). This representation clearly shows the typical discontinuous change of pK_a versus α, contrary to PAA. Such behavior, which indicates a transition from a rather compact state to an extended one is obviously reinforced when the number n of Phe residues increases. It must be also noted that the range of ionization where the conformational transition takes place is shifted to large values of α with the increasing size of the peptide chain.

The thermodynamic parameters for the transition process are related to the potentiometric results. The free energy change ΔG_0^t can be evaluated for the hypothetical process:

$$(\text{PAA.Phe}_n, \text{ hypercoil, } \alpha=0) \rightarrow (\text{PAA.Phe}_n, \text{ extended, } \alpha=0)$$

According to the Nagasawa's method,[28] ΔG_0^t is obtained by its proportionality with the area A between the observed titration curve for PAA.Phe$_n$ and a theoretical one corresponding to the ionization of the expanded coil from α = 0 to α = 1:

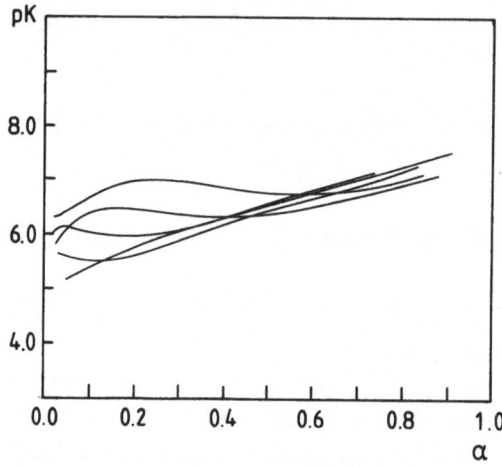

Figure 1. pK_a's variation with the ionization degree α. Counterion: K$^+$; supporting salt: KCl 2.10^{-3}M; Cp = 2.10^{-3}M; 0:PAA; 1:PAA.Phe$_1$; 2:PAA.Phe$_2$; 3:PAA.Phe$_3$; 4:PAA.Phe$_4$.

$$\Delta G_0^t/N = (RT/0.434)xA$$

($\Delta G_0^t/N$ is the free energy change per monomeric unit, each of them carrying a carboxylic group).

For a rough evaluation, the pK_a's variation versus α for PAA was chosen as the reference curve; as the grafting values of PAA.Phe$_n$ are relatively low, the assumption of close values of pK_0 (the intrinsic dissociation constant) for PAA and PAA.Phe$_n$ must be realistic.

The values reported in Table II indicate that the strength of cohesive interactions is readily increased with the length of the peptidic side chains. This is in qualitative agreement with results previously obtained for 1-1 copolymers of maleic acid and alkylvinyl ethers[9,11] although no proportionality between n (number of Phe groups per side chain) and $\Delta G_0^t/N$ can be proposed for PAA.Phe$_n$, but rather a cooperative behavior. This suggests, especially for PAA.Phe$_4$, an intramolecular aggregation of the aromatic residues and therefore the formation of micelle-like domains.

Table II

Free energy change for the transition from compact to extended

coil for PAA.Phe$_n$:

POLYELECTROLYTE	$\Delta G_0^t/N$ [a]
PAA.Phe$_1$	$\simeq$ 100 cal.
PAA.Phe$_2$	$\simeq$ 230 cal.
PAA.Phe$_3$	$\simeq$ 700 cal.
PAA.Phe$_4$	$\simeq$ 1100 cal.

[a] (for a PAA.Phe$_1$ sample with a grafting degree X = 0.065, the

conformational behavior was found -from potentiometric data-

normal as compared to PAA)

Viscometric Titrations

The hydrodynamic behavior, as evidenced from viscosity data, is in good agreement with the conclusions drawn from potentiometric experiments. Viscometric titrations (Figure 2) were performed in experimental conditions (counterion, added salt, polymer concentration) identical to those chosen for potentiometry. The reduced viscosity η sp/c is conventionally given in dl g^{-1}; as in the series PAA.Phe$_n$ because the mean molecular weight per monomeric unit is not identical, the absolute values are not to be compared. For PAA, the expansion of the polyelectrolyte chains occurs as soon as the ionization increases, contrary to what is observed for PAA.Phe$_n$. From a comparison among these curves there is evidence that such an expansion is delayed for PAA.Phe$_n$, this trend being reinforced from PAA.Phe$_1$ to PAA.Phe$_4$. Furthermore, the viscosity is drastically reduced at low ionization, as compared with PAA. This presumably also reflects the hypercompact state of the grafted polyelectrolytes.

Dye-Polyion Interactions

We now wish to report some complementary results dealing with the binding of dyes to PAA.Phe$_n$ polyions. As mentioned above, among the factors playing an efficient role in the interaction process dye-polyion, it appears that the overall conformational state and/or microscopic domains of the polyelectrolyte chains are of major importance. In this respect, examination of some results of dialysis experiments may provide us with meaningful observations, in connection with the above given potentiometric and viscometric data.

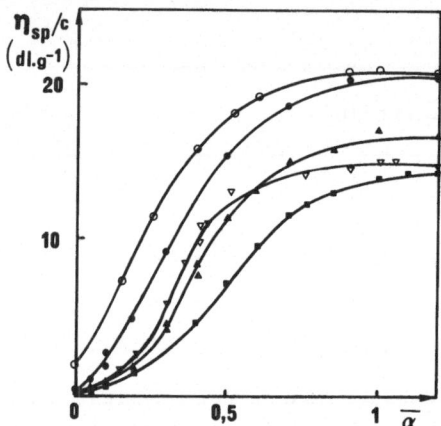

Figure 2.   Viscometric titrations of PAA.Phe$_n$.   Experimental conditions: same as in Figure 1;   o o PAA; $\bullet$ $\bullet$ PAA.Phe$_1$; ∇ ∇ PAA.Phe$_2$; $\blacktriangle$ $\blacktriangle$ PAA.Phe$_3$; $\blacksquare$ $\blacksquare$ PAA.Phe$_4$.

We previously quoted that a PAA.Phe$_1$ sample of low grafting value (X = 0.065) must be considered as normal and found similar to PAA potentiometry being taken into account. However, from a comparative study of the binding ability under similar essay conditions, a marked distinction between these two polyelectrolytes is evidenced.

The fraction of bound dye q (q = $C_{EB,b}/C_{EB,t}$ where $C_{EB,b}$ and $C_{EB,t}$ are the bound and total dye concentration, respectively, for the cationic dye Ethidium Bromide (EB) was first investigated throughout the neutralization range ($\bar{\alpha}$, the neutralization degree is varying from 0 to 1) at low ionic strength (Figure 3). Despite some scattering of the points, due to the rather low reproducibility of equilibrium dialysis experiments, these curves exhibit the following features:
- at low ionization, the extent of binding is much higher for PAA.Phe$_1$ than for PAA
- with increasing ionization, the binding increases slowly for PAA and decreases for PAA.Phe$_1$; the fraction of bound dye tends to a constant value of the same magnitude for the two polyelectrolytes.

Obviously, these differences in the extent of binding cannot be attributed to peculiar charge effects but rather to hydrophobic interactions. As in addition PAA.Phe$_1$, at low ionization, does not appear as hypercoiled, we tentatively explain these data as follows: for the unneutralized PAA.Phe$_1$, the presence of aromatic rings in the side chains (due to phenylalanine residues) may induce the formation of some hydrophobic intramolecular regions. These domains act as a local organic microphase and therefore must provide a more favorable environment for the binding of EB. Furthermore, in our opinion, the presence of relatively aggregated aromatic rings within these sites should lead to a strong binding process, as observed for EB intercalated between base pairs of DNA. Increasing neutralization of PAA.Phe$_1$ involves a progressive disorganization of these domains and for higher $\bar{\alpha}$ values the polymers chains should assume whole open conformations, as also observed for PAA. Then the binding of the dye becomes a mainly electrostatically governed process and remains therefore identical for the two polyelectrolytes.

Inspection of the results given in Figure 4 indicates also that the observed differences should be attributed to hydrophobic interactions. For both polymers the extent of binding diminishes with increasing ionic strength, but at low ionization the binding of EB to PAA.Phe$_1$ is less sensitive to such changes: indeed, the fraction of bound EB remains strongly enchanced in the presence of PAA.Phe$_1$ and the binding process may be identified as a "non-competitive" one.

Spectral data can provide some information about the environment of the bound dye molecules.

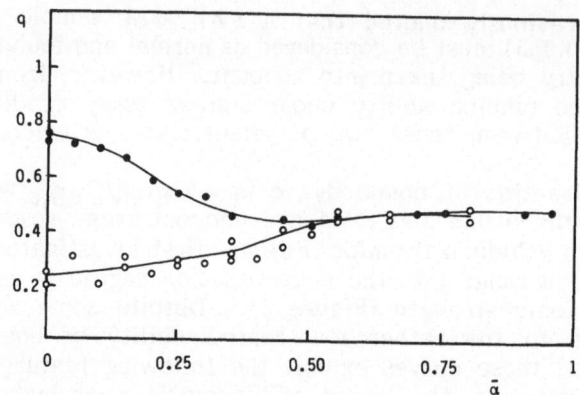

Figure 3.   Fraction of bound EB (q) dependence on $\bar{\alpha}$. C_{EB}:5.10^{-5}M;
Cp:2.10^{-3}M (P/D = 40); supporting salt: KCl 2.10^{-3}M.
o o PAA;   ● ● PAA.Phe$_1$.

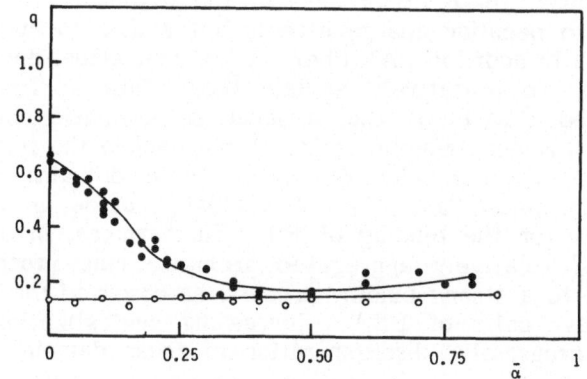

Figure 4.   Fraction of bound EB dependence on $\bar{\alpha}$. Same symbols and
conditions as in Figure 3 but KCl: 2.10^{-2}M.

    It is known that Ethidium Bromide shows a tendency to dimerize in
aqueous solution. In connection with the dimerization process a shift of
the absorption spectra is observed (maximum absorption of EB monomeric form: 480 nm, maximum absorption of EB dimeric form: 520 nm).
This red shift is also observed for the spectrum of EB intercalated within
DNA[29] and also when this dye is bound to some synthetic polyelectrolytes.[1,18] Such a feature can be explained by a stacking of EB
(dimers or higher aggregates) along the macroions, or alternatively by a
change of environment of the bound monomeric dye.   These data are
conveniently normalized[29] in order to evidence the absorption maximum
shift, as the optical density difference per total dye concentration (see
Figures 5-6).

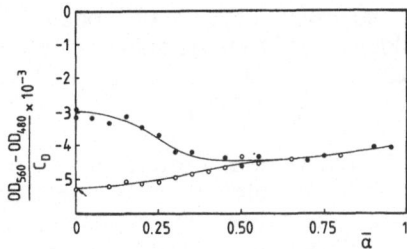

Figure 5. Reduced Optical Density Difference of total EB dependence on $\bar{\alpha}$. Same symbols and conditions as in Figure 3.

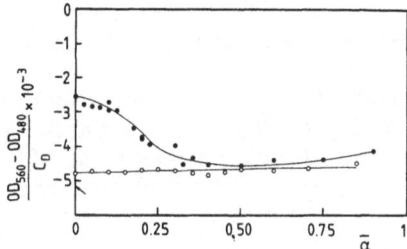

Figure 6. Reduced Optical Density Difference of total EB dependence on $\bar{\alpha}$. Same symbols and conditions as in Figure 4.

It should be noted that these changes recorded over the entire neutralization range are in marked correlation with the binding curves $q = f(\bar{\alpha})$ reported Figures 3-4, for identical experimental conditions.

For sufficiently high polymer-to-dye ratio (P/D), it is generally considered that the bound EB is essentially in monomeric form. Since P/D = 40 in these experiments, and the close agreement between the two sets of results being taken into account, the observed spectral changes must be related to environment effects. In our opinion, they are consistent with the formation of local hydrophobic domains along the macromolecular chains of PAA.Phe$_1$ (for $\bar{\alpha} < 0.5$), as considered above.

An alternative assumption could be considered, not inconsistent with this interpretation. At low ionization, only a few binding sites are available and the flat aromatic rings of EB may preferentially bind to the phenylalanine moieties grafted along PAA.Phe$_1$ chains. This kind of interaction can also explain the spectral displacement as found for EB-polynucleotides systems. Under higher ionization the number of binding sites greatly increases and the dye is likely to be more statistically distributed along the macromolecular chains.

Although these spectral data may appear still inconclusive by themselves, we consider that they provide supplementary evidence in establishing the existence of overall and/or local conformational states of polyelectrolytes such as PAA.Phe_1.

As already mentioned, various factors are to be taken into account to provide us with a coherent picture of a polyelectrolyte-dye system. Among them, we would now place the emphasis on the polymer-to-dye ratio. In the cases previously considered (Figures 3 to 6) a relatively high excess of the polyelectrolyte with respect to the dye was used (P/D = 40). The results of similar equilibrium dialysis experiments (and related spectral data) are now given Figures 7-8, for P/D = 10.

In view of these curves, somewhat surprising features show. First, contrary to the difference established from preceeding data, the behavior of PAA.Phe_1 and PAA become identical in regard to their binding ability for EB. Secondly, the fraction of bound dye remains very high and constant over the entire neutralization range (with the exception of the low range of $\bar{\alpha}$, where a partial precipitation cannot be avoided). Thirdly, the absorption spectrum of Ethidium Bromide is found much more displaced to longer wavelength than when P/D = 40 and this tendency appears to be stronger at low ionization.

In our opinion, these results may be consistently explained by asserting that at low P/D the conditions are such that the binding of EB becomes a cooperative process. Obviously, when P/D decreases, the tendency of EB to bind next to a dye already bound to the polyelectrolyte is enhanced (stacking process).[1,3] This stacking is in turn favorable to further binding: the growth of such aggregates along the macromolecular chains results in micellar-like regions, which are expected to "solubilize" the dye to a greater extent than for a non-hypercoiling polyelectrolyte. To sum up this picture, at low P/D, Ethidium Bromide does not discriminate between the two kinds of polyelectrolytes:
- PAA.Phe_1 which is "intrinsically" hydrophobic (due to hydrophobic moieties covalently grafted)
- PAA which may become "extrinsically" hydrophobic (through hydrophobic moieties electrostatically bound)

Lastly, for the PAA.Phe_4 sample (the overall conformation of which was shown from potentiometric and viscometric data to be the more compact in the PAA.Phe_n series), we now report results of dialysis experiments concerning this polyelectrolyte (Figure 9). For a better comparison, measurements dealing with PAA and PAA.Phe_1 are given again.

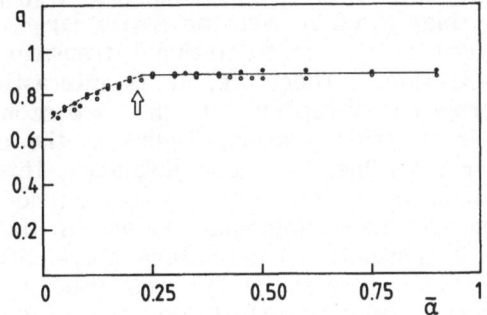

Figure 7.　Fraction of bound EB dependence on $\bar{\alpha}$.　$C_{EB}:5.10^{-3}M$; $C_P:5.10^{-4}M$ (P/D = 10); supporting salt: KCl $2.10^{-3}M$. o PAA; ● ● PAA.Phe$_1$.　(arrow: beginning of partial precipitation)

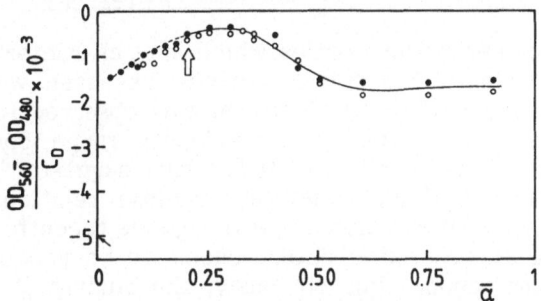

Figure 8.　Reduced Optical Density Difference of total EB dependence on $\bar{\alpha}$.　Same symbols and conditions as in Figure 7.

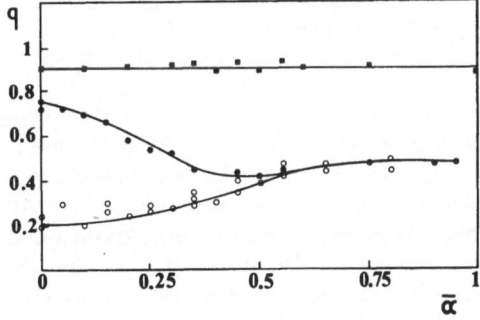

Figure 9.　Fraction of bound EB (q) dependence on $\bar{\alpha}$.　C_{EB}, C_P, counterion, supporting salt concentration: same conditions as in Figure 3.　o o PAA;　● ● PAA.Phe$_1$;　■ ■ PAA.Phe$_4$.

Inspection of these data reveals that the fraction of bound EB remains constantly high ($q > 0.9$) over the entire range of neutralization. However (see Figure 1) the conformational transition is initiated at relatively low ionization.   Therefore, it is interesting to note that despite the progressive disruption of the hypercompact coil (also evidenced from viscometric titration, Figure 2) the binding ability is found constant for PAA.Phe$_4$.   In our judgment, this result must be considered as conclusive evidence of the persistence of hydrophobic microdomains along the macromolecular chains.   In fact, the conformational transition of synthetic hypercoiling polyelectrolytes must be generally described as a process of low cooperativity as compared to highly ordered polymers (such as poly-L-glutamic acid).   Regarding this, for PAA.Phe$_4$ the disappearance of intramolecular micelle-like sites is in all probability not fully achieved even at full neutralization, and the remaining hydrophobic domains can be held responsible for the high binding ability of this polyelectrolyte.

Induced Circular Dichroism of Bound Acridine Orange

Among the spectral properties which may characterize the binding of dyes to biopolymers, the optical activity has been widely used.   In particular, an induced circular dichroism was observed (especially with Acridine Orange (AO)) within the wavelength range corresponding to absorption bands of the bound dye. It has been suggested[31-32] that such an induced optical activity could be in close relationship with the conformational state of biopolymers, although we recently showed[33] that the existence of an ordered structure seems to be unnecessary to give rise to this phenomenon.   But obviously, the binding of dye molecules must occur in an asymmetric environment.   In this connection, as a result of grafted   L-phenylalanine moieties along PAA.Phe$_1$ macromolecular chains, a chiral carbon can be found close to the aromatic ring which is a part of this group.   Such a feature may be a favorable opportunity: if in interacting with PAA.Phe$_1$, AO lies in the vicinity of this aromatic ring, the process can give rise to an induced circular dichroism and will allow us to distinguish this type of binding.

As shown in Figure 10, this effect was actually found.   The observed CD signal is strongly dependent on P/D when this ratio is close to 1.   This relation of the extrinsic Cotton effect to P/D suggests that in this range the mode of binding becomes more operative in the vicinity of the Phe moieties.   Moreover (results not given here) at constant P/D the induced optical activity is also notably modified by the ionization of the polyelectrolyte.   These results will be examined at greater length later but in our opinion these variations can be correlated with the aggregation state of AO on its binding site (dimers or higher aggregates) as previously shown for another synthetic polyelectrolyte.[33]

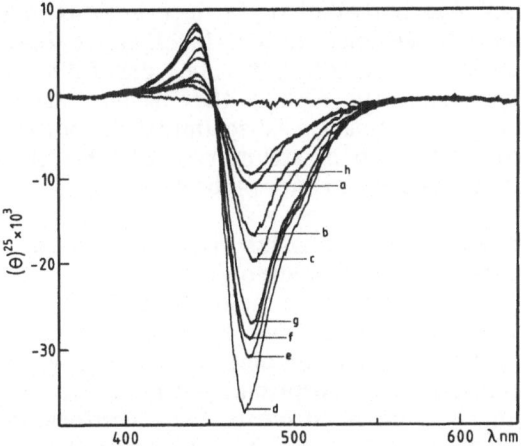

Figure 10.   Induced CD Spectra of AO (C_{AO} = 2.10^{-5}M) in the presence
of unneutralized PAA.Phe_1. $(\theta)^{25}$:   measured ellipticity in
degree. Optical path = 0.2 dm; supporting salt KCl 2.10^{-3}M.
P/D: a = 0.50;   b = 0.66;   c = 0.80;   d = 0.92;   e = 1.05;
f = 1.20;   g = 1.30;   h = 1.65.

CONCLUSION

     In this survey our aim was to provide a reliable picture of some
effects connected with hydrophobic interactions, with our attention
focussed on the peculiar contribution of the aromatic ring, when such a
group is present within the side-chain of a water-soluble polyelectrolyte.
With the aid of the PAA.Phe_n model, several aspects appear to be worth
noting.

     First, as also evidenced from previous studies dealing with
hypercoiling polyelectrolytes carrying alkyl side-chains, the most characteristic effect is the ability of the aromatic group to act in stabilizing
hypercompact conformational states and/or local micellar-type domains.
However, as obvious from potentiometric and viscometric data, the
formation (through an increase of the length of grafted hydrophobic
moieties) and disruption (through an increase of the ionization degree) of
such structures is relevant to processes of relatively low cooperativity.

     Secondly, in the light of dialysis experiments (and related spectral
data) the dye binding process is likely to depend notably on the presence
of the aromatic ring. This may involve this group directly by some
ability to interact preferentially with dye molecules (as also implied
from induced circular dichroism data) and/or indirectly as inducing
micelle-like domains along the macromolecular chains. But it has to be

kept in mind that dye-polyelectrolyte interactions have a somewhat intricate character and various factors (such as ionization state, ionic strength, polymer-to-dye ratio) must be carefully inspected. For example, it appears from our results that when P/D is large enough, Ethidium Bromide may clearly "discriminate" between hydrophobic microdomains and whole open structures of the investigated polyelectrolytes, whereas this is no more observed at low P/D.

Finally, we wish to emphasize that a general picture of the conformational state of a polyelectrolyte is connected to the experimental method used in order to elucidate it. Regarding this, potentiometric and viscometric data may led us to ascertain the overall conformational state of the macromolecular chains, whereas dye binding studies (equilibrium dialysis, absorption spectroscopy, fluorescence, optical activity) are much more sensitive to local conformational states. In our opinion, these latter methods can be conveniently employed in ascertaining hydrophobic microdomains.

ACKNOWLEDGMENT

The support of this research by a grant from the Alexander Von Humboldt-Stiftung is gratefully acknowledged. One of the authors (J.B.) is indebted to Professor E. Selegny for his interest and helpful discussions in the development of these investigations.

REFERENCES

1. See (a) A. Blake, and A. R. Peacocke, Bipolymers, 6, 1225 (1968); (b) V. Crescenzi, and F. Quadrifoglio in Polyelectrolytes and their applications, A. Rembaum and E. Selegny, Eds., Reidel, Dordrecht, Holland, 2, p. 217 (1975); (c) J. Pauluhn, and H. Zimmermann, Ber. Bunsenges. Phys. Chem., 82, 1265 (1978); (d) J. Pauluhn, and H. Zimmermann, Ber. Bunsenges. Phys. Chem., 83, 768 (1979); (e) W. Burkart, and H. Klump, Ber. Bunsenges. Phys. Chem., 85, 371 (1981); (f) E. R. Lochmann, and A. Micheler in Physico-Chemical Properties of Nucleic Acids, J. Duchesne, Ed., Acad. Press, 1, 223 (1973)
2. F. Brun, J. J. Toulme, and C. Helene, Biochemistry, 14, 558 (1975)
3. V. Von Tscharner, and G. Schwarz, Biophys. Struct. Mechanism, 5, 75 (1979)
4. M. Mandel, J. C. Leyte, and M. G. Stadhouder, J. Phys. Chem., 71, 603 (1967)
5. V. Crescenzi, Adv. Polym. Sci., 5, 358 (1968)
6. V. Bottiglione, M. Morcellet, and C. Loucheux, Makromol. Chem., 181, 485 (1980)
7. P. Dubin, and U. P. Strauss, J. Phys. Chem., 74, 2842 (1973)
8. P. Dubin, and U. P. Strauss, ibid., 77, 1427 (1973)

9. P. Dubin, and U. P. Strauss, in Polyelectrolytes and their Applications, A. Rembaum and E. Selegny, Eds., Reidel, Dordrecht, Holland, 2, p. 3 (1975)
10. P. J. Martin, and U. P. Strauss, Biophys. Chem., 11, 397 (1980)
11. P. J. Martin, L. R. Morss, and U. P. Strauss, J. Phys. Chem., 84, 577 (1980)
12. See (a) C. Villiers, and C. Braud, Nouv. J. Chim., 2, (1), 33 (1977); (b) J. C. Fenyo, F. Delben, S. Paoletti, and V. Crescenzi, J. Phys. Chem., 81, 1900 (1977)
13. C. Braud, Europ. Polym. J., 11, 421 (1975)
14. G. Muller in Polyelectrolytes, E. Selegny, Ed., Reidel, Dordrecht, Holland, p. 195 (1974)
15. W. H. Stork, J. A. M. Van Boxsel, A. F. P. M. De Goeij, P. L. De Haseth, and M. Mandel, Biophys. Chem., 2, 127 (1974)
16. B. Erny, and G. Muller, J. Polym. Sci., Polym. Chem. Ed., 17, 4011 (1979)
17. M. C. Vandevelde, and J. C. Fenyo, Europ. Polym. J., 15, 431 (1979)
18. J. C. Fenyo, L. Mognol, F. Delben, S. Paoletti, and V. Crescenzi, J. Polym. Sci., Polym. Chem. Ed., 17, 4069 (1979)
19. G. Muller, and J. C. Fenyo, J. Polym. Sci., Polym. Chem. Ed., 16, 327 (1978)
20. J. C. Fenyo, C. Braud, J. Beaumais, and G. Muller, J. Polym. Sci., Polym. Lett. Ed., 13, 669 (1975)
21. G. Muller, and J. C. Fenyo, J. Polym. Sci., Polym. Chem. Ed., 16, 77 (1978)
22. J. C. Fenyo, J. Beaumais, and E. Selegny, J. Polym. Sci., Polym. Chem. Ed., 12, 2659 (1974)
23. G. Muller, J. C. Fenyo, J. Beaumais, and E. Selegny, J. Polym. Sci., Polym. Chem. Ed., 12, 2671 (1974)
24. J. Beaumais, J. C. Fenyo, and G. Muller, J. Polym. Sci., Polym. Chem. Ed., 13, 2305 (1975)
25. V. Vitagliano, L. Constantino, and A. Zagari, J. Phys. Chem., 77, 204 (1973)
26. V. Vitagliano in Chemical and Biological Applications of Relaxation Spectrometry, E. Wym-Jones, Ed., Reidel, Dordrecht, Holland, p. 437 (1975)
27. J. Beaumais, and Th. Ackermann, to be published
28. M. Nagasawa, and A. Holtzer, J. Am. Chem. Soc., 86, 538 (1964)
29. M. J. Waring, J. Mol. Biol., 13, 269 (1965)

30. V. Crescenzi, and F. Quadrifoglio, Europ. Polym. J., 10, 329 (1974)
31. E. R. Blout, and L. Stryer, Proc. Nat. Acad. Sci. US, 45, 1591 (1959)
32. L. Stryer, and E. R. Blout, J. Am. Chem. Soc., 83, 1411 (1961)
33. J. Beaumais, J. C. Fenyo, and G. Muller, J. Polym. Sci., Polym. Chem. Ed., 18, 1367 (1980)

(ACID-BASE)-DEPENDENT GLOBULAR STRUCTURES OF PARTIALLY

N-ALKYLATED POLY(TERTIARY AMINES)

Jovanka Huguet and Michel Vert
Laboratoire des Substances Macromoleculaires
ERA CNRS 471
INSCIR, BP 8, 76130 Mont-Saint-Aignan,
FRANCE

SUMMARY

A new type of bifunctional polybase is presented which is obtained by partial N-alkylation of poly(tertiary amines) with pendant amino groups. The ability of partially N-methylated poly[thio-1-(N-N-diethyl-aminomethyl)ethylenes], Q-P(TDAE)X, to form hydrophobic micro-domains in aqueous medium is discussed in regard to the amounts of N-alkylated repeating units and to acid-base reactions at residual tertiary amine sites. Emphasis is on deprotonated compounds with low contents of quaternary ammonium groups, as they can take on polysoap-like globular structures, thus explaining the difference between N-alkylation and N-protonation to form water soluble poly(tertiary amine) macro-molecules. The acid-base dependency of these globular structures is discussed and it is shown that cooperative chemical and conformational transitions can be observed for some compounds. The effect of the nature of polymer backbones is investigated by comparing two polymers with the same degree of N-methylation but different polymer backbones, of the poly(thioether) and the poly(olefin) types, respectively. Potentio-metric, viscometric and polarimetric titrations, and solubilization of progesterone, (a lipophilic water-insoluble compound), are used to char-acterize the physico-chemical states of these bifunctional polybases in water.

INTRODUCTION

Most of the polybases of the polytertiary amine type with pendant amino groups are water-insoluble at neutral pH. However, it is well known that water-solubility can be obtained by introducing electric charges on the polymer chains either by adding strong acids that cause reversible N-protonation of tertiary amine groups or by partial or total N-alkylation that generate stable quaternary ammonium groups.[1]

water-insoluble   water-soluble

Surprisingly enough, we have found in several cases where tertiary amine groups are located in side-chains, that a rather high proportion of N-protonated repeating units is necessary to effectively solubilize all the macromolecules while a very small proportion of N-alkylated groups is enough to retain even deprotonated macromolecules in solution.

A typical example has been recently reported for poly[thio-1-(N-N-diethyl) aminomethyl ethylene] (TDAE), a polybase of the polytertiary amine type with a polythioether backbone and tertiary amine pendant groups.

P(TDAE)

For P(TDAE) itself, the presence of ca. 90% of protonated units is necessary to maintain all the macromolecules in salt-free water.[2]

In contrast, for partially N-methylated poly[thio-1-(N-N-diethyl amino methyl) ethylene], Q-P(TDAE)X, with X = percent of methylated units, a very few amounts of charged units of the tetralkyl ammonium type, ca. 4%, has been found sufficient to solubilize all the macro-molecules in aqueous solution, regardless of the degree of protonation of residual tertiary amine repeating units.[2]

In order to account for the difference of efficiency between N-protonation and N-methylation to solubilize P(TDAE) polymer chains, investigations have been undertaken of the solution properties of different partially quaternized P(TDAE) with various proportions of tetralkyl ammonium and tertiary amine repeating units.

At the moment, our understanding of this difference is based on the evidence that for partially alkylated compounds with low content of quaternary ammonium groups, polymers do precipitate in water as non-alkylated poly(tertiary amines), but the precipitated phase remains dispersed in water because of the electric charges due to permanently ionized quaternary ammonium groups. Macromolecules take on globular structures which can be monomolecular, or plurimolecular because of aggregation, depending on the degree of N-alkylation. The inner part of these globules forms polysoap-like hydrophobic microdomains where the deprotonated tertiary amine repeating units cling together in a water-free medium.[2]

In so far as microdomains in water are concerned, partially quaternized polytertiary amines with pendant tertiary amine groups possess a property of special interest. The deprotonated microdomains can be readily destabilized by adding strong acids, which cause protonation of tertiary amine residues and thus solubilize the polymer chains conventionally. The destabilization occurs at a constant pH through cooperative chemical and conformational transitions of the all-or-none type for a large range of the degree of protonation.[2]

The destabilization is absolutely reversible and the microdomains reform immediately by adding bases which cause deprotonation of the formerly protonated tertiary amine residues. Such behavior is of special interest if temporary trapping of hydrophobic water-insoluble species by solubilization in the microdomains is desired. Indeed, the trapped species should be releasable instantaneously and at constant pH by adding at most a stoichiometric amount of strong acid.

In this contribution we first describe previous work dealing with the X-dependency of the physico-chemical properties of Q-P(TDAE)X compounds. Special emphasis will be given to the acid-base reaction-dependency of the dispersed microphase formed by the compounds with low X values. In a second part, we report recent work undertaken to evaluate the influence of the polymer backbone on acid-base reactions of partially quaternized polytertiary amines. Comparison will be made between a monomolecular globule-forming Q-P(TDAE)X and partially quaternized poly[(N-sec-butyl N-methyl aminomethyl)ethylene] Q-P(BM-AE)X, a compound with the same degree of methylation but with a backbone of the poly(olefin) type.

$$\left[\begin{array}{c} CH-CH_2 \\ | \\ CH_2 \\ | \\ CH_3-N \\ | \\ C_2H_5 \end{array} \quad\quad \begin{array}{c} CH-CH_2 \\ | \\ CH_2 \\ | \\ CH_3-N^+-CH_3,X^- \\ | \\ C_2H_5 \end{array}\right]$$

Q-P(BMAE)X

In a third part, we compare the ability of both compounds to solubilize progesterone (a hydrophobic steroid hormone virtually insoluble in water) and to release the trapped progesterone suddenly and at constant pH.   Possible biomedical uses of such systems will be commented.

X- AND PROTONATION DEPENDENCIES OF Q-P(TDAE)X POLYELEC-TROLYTE PROPERTIES

Free tertiary amine groups and positively charged groups are present in Q-P(TDAE)'s.   The behavior of positively charged groups depends on the counterion.   When neutralized by anions derived from strong acids like HCl, HBr, etc., quaternary ammonium groups behave as neutral salts, though they are strongly basic when counterions are OH^- ions.   In the latter case, which is obtained after percolating protonated copolymer solutions through an anion-exchange column (OH^- form), the neutralization by strong acids (HX) can be represented in two steps as follows:

Step 1:   Ion-exchange on quaternary ammonium groups

$$\left(\begin{array}{c} -S-CH-CH_2-S-CH-CH_2- \\ | \quad\quad\quad | \\ CH_2 \quad\quad CH_2 \\ | \quad\quad\quad | \\ N^+CH_3,OH^- \quad N \\ / \quad \backslash \quad\quad / \quad \backslash \\ CH_2 \; CH_2 \; CH_2 \; CH_2 \\ | \quad | \quad | \quad | \\ CH_3 \; CH_3 \; CH_3 \; CH_3 \end{array}\right) + H^+X^- \rightleftharpoons \left(\begin{array}{c} -S-CH-CH_2-S-CH-CH_2- \\ | \quad\quad\quad | \\ CH_2 \quad\quad CH_2 \\ | \quad\quad\quad | \\ N^+CH_3,X^- \quad N \\ / \quad \backslash \quad\quad / \quad \backslash \\ CH_2 \; CH_2 \; CH_2 \; CH_2 \\ | \quad | \quad | \quad | \\ CH_3 \; CH_3 \; CH_3 \; CH_3 \end{array}\right)$$

Step 2: Protonation of tertiary amine groups

$$\left[\begin{array}{c}\text{S-CH-CH}_2\text{—S—CH-CH}_2\\ \qquad\text{CH}_2\qquad\qquad\text{CH}_2\\ \text{N}^+\text{CH}_3,\text{X}^-\quad\text{N}\\ \text{CH}_2\ \text{CH}_2\quad\text{CH}_2\ \text{CH}_2\\ \text{CH}_3\ \text{CH}_3\quad\text{CH}_3\ \text{CH}_3\end{array}\right] + \text{H}^+\text{X}^- \underset{K_b}{\overset{K_a}{\rightleftharpoons}} \left[\begin{array}{c}\text{S-CH-CH}_2\text{—S-CH-CH}_2\\ \qquad\text{CH}_2\qquad\qquad\text{CH}_2\\ \text{N}^+\text{CH}_3,\text{X}^-\quad\text{N}^+\text{H},\text{X}^-\\ \text{CH}_2\ \text{CH}_2\quad\text{CH}_2\ \text{CH}_2\\ \text{CH}_3\ \text{CH}_3\quad\text{CH}_3\ \text{CH}_3\end{array}\right]$$

where K_a is the dissociation constant of protonated amino-groups and K_b is the protonation constant of the tertiary amine ones.

During step 1 (ion-exchange), the number of ionized repeating units defined by the degree of quaternization ζ, remained constant. In contrast, the protonation of tertiary amine groups (step 2) increased the charge density of macromolecules. Given the presence of two types of basic units, the following parameters were considered to describe the action of HX on the polybasic form of partially quaternized P(TDAE).

(i) ε is the degree of neutralization of the hydroxylated quaternary ammonium groups.

(ii) β is the degree of protonation of tertiary amine groups. These basic groups being of medium strength, β was considered as equal to the normality of the strong acid added beyond the neutralization of strong basic groups divided by the normality of tertiary amine repeating units ($\bar{\beta}$).

(iii) δ is the degree of neutralization of all the basic repeating units in the dibasic form of Q-P(TDAE)'s $\delta = [\text{HCl}]/T_n$.

(iv) γ is the mole fraction of the ionized monomeric units $\gamma = \tau + \beta (1 - \tau)$ which can be assimilated to a degree of ionization if all the positive charges are assumed equivalent.

In a first approach, we have studied the polyelectrolyte properties of a series of Q-P(TDAE)X, with X = 7, 17, 22, 35 and 48. The conformational and the ionization behaviors of these compounds were investigated by potentiometry, viscometry, laser light scattering and by Optical Rotatory Dispersion (ORD) and Circular Dichroism (CD) for some optically active derivatives.[2-4]

Q-P(TDAE)48 was found to behave as expected for a typical hydrophilic polyelectrolytic macromolecule bearing both strongly basic and weakly basic groups. Q-P(TDAE)35 behaved conventionally too,

however, potentiometric and viscometric titration curves suggested that
the lower degree of N-alkylation, and thus the relatively larger
hydrophobic character, allow the macromolecules to take on compact-
coil conformations when largely deprotonated, in a manner quite similar
to the now well-known amphiphilic polyacids.[5-6]   In contrast, unusual
characteristics, in regard to classical monofunctional polyelectrolytes,
were found for copolymers with lower X. Indeed, Q-P(TDAE)22 and Q-
P(TDAE)17 gave potentiometric titration curves with high buffering
regions similar to that observed for precipitating non-alkylated
P(TDAE).[7]

In order to account for the presence of such high buffering regions
in spite of the apparent water-solubility, one may suggest the presence
of a very dispersed microphase in which deprotonated tertiary amine
residues would cling together in a water-free environment, i.e. in a
state similar to that found in the macroscopic precipitate of non-
methylated P(TDAE).

In agreement with the very low viscosity of solutions of deprotonated Q-P(TDAE)22 and 17, the microphase was supposed to be formed
of very small globules stabilized in a polysoap-like structure by the
quaternary ammonium electric charges arranged at the surface. Furthermore, a non-conventional protonation-deprotonation mechanism has been
proposed for the tertiary amine residues buried in the inner part of the
globules. This mechanism is based on the macroscopic phase equilibrium
which occurs in the case of non-quaternized precipitating poly(tertiary
amine) in the high buffering zone of the potentiometric titration curves.
The dispersed macromolecules are distributed in two populations at
equilibrium, respectively composed of solvated and highly protonated
macromolecules and of "micro-precipitated" deprotonated macromolecules, forming a two-phase system in so far as tertiary amine units are
concerned (scheme I).

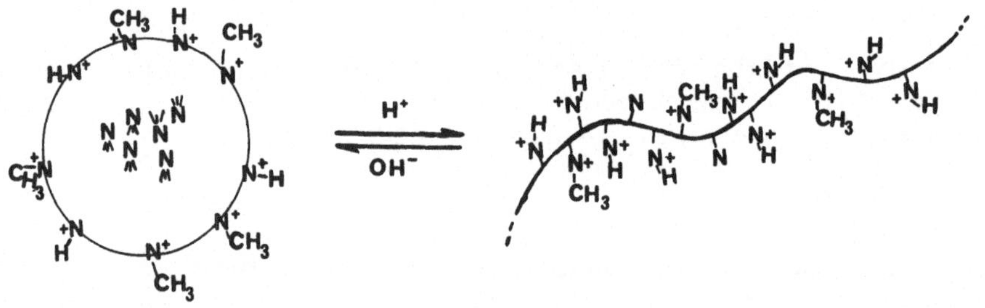

Scheme I

Scheme I.   Schematic representation of macromolecules in the water-
             free globular conformation and of those in the highly
             protonated extended state at equilibrium in the two-phase
             system.

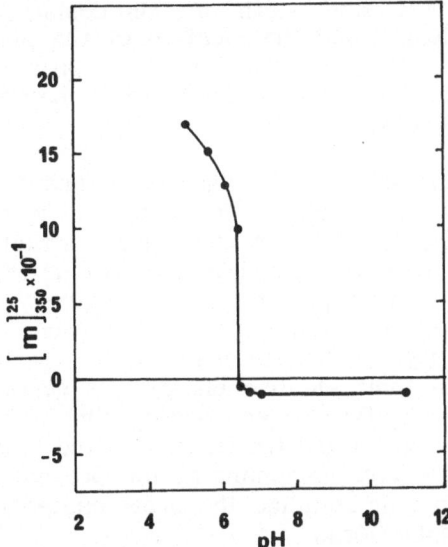

Figure 1.   Variation of mean residue optical rotatory power [m] with pH for optically active partially quaternized poly[N-N-diethyl aminomethyl thiirane] with 17% N-methylated units in salt-free water ($T'_N = 1 \times 10^{-2}$ N; titrating reagent 1 M HCl).

The respective proportions of both populations depend linearly on the amount of acid added in regard to the content in tertiary amine residues, i.e., on the theoretical averaged degree of protonation, , of all the macromolecules present in the two phases. Such a mechanism, which is equivalent to simultaneous chemical and conformational cooperative transitions,[2] well accounts for the high buffering region for water-soluble macromolecules. It also agrees with the linear variation of the reduced viscosity observed in viscometric titration curves and the sharp variations of optical activity found for an optically active sample of Q-P(TDAE)17*, in particular when acid-induced optical activity changes are plotted vs pH, (figure 1) as usually done for polypeptides and proteins.

For Q-P(TDAE)7, we have found that deprotonated globules are no longer monomolecular. The larger hydrophobic character and the lower content in quaternary ammonium groups allow several macromolecules to aggregate as shown by laser light scattering.[4]

Based on the characteristics shown by the five Q-P(TDAE)X above, a kind of phase-diagram has been established considering the degree of methylation, $\tau = X/100$, and the molar fraction of charged units, $\gamma = \tau + \beta(1 - \tau)$, as the variables (Figure 2). In this phase diagram, domain 3 corresponds to the zone where a plateau is observed in potentiometric titration curves.   The size width of this domain was approximated deduced from the length and the location of the plateau in regard to values. The limits of the other domains of the phase-diagram have been defined approximately, too, so as to include the characteristics found for the various Q-P(TDAE)X.[2-4]

The equilibrium between globular and extended macromolecules is obviously controlled by electrostatic repulsions and entropic effects involving hydrophobic interactions and solvent structuration. However, the main factor is the ability of hydrophilic protonated tertiary amine to deprotonate easily to hydrophobic tertiary amine. Therefore, the difference between N-protonation and N-methylation in their ability to impart water-solubility to P(TDAE) chains is because the non-methylated chains can expel all their electric charges by deprotonation and thus macroscopically precipitate when hydrophobic interactions become greater than electrostatic repulsive forces. In the case of N-methylated molecules, stable tetralkyl ammonium groups prevent the loss of all the electric charges and thus stabilize the deprotonated macromolecules in their dispersed globular form.

The characteristics (pH and length of the plateau in potentiometric titration curves) have already been shown to depend on factors that usually affect both the globules and the extended macromolecules (polymer and salt concentrations, presence of organic solvent, nature of the alkyl substituent of nitrogen atoms).[2-4]   Another factor which might be important is the nature of the polymer backbone.

EFFECT OF POLYMER BACKBONE: COMPARISON BETWEEN Q-P(TDAE)19 AND   Q-P(BMAE)19.

Synthesis

Q-P(TDAE)19 was obtained from P(TDAE) using the same procedure as for the other Q-P(TDAE)X.[2] N-methylation of P(TDAE) (MW = 1×10^6, 50% protonated) was carried out in 40-60 V/V methanol/benzene using methyl sulfate (Me_2SO_4/amine = 1) as methylating reagent. Q-P(TDAE)19 (MW - 12×10^4, H^+ form, by laser light scattering in 0.1 N KCl) was recovered after 24 hrs. reaction.   The MW decrease is due to chain degradation occuring at sulfur sites during N-methylation.[2]

Q-P(BMAE)19 was obtained according to the following route, based on catalytic hydrogenation of poly(N-sec-butyl-N-methyl acrylamide)[7] and partial methylation of the resulting polytertiary amine using ICH_3 in benzene.

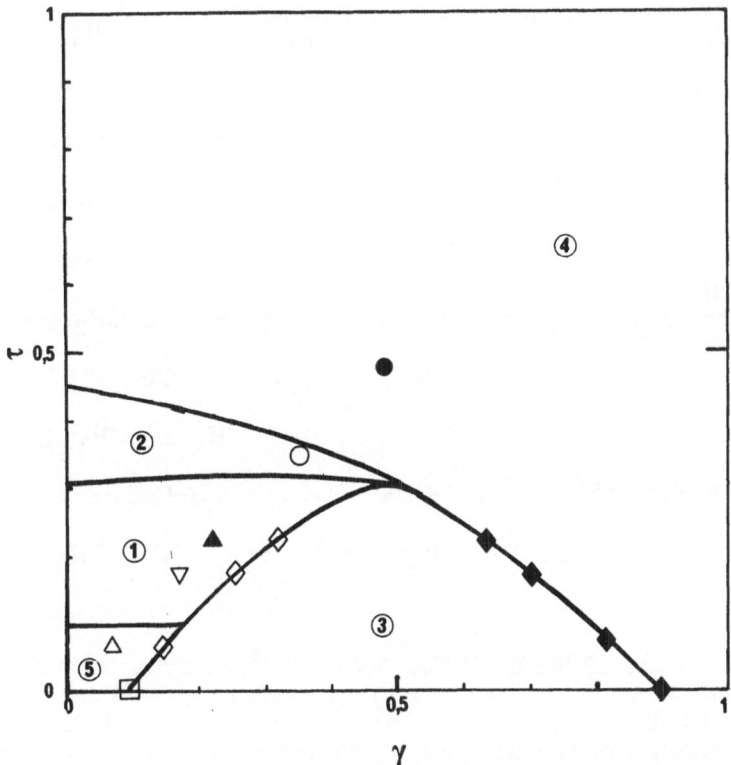

Figure 2.   Polyelectrolyte behavior of Q-P(TDAE) with regard to the
            degree of quaternization τ and to the mole fraction of
            charged units γ, as defined from γ_1 and γ_2 deduced from
            extremities of the plateau of pH observed in potentio-
            metric titration curves.

            ① water-free globules, ② compact coils, ③ two-phase
            system, ④ conventional polyelectrolyte, ⑤ aggregates

            (experimental data: □ P(TDAE), △ Q-P(TDAE)7,
             ▽ Q-P(TDAE)17, ▲ Q-P(TDAE)22, ○ Q-P(TDAE)35,
             ● Q-P(TDAE)48, γ_1 ◇ , γ_2 ◆)

$$\left[\begin{array}{c}CH\text{-}CH_2\\ |\\ C=O\\ |\\ CH_3\text{-}N\\ |\\ H\text{-}C\text{-}CH_3\\ |\\ CH_2\text{-}CH_3\end{array}\right] \quad \xrightarrow[\substack{\text{2 days in}\\ \text{anhydrous}\\ \text{dioxane}}]{AlLiH_4} \quad \left[\begin{array}{c}CH\text{-}CH_2\\ |\\ CH_2\\ |\\ CH_3\text{-}N\\ |\\ H\text{-}C\text{-}CH_3\\ |\\ CH_2\text{-}CH_3\end{array}\right]$$

P(BMAE)

$$\xrightarrow[\text{benzene}]{ICH_3} \quad \begin{array}{c}\text{—}CH\text{-}CH_2\text{—} \cdots \text{—}CH\text{-}CH_2\text{—}\\ |\qquad\qquad\qquad\qquad |\\ CH_2\qquad\qquad\qquad CH_2\\ |\qquad\qquad\qquad\qquad |\\ CH_3\text{-}N\qquad\quad CH_3\text{-}N^+\text{-}CH_3,\ X^-\\ |\qquad\qquad\qquad\qquad |\\ H\text{-}C\text{-}CH_3\qquad\quad H\text{-}C\text{-}CH_3\\ |\qquad\qquad\qquad\qquad |\\ CH_2\text{-}CH_3\qquad\quad CH_2\text{-}CH_3\end{array}$$

Q-P(BMAE)

Main chains of P(BMAE) ($M_w = 15. \times 10^4$) carrying no sulfur atom, N-methylation occured without significant chain degradation. The resulting Q-P(BMAE)19 ($M_w = 15. \times 10^4$, H^+ form in 0.1N KCl) is water-soluble regardless of the pH, contrary to its precursor P(BMAE) which is soluble in acidic medium only.

POTENTIOMETRIC AND VISCOMETRIC TITRATIONS OF Q-P(TDAE)19 AND Q-P(BMAE)19.

Figure 3 shows potentiometric titration curves of Q-P(TDAE)19 and Q-P(BMAE)19 at the same concentration in salt-free water. Both curves show the same general characteristics, particularly a buffering-region in the $0.1 < \beta < 0.5$ range. However, the pH, which remains almost constant for Q-P(TDAE)19 (Δ pH $\simeq$ 0.02 units), varies between 6.2 and 6.0 in the same range for Q-P(BMAE)19. A difference in the behavior of the two compounds is also observed when comparing viscometric titration curves (figure 4). While the viscosity of Q-P(TDAE)19 in the ion-exchange zone ($0 < \varepsilon < 1$) is low ($\eta_{sp}/c \simeq 0.2$ 100 cm^3.g^{-1}) and remained almost constant (as observed for Q-P(TDAE)17) Q-P(BMAE)19 shows a higher initial viscosity ($\eta_{sp}/c \simeq 0.4$ 100 cm^3.g^{-1}) for similar molecular weights. Furthermore, η_{sp}/c, rises immediately as ε increases from 0. Therefore, the compact conformation taken on by deprotonated Q-P(BMAE)19 macromolecules seems to be different and less hydrophobic than the water-free globular structure of Q-P(TDAE)19.

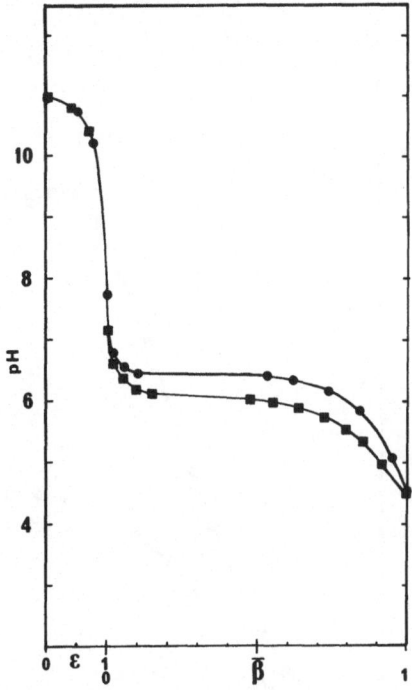

Figure 3.   Potentiometric titration curves of partially quaternized
poly[N-N-diethyl aminomethyl thiirane]   and poly[N-sec—
butyl N-methyl aminomethyl ethylene] with 19% N-methylated units in salt-free water ($T'_N = 1 \times 10^{-2}$ N; titrating
reagent 1 M HCl).
● Q–P(TDAE)19
■ Q–P(BMAE)19

In other words and from a qualitative point of view, Q–P(BMAE)19
behaves much more like Q–P(TDAE)X with X values corresponding to
domain 2 of Figure 2 than like globule-forming Q–P(TDAE)17 or 19
(domain 1 of Figure 2).

ABILITY OF Q–P(TDAE)19   AND Q–P(BMAE)19   TO SOLUBILIZE
PROGESTERONE.

Solutions of Q–P(TDAE)19 ($T_N = 2.5 \times 10^{-2}$ N) and of Q–P(BMAE)19
($T_N = 3.4 \times 10^{-2}$ N) were adjusted to different pHs corresponding to
various states of protonation in the range : $0 < \bar{\beta} < 1$.   Powdered
progesterone (3 mg) was added to 5 ml aliquots of the resulting solutions
and the mixtures were sonicated for 5 min.

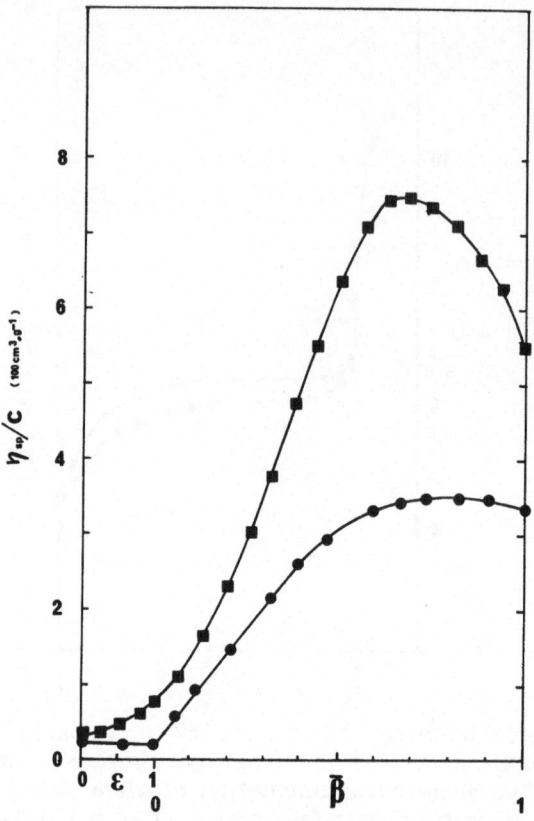

Figure 4.   Viscometric titration curves of Q-P(TDAE)19 and Q-
            P(BMAE)19 in salt-free water (T_N = 1 x 10^{-2} N; titrating
            reagent 1 M HCl).
            ● Q-P(TDAE)19
            ■ Q-P(BMAE)19

        After filtration, the filtrate (1 ml) was diluted to 5 ml by the
addition of pure ethanol. CD of the diluted solutions was measured and
compared with the optical activity of standard solutions of progesterone
in the same solvent medium.   The amount of progesterone dissolved in
the filtrate thus obtained has been plotted vs $\bar{\beta}$ for both compounds
(Figure 5).   The larger amount of progesterone initially dissolved in the
presence of deprotonated Q-P(TDAE)19 is in good agreement with the
greater hydrophobic character of the microdomains formed by this

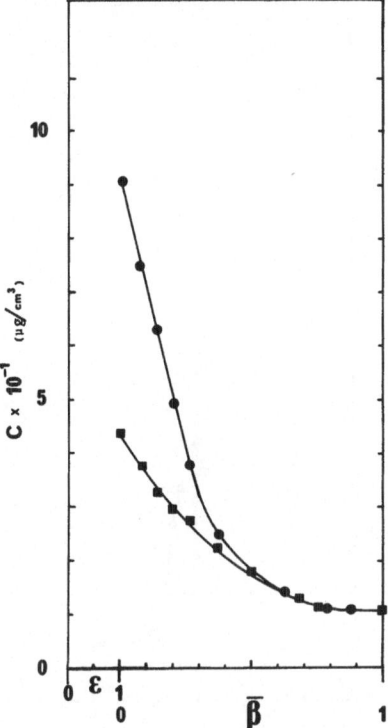

Figure 5.   Variation of the amount of solubilized progesterone in regard to   for Q-P(TDAE)19 and Q-P(BMAE)19 in salt-free water.

● Q-P(TDAE)19 (T_N = 2.5 x 10^{-2} N)
■ Q-P(BMAE)19 (T_N = 3.4 x 10^{-2} N)

compound in regard to Q-P(BMAE)19, as deduced from potentiometric and viscometric titration curves. The range corresponding to the ion-exchange at quaternary ammonium sites ($0 < \varepsilon < 1$) was not explored because of the too high pH values. In the $0 < \bar{\beta} < 0.5$ range corresponding to protonation at tertiary amine sites, the amount of dissolved lipophilic compound decreased drastically for both compounds. The same data plotted in regard to pH (Figure 6) show the sharpness of the acid-induced transition which occured at pH $\simeq$ 7 for Q-P(TDAE)19 under the selected conditions. Comparatively, the drop observed for Q-P(BMAE)19 has a smaller magnitude and is less sharp, in agreement with the less

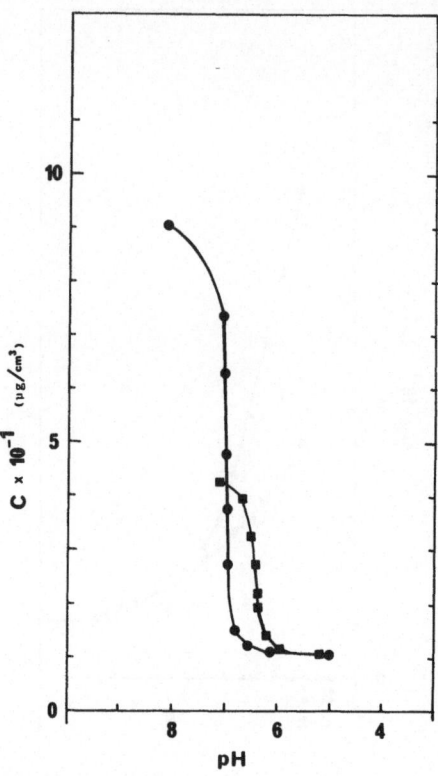

Figure 6.   Variation of the amount of solubilized progesterone in
             regard to pH for Q-P(TDAE)19 and Q-P(BMAE)19 in salt-
             free water.
             ● Q-P(TDAE)19 (T_N = 2.5 x 10^{-2} N)
             ■ Q-P(BMAE)19 (T_N = 3.4 x 10^{-2} N)

hydrophobic character assigned to Q-P(BMAE)19 on the basis of viscometric titration curves. At low pH values, both systems lead to the same
limit which corresponds to the solubility of progesterone in water, thus
showing that Q-P(TDAE)19 and Q-P(BMAE)19 in the protonated form do
not promote significant progesterone solubilization.

On the other hand, Figure 6 shows that Q-P(TDAE)19 becomes more efficient than Q-P(BMAE)19 to solubilize progesterone in the high buffering region of the potentiometric titration curve. We believe that this finding, together with the linear variation of the concentration of the dissolved progesterone in regard to $\bar{\beta}$ in the $0 < \bar{\beta} < 0.4$ range (Figure 5) can be regarded as agreeing well with the existence of the particular organic microphase formed by deprotonated Q-P(TDAE)19 macromolecules whose concentration has to vary linearly according to the two-phase model (scheme I and fig. 2).

CONCLUSIONS

In a previous paper, we have shown that Q-P(TDAE)X, whose side chains consist in N-N-diethylaminomethyl groups, and partially quaternized poly [thio-1 (N-sec-butyl--N-methyl amino methyl) ethylene], whose side chains are identical to those of Q-P(BMAE)X, behave similarly.[4] Therefore, it was concluded that changes in nitrogen substituents has no drastic effect on the existence of globular microdomains and or on the acid-base-induced cooperative transitions. The degree of methylation is the most important parameter.[4]

The results reported herein show that another important factor is the nature of polymer main chain. Many subfactors can be involved in main chain contribution : structural ones, such as the nature of the atom present in the main chain repeating units ($-(S-CH-CH_2-)_n$ for Q-P(TDAE)19 and $-(CH-CH_2)-$ for Q-P(BMAE)19), the interaction of these atoms with water and aqueous solution components, etc., and polyelectrolyte ones, such as differences in the linear charge parameter for a given ionization state, dimension of the polyion, etc.

Based on these conclusions, a decrease of the degree of methylation of Q-P(BMAE)X should increase its hydrophobic character and, thus, characteristics exhibited for Q-P(TDAE)19 should become observable. Work along these lines is under way together with investigations of the potential of these monomolecular microphases for drug transportation and delivery in body fluids. These applications would take advantage of the sharp acid-induced conformational transitions at constant pH.

ACKNOWLEDGEMENTS

This work has been supported by the French Ministry of Education through its C specific program.

REFERENCES

1. M. Fred Hoover, J. Macromol. Sci. Chem., A4, 1327 (1970).
2. D. Vallin, J. Huguet and M. Vert, Polymer J., 12, 113 (1980).
3. D. Vallin, J.Huguet and M. Vert, "Polymeric Amines and Ammonium Salts," IUPAC, E. J. Goethals, Ed. Pergamon Press Oxford, p. 219, (1980).
4. J. Huguet, D. Vallin and M. Vert, Polymer J., 14, 335 (1982).
5. J. C. Fenyo, J. Beaumais and E. Selegny, J. Polym. Sci., Polym. Chem. Ed., 12, 2659 (1974).
6. M. Mandel and J. C. Leyte, J. Polym. Sci., A, 2, 3771 (1964).
7. J. Huguet, Nouv. J. Chimie, 3, 293 (1979).
8. F. Danusso and P. Ferruti, Chimica e Industria (Milan) 50, 71 (1968).

HYDROPHOBIC REGION OF POLY(STYRENESULFONIC ACID) AS STUDIED BY ELECTRIC DICHROISM MEASUREMENTS

Akihiko Yamagishi
Department of Chemistry
Faculty of Science
Hokkaido University
Sapporo 060, Japan

ABSTRACT

The nature of the hydrophobic region in poly(styrenesulfonic acid) was studied by the electric dichroism measurements when the polymer intereacted with a metal chelate cation. The chelates used were bis(2-(2-pyridylazo)-1-naphthol)cobalt(III), bis(1-(2-pyridylazo)-2-naphthol)cobalt(III) and bis(10-(2-pyridylazo)-9-phenanthrol)cobalt(III). For all the systems investigated, the bound chelates were concluded to have their aromatic ligands nicely faced with the phenyl rings of the polyelectrolyte residues. The results are considered to present the molecular basis for the hydrophobicity of the concerned polymer.

INTRODUCTION

The presence of hydrophobic region in the polystyrene-type polymers has been noted by several investigators.[1] Such hydrophobicity plays an essential role in the binding and activation of a small molecule by the polymers. In spite of a number of thermodynamic and kinetic experiments, however, no distinct molecular model has been presented in interpreting the phenomena. In fact, it is a difficult task to make clear the microenvironments of a polymer especially when the polymer is dissolved in a homogeneous solution.

This communication reports that the electric dichroism measurements give an important clue to understanding the hydrophobic interaction of a polymer with a small molecule. Electric dichroism of a polymer solution is caused by the orientation of a polymer chain in the direction of an intense electric field. If a polymer binds a small chromophore, the observed dichroism allows us to determine the orientation of the transition moment of the chromophore.[2] Based on the results, it is possible to construct the molecular models of the polymer-small molecule complexes in a solution. The resultant models are regarded as presenting a molecular basis for the hydrophobicity of the polymer.

The systems investigated were an aqueous solution of poly-(styrenesulfonic acid)(PSS) and three kinds of Co(III) chelates: bis(2-(2-pyridylazo)-1-naphthol)cobalt(III)(Co(αPAN)$_2^+$), bis(1-(2-pyridylazo)-2naphtholcobalt(III)(Co(βPAN)$_2^+$) and bis(10-(2-pyridylazo)-9-phenanthrol)cobalt (III)(Co(PAPhen)$_2^{2+}$). As shown in Fig. 1, these chelates are all wrapped with the planar aromatic ligands. They were expected to be bound with PSS through both the electrostatic and hydrophobic interactions. The detailed structures of such a polymer-metal chelate complex show how the hydrophobic region of a polymer chain plays its role in binding a small molecule.

EXPERIMENTAL

Co(αPAN)$_2$Cl, Co(βPAN)$_2$Cl and Co(PAPhen)$_2$Cl were synthesized by mixing individual ligands with CoCl$_2$ and oxidizing the Co(II) chelates with air. PSS(m.w. 3×10^5) was used as a potassium salt.

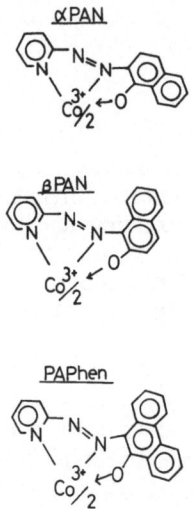

Figure 1. The structures of the Co(III) chelates used in this work.

Electronic spectra were measured with a EPS-3T spectrophotometer at 20°C. Electric field pulse for the measurements of electric dichroism was generated with a high field pulse generator manufactured by Denkenseiki Co., Ltd. (Kanagawa, Japan). A pulse of 4 kvolt and 1 msec was imposed on a cell with two gold electrodes (0.4 cm x 0.50 cm^2). The rise and decay of an electric field occurred within 10 µsec. Another pulse was obtained with a temperature-jump apparatus (Union Giken, Osaka, Japan). The height of a pulse was 13-25 kvolt. The electric field decayed exponetially with the half-life of 20-100 µsec, depending on the conductivity of a solution. The electric dichroism was monitored by a light polarized linearly with a linear polarizer (Nikon Kogaku).

Resolution of $Co(\beta PAN)_2^{2+}$ was performed chromatographically as described elsewhere.[3] The molecular rotation of $(-)_D$-$Co(\beta PAN)_2^{2+}$ was determined to be -35000.

RESULTS

Figs. 2-4 show the electronic spectra of the cobalt chelates in the presence of various amounts of PSS. When the polymer-to-chelate ratio (P/M) was less than 1, the absorbances in the wavelength of 340 - 650 nm decreased with the simultaneous increase above 650 nm. This corresponded to the quantitative addition of a free metal chelate (M^+) on the styrenesulfonic residue (SS^-) of PSS:

$$M^+ + SS^- \rightleftharpoons M^+ \cdot SS^- \quad ; \quad K_1 \qquad (1)$$

K_1 was estimated to be larger than 10^{-5} M. For P/M larger than 1, the spectra exhibited small change with different isosbestic points. In this process, the absorbance above 650 nm decreased. The latter change corresponded to the dilution process of adsorbed chelates along the empty sites of the polymer chain. When P/M exceeded 10, the spectra showed no further change. In these states, the bound chelates were an isolated species interacting with no neighboring chelates. These behaviors were analogous to the binding of an organic dye cation rather than to that of a hydrated metal ion. In the latter case, a metal ion like alkali metal ion is more loosely bound with PSS and the ion is in an association-dissociation equilibrium in the range of P/M = 0 - 10^2.[1]

Electric dichroism was measured for a solution of PSS and the metal chelates at the various P/M ratios. When the square electric field pulse (1 ms, 4 kvolt) was imposed on the solution, the transient change of transmittance was observed as shown in Fig. 5. The amplitude of the signal depended on the angle (θ) between the polarization of a monitoring light and the electric field,[5]

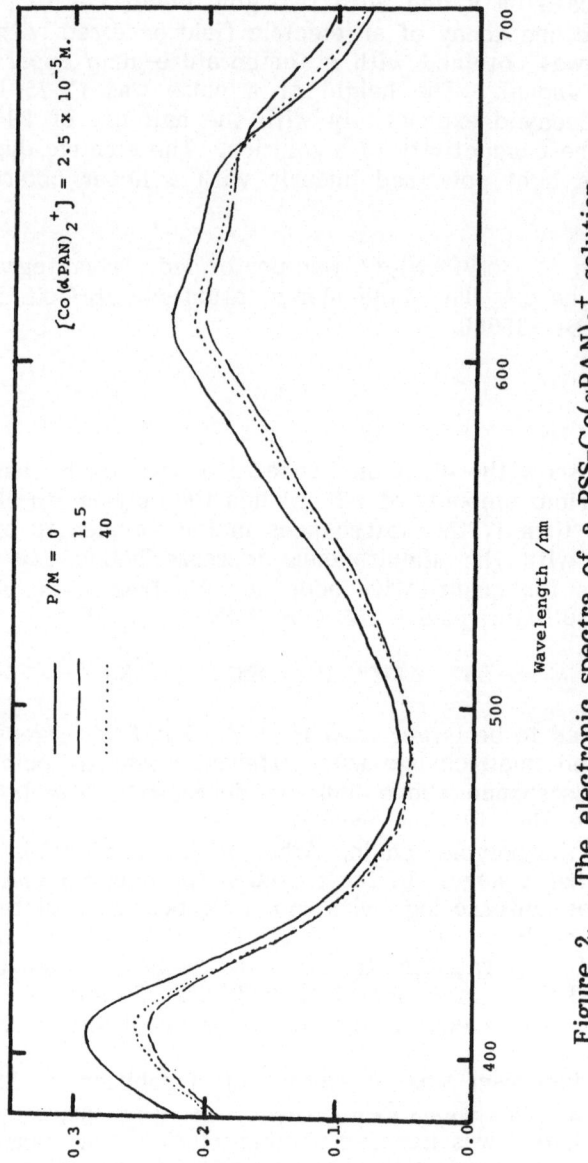

Figure 2. The electronic spectra of a PSS–Co(αPAN)₂⁺ solution.

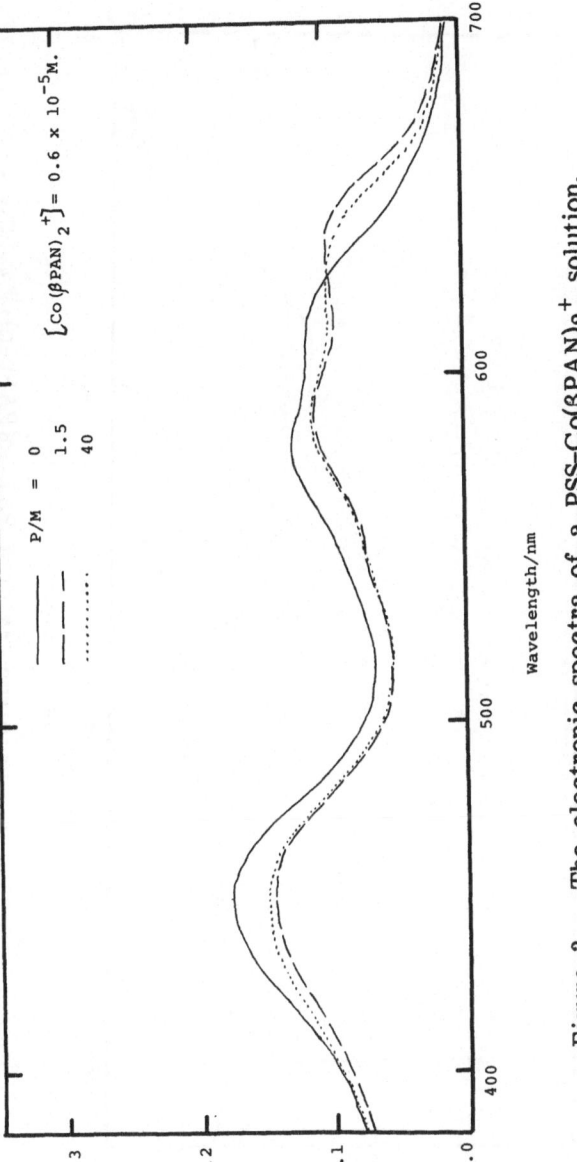

Figure 3.   The electronic spectra of a PSS–Co(βPAN)$_2^+$ solution.

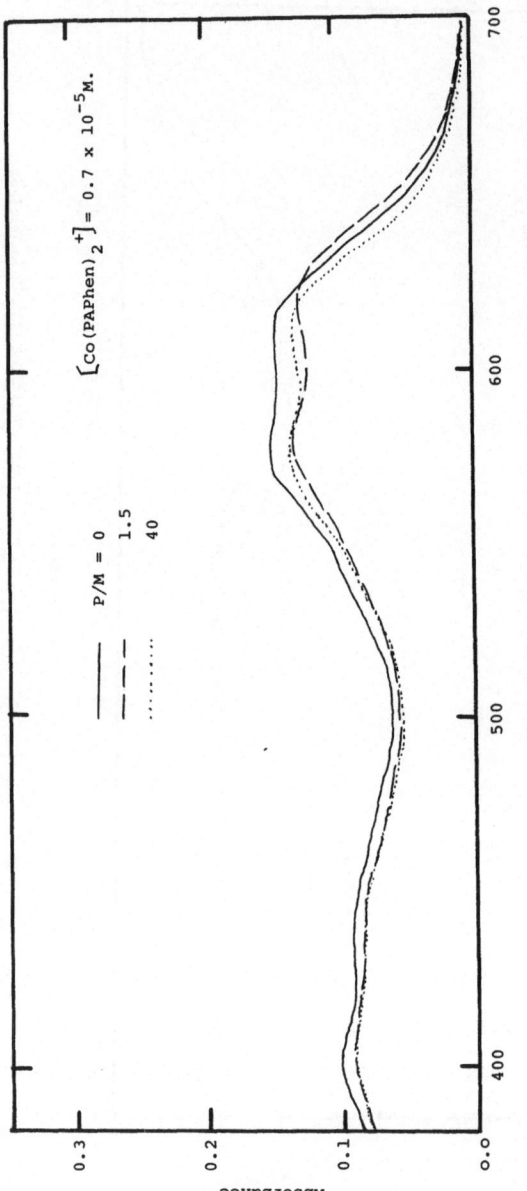

Figure 4.   The electronic spectra of a PSS–Co(PAPhen)$_2^+$ solution.

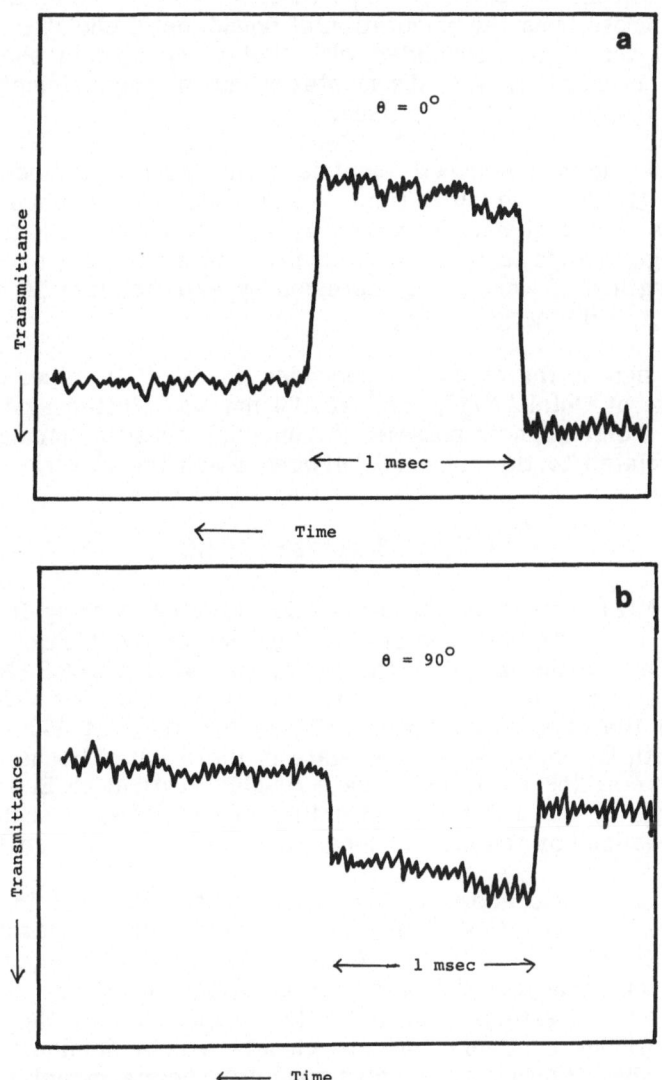

Figure 5. (a) The transient transmittance at 580nm and $\theta = 0^o$, when a square electric field pulse was imposed on a PSS-Co(αPAN)$_2^+$ solution at P/M = 1.5. (b) The same signal except for $\theta = 90^o$.

$$\Delta A/A = (\rho/6)(1 + 3 \cos 2\theta) \tag{1}$$

in which $\Delta A/A$ and ρ are the relative absorbance change and the reduced linear dichroism, respectively. The results confirmed that the observed dichroism arose from the orientation of bound metal chelates. The time-courses of the signals coincided with that of an applied electric field. Thus the orientation and disorientation of a polymer-metal chelate complex completed within 10 μsec.

When the spike-shaped electric field from a temperature-jump apparatus (13-25 kvolt, life time ~ 100 μsec) was applied, the peak amplitudes of the signals increased by less than 10% with the increase of discharge voltage from 13 to 25 kvolt. The amplitudes at the infinite field strength, ΔA^{∞} were safely obtained by extrapolating ΔA against the square of the discharge voltage.

The sign of the dichroism depended on the P/M values drastically. In the case of $Co(\alpha PAN)_2^+$, ΔA^{∞} at 580 nm was plotted against P/M in Fig. 6. If a chromophore posseses the uniaxial transition moment $(\vec{\mu})$, ρ in eq.(1) is related to the angle (θ) between $\vec{\mu}$ and the electric field ($\vec{E}$) as below[5]

$$\rho = (3/4)(1 + 3 \cos 2\phi) \ \Phi(\vec{E}) \tag{2}$$

in which $\Phi(\vec{E})$ represents the orientation function ranging from 0 to 1. According to eq.(2), the inversion of the sign of the dichroism in Fig.6 was a definite indication that the orientation of $\vec{\mu}$ in $Co(\alpha PAN)_2^+$ varied its direction with respect to $\vec{E}$ with the increase of P/M. That is, the chelate in the aggregated form (P/D ~1) had its $\vec{\mu}$ at 580 nm roughly parallel with $\vec{E}$, while the same chelate in the isolated bound state (P/D > 4) had its $\vec{\mu}$ at the same wavelength roughly vertical to $\vec{E}$. As long as $\vec{\mu}$ was fixed in the octahedral structure of $Co(\alpha PAN)_2^+$, such a change in $\vec{\mu}$ was realized by the orientational change of the bound chelate itself.

When KCl was added to the solution at P/D = 6.5 in Fig.6, the amplitude of the negative dichroism decreased until it became positive again at [KCl] = 2×10^{-4} M. Added K^+ ions might compete with Co-$(\alpha PAN)_2^+$ for occupying the free styrenesulfonic residues. Accordingly the observed KCl effects lead us to the conclusion that the number of polymer residues available for one chelate was a dominant factor in determing the structure of a polymer-metal chelate complex.

Figures 7 and 8 show the wavelength dependences of the amplitudes, ΔA^{∞}, at θ = 0 deg for P/M = 1.5 and 40 in the case of $Co(\alpha PAN)_2^+$, respectively. The former and latter curves corresponded to the aggregate and isolated forms of the chelate, respectively. For the aggregate form, both of the amplitudes for the first transition (350 - 450 nm) and the second one (500 - 700 nm) (denoted by ΔA_1 and ΔA_2, respectively) had a positive value. To the contrary of this, for the

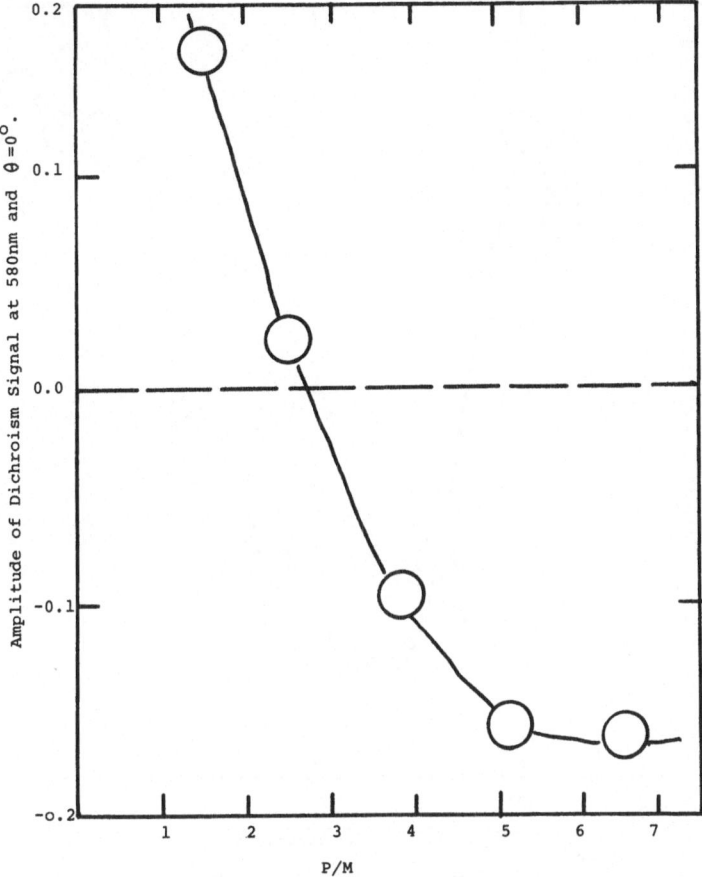

Figure 6. The dependence of the amplitude of the dichroism signal at 580nm, $\theta = 0^\circ$ and $E = \infty$ on P/M.

isolated form, ΔA_1 had a negative small value, while ΔA_2 had a positive large value. Figures 9 and 10 are the similar plots for the system of PSS and $Co(\beta PAN)_2^+$. Analogous tendency was observed for both P/D conditions, although the relative amplitudes of ΔA_1 and ΔA_2 were different from those of $Co(\alpha PAN)_2^+$. The figures also include the wavelength dependences for optically active $(-)_D$-$Co(\beta PAN)_2^+$. It is noteworthy that, in the aggregate form, ΔA_1 was smaller for $(-)_D$-Co $(\beta PAN)_2^+$ than for racemic $Co(\beta PAN)_2^+$, while the tendency was reversed for the magnitude of ΔA_2. Thus there were at least stereospecific effects in forming the bound aggregates of this kind of chelate on a PSS chain. In the isolated form, there was no difference in both ΔA_1 and ΔA_2 between the racemic and enantiomeric $Co(\beta PAN)_2^+$. This was reasonable because PSS itself was optically inactive. Thus no stereospecificity

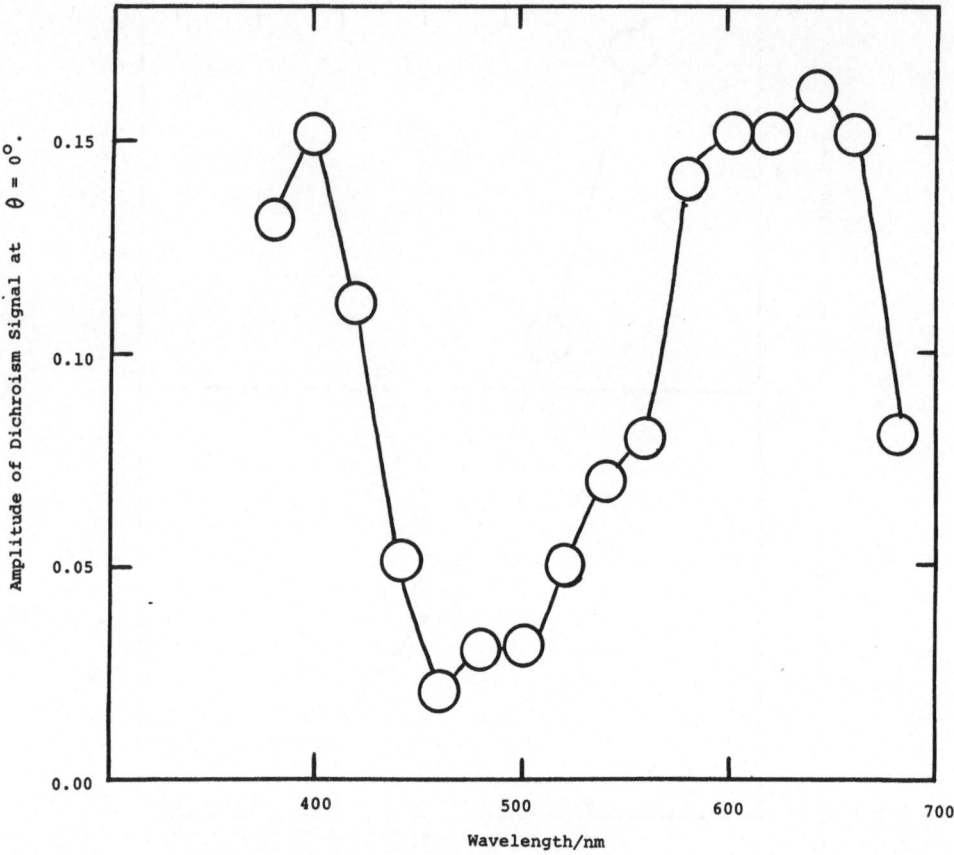

Figure 7. The wavelength dependence of ΔA^{∞} for a PSS-Co(αPAN)$_2^+$ solution at P/M = 1.5.

arose when the chelate existed as an isolated species. In case of Co-(PAPhen)$_2^+$, no transmittance change was observed at P/D = 1.5. Thus only ΔA^{∞} at P/D = 40 was shown in Fig.11. The spectrum was roughly analogous to those of Co(αPAN)$_2^+$ and Co(βPAN)$_2^+$. The structural implications of the above results are discussed in the next section.

DISCUSSION

The observed high affinity of the cobalt chelates toward PSS is apparently originated from the attractive interaction between the aromatic ligands in the chelates and the phenyl groups in PSS. The

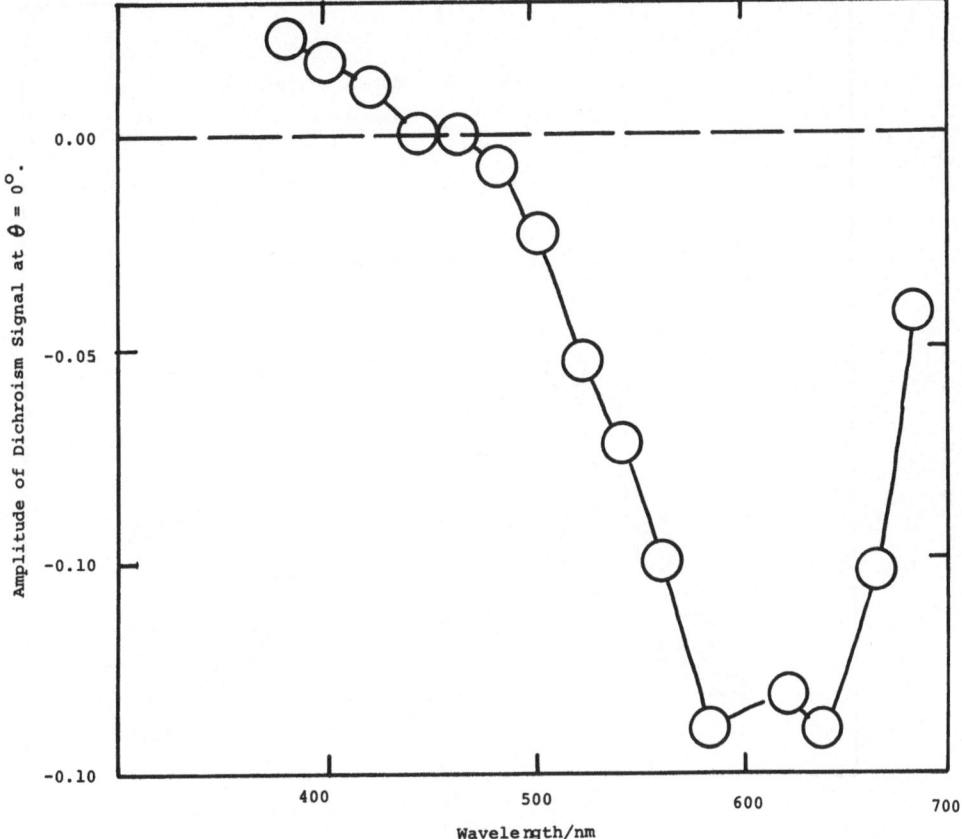

Figure 8.   The wavelength dependence of ΔA^{∞} for a PSS-Co(αPAN)$_2^+$
solution at P/M = 40.

detailed structures of the polymer-chelate complexes were not evident
from the electronic spectra only. We discuss below the possible
structures of the bound states of the chelates on the basis of the electric
dichroism results.

According to the previous theoretical calculations, the neutral
form of αPANH is predicted to have three observable transitions in the
visible region (Table 1).[6] The coordinated αPAN molecules are under the
perturbation of metal ion (Co^{3+}) instead of H$^+$, resulting in the red-shift
of the whole spectrum. We assign that the second and first transitions
in Table I correspond to the observed absorption bands at 400 and 600
nms in the spectrum of Co(αPAN)$_2^+$, respectively (Fig.2). If so, the

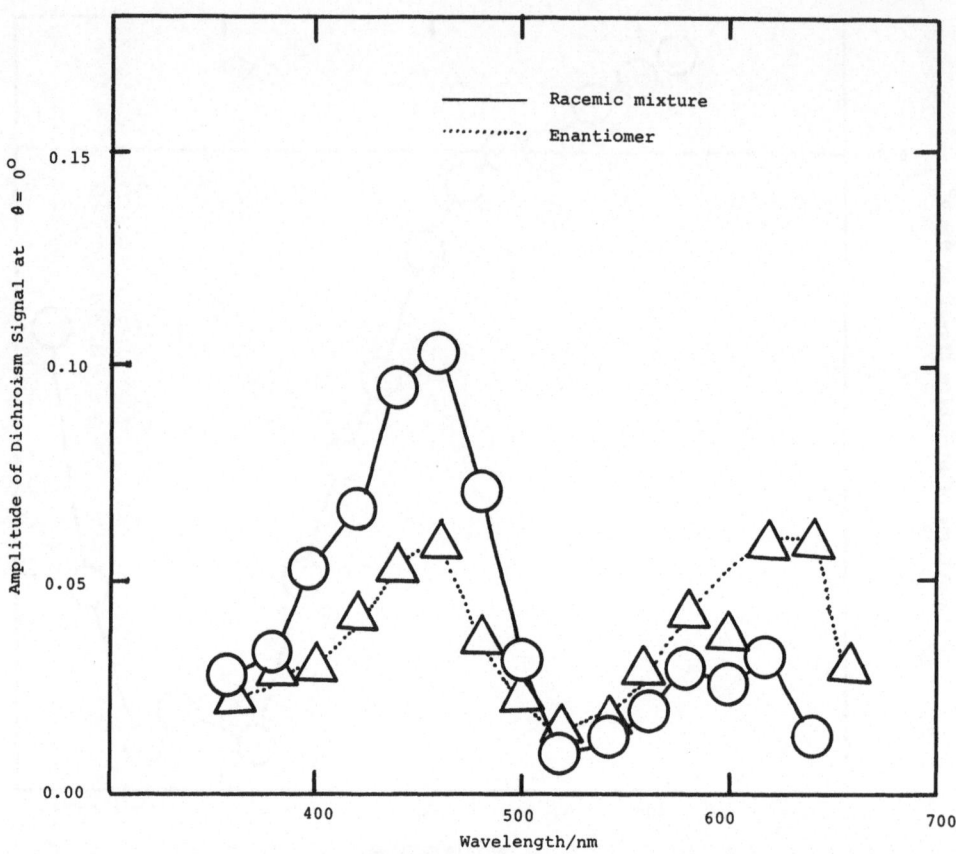

Figure 9. The wavelength dependence of ΔA^{∞} for a PSS–Co(βPAN)$_2^+$ solution at P/M = 1.5. The dotted curve is for (−)$_D$–Co(βPAN)$_2^+$.

transition moments for these excitations lie in the directions of the short and long axes of the αPAN molecule, respectively (Fig.12). They are denoted by $\vec{\mu}_i{}^a$ and $\vec{\mu}_i{}^b$ (i = 1 and 2), respectively, in which a and b indicate the two ligands and 1 and 2 the transitions at 400 and 600 nms, respectively.

Since $\vec{\mu}_1{}^a$ and $\vec{\mu}_1{}^b$ both lie on the same straight line (Fig.13), the total transition moment for the first absorption (400 nm) of Co(αPAN)$_2^+$ is oriented in the direction of the x-axis in the figure. On the other hand, $\vec{\mu}_2{}^a$ and $\vec{\mu}_2{}^b$ are directed along the y- and z- axes, respectively. Assuming that the two ligands negligibly interact with each other,[6] the total transition moment for the second absorption (600 nm) of Co-(αPAN)$_2^+$ is concluded to take any direction on the plane determined by two vectors, $\vec{\mu}_2{}^a$ and $\vec{\mu}_2{}^b$, or on the y–z plane in Fig.13.

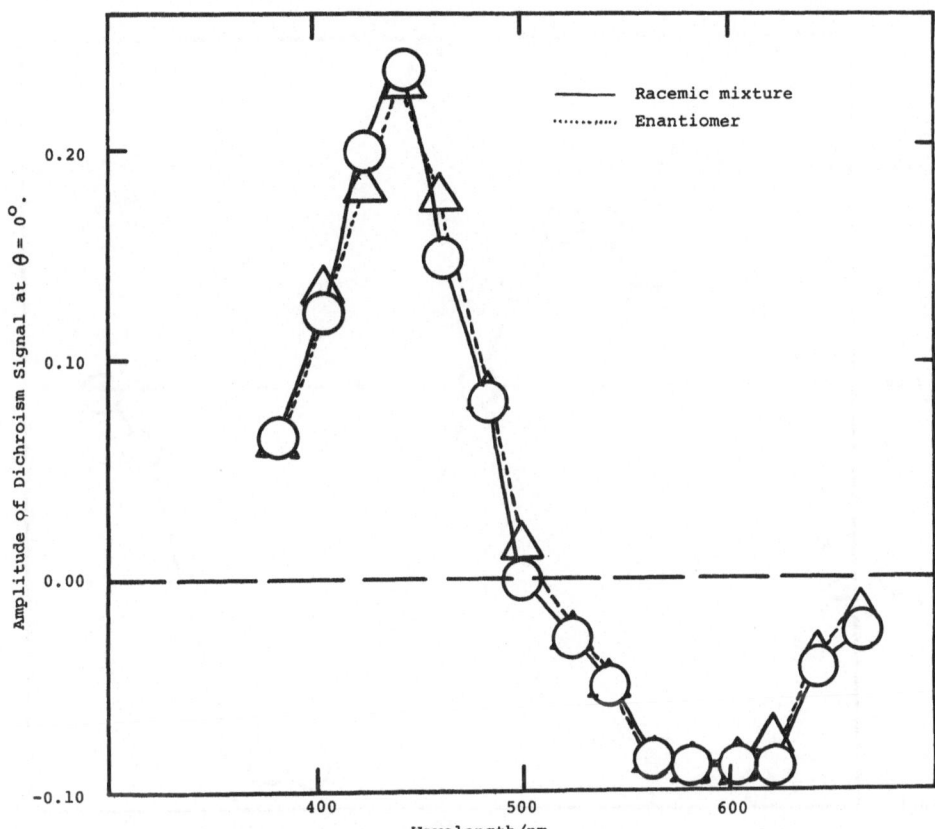

Figure 10.   The wavelength dependence of ΔA^{∞} for a PSS-Co(βPAN)$_2{}^+$ solution at P/M = 40.   The dotted curve is for $(-)_D$-Co(βPAN)$_2{}^+$.

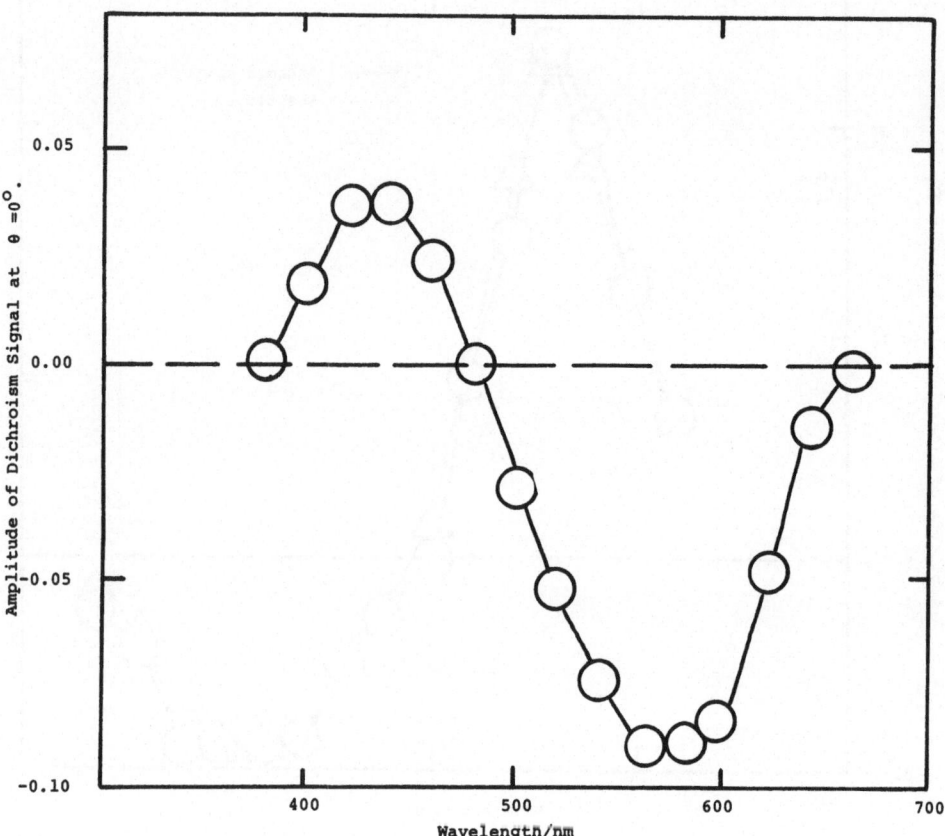

Figure 11.   The wavelength dependence of ΔA^{∞} for a PSS–Co(PAPhen)$_2{}^+$ solution at $P/M = 40$.

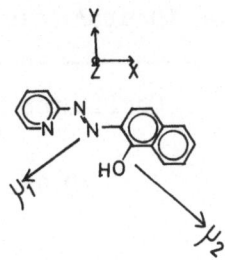

Figure 12. The directions of $\vec{\mu}_1$ and $\vec{\mu}_2$ for PANH.

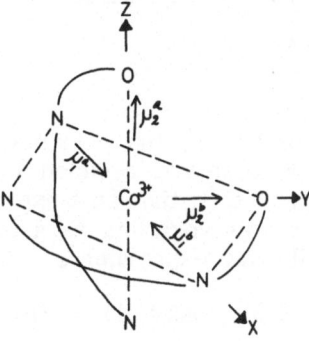

Figure 13. The directions of $\vec{\mu}_1{}^a$, $\vec{\mu}_1{}^b$, $\vec{\mu}_2{}^a$ and $\vec{\mu}_2{}^b$ for the Co(III) chelate.

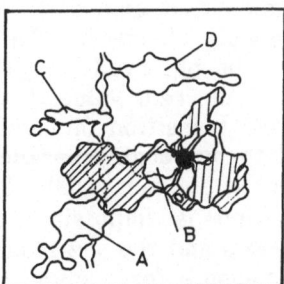

Figure 14. The proposed structure of a PSS–Co(αPAN)$_2{}^+$ complex at P/M larger than 4. Each coordinated PAN molecule is faced with two phenyl groups located in the head and tail of the ligand. As a result, four residues are used in binding the chelate.

Table 1. Molecular Orbital Calculation of αPANH.

M_x	M_y	M_z	Oscillator Strength	Energy/nm
0.50	-0.18	0.0	0.065	477
-0.79	-0.50	0.0	0.227	418
0.53	-0.48	0.0	0.137	408

(*) M_i denotes the transition moment in the direction of the axis in Figure 12.

As for the first absorption, the previous expression for (eq(2)) is applicable in relating the observed reduced linear dichroism with the angle, ϕ. To the contrary of this, the same expression was not applicable because the chromophore was regarded as a circular absorber for this transition. Thus the following expression should be used instead of (2);

$$\rho = (3/8)(3\cos 2\phi - 1) \; \Phi \; (E) \tag{3}$$

in which ϕ represents the angle beween the y-z plane and the electric field direction. Within the approximations stated above, the first transition moment (x-axis) is always perpendicular to the second transition moment (y-z plane). The same situations are assumed to hold also for the first and second transitions of $Co(\beta PAN)_2^+$ and $Co(PAPhen)_2^+$.

Table 2 gives the calculated angles, ϕ, for the first and second transitions of the bound cobalt chelates. The sign, < or > , was added before each value, because the observed ρ values were considered to be the upper or lower limits of the real values.[7] As a general conclusion, both of the first and second transition moments for the chelate in the aggregate form are found to be roughly parallel with the electric field direction (or with the oriented polymer axis). This conclusion is peculiar because such a conformation is impossible due to the geometrical restriction. That is, the x-axis and the y-z plane cannot be parallel with each other. One counter-measure for it might be that the bound chelate in the aggregate form was not fixed rigidly on a PSS chain, but there remained some freedom of rotation. For example, if the chelate rotates freely around the axis of y = z (x = 0), ρ values for both the first and second transitions can take a positive value. The proposed picture is

Table 2.　Observed Reduced Linear Dichroism and Calculated Angle, ϕ.

Co(III) Chelate	Bound State (P/M)	First Transition(~400nm)		Second Transition(~600nm)	
		ρ	ϕ	ρ	ϕ
Co(αPAN)$_2^+$	Aggregate (1.5)	+0.50	<48°	+0.50	<19°
	Isolated Monomer (40)	+0.07	<54°	-0.66	>52°
Co(βPAN)$_2^+$	Aggregate (1.5)	+0.66	<46°	+0.20	<30°
	Isolated Monomer (40)	+1.00	<42°	-0.45	>67°
Co(PAPhen)$_2^+$	Isolated Monomer (40)	+0.54	<48°	-0.90	>59°

reasonable, because the bound chelates at low P/D interact with only one or two residues in PSS. Under such situations, the chelate was bound with the polymer mainly through the electrostatic force. Thus the chelate was likely to be allowed some residual freedom.

As for the results at P/D = 40 or in the isolated states, the transition moment of the first transition (or the x-axis) was roughly parallel with the polymer chain, while the transition moment of the second transition (or the y-z plane) was roughly vertical to it. Such a conformation is realizable, because the x-axis and the y-z plane are perpendicular to each other. In comparison with the proposed structure of the aggregate form of the chelates (P/D 1.5), the chelates in the isolated form (P/D = 40) are concluded to be fixed with respect to the polymer chain. Since the isolated form of the bound chelate became predominant above P/M = 4 (Fig.6), at least four polymer residues were participating in that bound state. Based on this, one of the possible structures is displayed in Fig. 14. In that structure, each ligand is stacked with two phenyl groups in PSS, leading to the rigid bound state. It is apparent that such a structure is realized due to the attractive interaction of the aromatic planar ligands with the phenyl groups. The hydrophobicity of PSS is interpreted in this way in terms of molecular interactions.

The results in Figs. 9 and 10 confirm that some stereospecificity exists in forming the bound aggregate of $Co(PAN)_2^+$ on a PSS chain. Since both racemic and $(-)_D\text{-}Co(PAN)_2^+$ gave the similar electronic spectra in the presence of PSS, the present finding demonstrates the utility of electric dichroism for such phenomena. Based on the foregoing discussions, it is inferred that the absolute configurations of the neighboring bound chelates affect the rotational motions of them.

REFERENCES AND NOTES

1. (a) Okubo, T.; Ise, N., J. Am. Chem. Soc., 1973, 95, 2993.
    (b) Tondre, C.; Kale, K. M.; Zama, R., Eur. Polym. J., 1978, 14, 139.
2. Yamagishi, A., Biopolymers, 1982, 21, 89.
3. Yamagishi, A.; Ohnishi, R., J. Chromatography, 1982, 245, 213.
4. Vitagliano, V.; Constantino, L.; Zagari, Z., J. Phys. Chem., 1973, 77, 204.
5. Dourlent, M.; Hogrel, J. F.; Helen, C., J. Am. Chem. Soc., 1974, 96, 3398.
6. Yamagishi, A., J. Phys. Chem., 1982, 86, 223.
7. With the increase of the PSS concentration, for example, ΔA^∞ was found to decrease apparently.

PART II

ASSOCIATION, AGGREGATION AND GELATION

ASSOCIATION AND COMPLEX FORMATION IN STEREOREGULAR PMMA SOLUTIONS

G. Rehage and D. Wagner
Institute of Physical Chemistry
Technical University Clausthal
3392 Clausthal-Zellerfeld/FRG

INTRODUCTION

Association phenomena in dilute solutions of atactic (a) and stereoregular (i,s) PMMA in suitable solvents have been extensively studied in the past.[1] The anomalies observed even in dilute solutions were explained by intermolecular association of complementary sequences of the stereoregular polymer chains, so-called stereocomplex formation. Besides the well known stereocomplex, remarkable effects in solutions of either i- or s-PMMA alone led to the assumption that associated structures can also be formed as a consequence of interactions between sequences of equal tacticity.[2] It has been found that the solvent exhibits a striking influence on the association of PMMA. This paper deals predominantly with rheological investigations of dilute solutions of a-, i- and s-PMMA and their mixtures, as well as investigations on a stereoblock polymer. Results of calorimetric[3] and electron microscopic[4] investigations will be taken into account.

EXPERIMENTAL

The data of the PMMA polymers used are summarized in Table 1. They were obtained with the aid of Röhm GmbH, Darmstadt, where these polymers have also been characterized by GPC and NMR techniques. The solvents used were dimethylformamide (DMF), toluene, o-xylene and chloroform and were purified in accordance with established procedures. Stock solutions were prepared by dissolving the polymer in sealed glass flasks at 60°C. From these stock solutions, series of less concentrated samples were made by adding solvent.

Table 1.

DATA OF POLYMERS USED :

PMMA :	M_N	M_W	M_Z	TRIAD CONTENT		
				I :	H :	S
A-1	92.000	182.000	283.000	9	43	48
A-2	53.000	116.000	177.000	11	42	47
A-3	89.000	172.000	255.000	6	38	56
A-4	35.7oo	69.2oo	118.000	6	38	56
A-5	47.000	75.1oo	1o6.000	15	55	3o
A-6	23.8oo	41.9oo	62.5oo	15	54	31
A-7	4.4oo.000	12.1oo.000	-	6	35	59
I-1	14.000	189.000	2.76o.000	91	7	2
I-2	12.000	398.000	2.831.000	89	6	5
S-1	2o.000	132.000	331.000	9	24	67
S-2	13.000	198.000	1.655.000	3	17	8o
SB-1	6.000	1o.000	2o.000	64	12	24

Stereocomplexes were prepared by systematically mixing i- and s-PMMA solutions of equal concentration at room temperature. They were kept in a thermostatically controlled bath at 25°C before the measurements were done.

The polymer interactions were investigated by measuring reduced specific viscosities ($\eta_{red} = \eta_{sp}/ c$) as a function of the shear rate D and the concentration c, using a Zimm-Crothers viscometer (Krannich, Göttingen) and a "low shear" rotatory viscometer (LS 30, Contraves, Zürich), that can be successfully used in a range of the shear rate from 3.5×10^{-3} to 2.5×10^2 s^{-1}. Time dependent measurements were also performed and ratio i/s was changed.

The heats of complex formation were measured in a Tian-Calvet-Calorimeter (Setaram, Lyon). The electron micrographs were obtained using an electron microscope Hitachi H 500.

RHEOLOGICAL INVESTIGATIONS

a) <u>Atactic PMMA</u>: Viscosities of six PMMA samples (a-1 to a-6) were measured in the concentration range 0.1 to 6 g/dl at 25°C. In Figure 1 it is seen that in the case of a-1 all curves are linear. The intrinsic viscosity [η] = lim η_{red} (D→0, c →0) increases with increasing solvent power. Therefore, we get the following order: o-xylene (worst solvent), toluene, dimethylformamide, chloroform (best solvent). Values and variation of the Huggins-constant k' with solvent is generally in

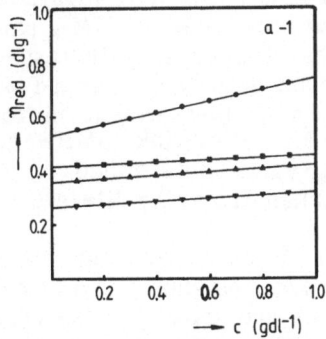

Figure 1.   Reduced viscosity η_{red} vs concentration c for a-1 PMMA at
25°C;   shear rate D = 0.01 s^{-1}.   ● chloroform;   ■ DMF;
▲ toluene;   ▼ o-xylene.

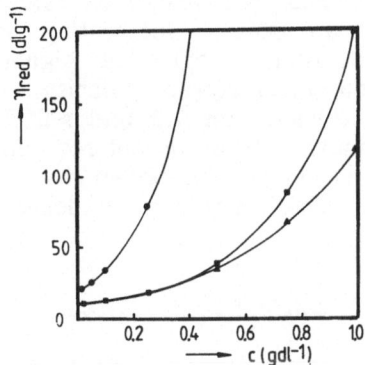

Figure 2.   Reduced viscosity η_{red} vs concentration c for a-7 PMMA at
25°C; D = 0.01 s^{-1}.   ● chloroform;   ■ DMF;   ▲ toluene.

reasonable agreement with those observed by others.   Similar curves
were also found for atactic PMMA samples a-2 to a-6.   The viscosities
of concentrated solutions (5 g/dl) did not show any time- or shear rate
dependence in the range of 3.5 x 10^{-3} to 2.5 x 10^{2} s^{-1}.   The results can
be interpreted in terms of highly flexible polymer chains which behave
like a Newtonian liquid in this region.

        The question whether a-PMMA showed intersegmental association
has been discussed since the discovery of stereocomplex formation
between stereoregular PMMA-molecules.   Association should be favored
in the case that the block length of stereospecific sequences is long
enough to allow intersegmental interactions. Therefore, measurements
were also done using an atactic PMMA of very high molecular weight
(a-7).

As shown in Figure 2, we found a strong concentration dependence of the reduced viscosity even at low concentrations, which in contrast to the findings with the low molecular a-PMMA samples is not linear. It is worthwhile to note that the order with respect to solvent power is obviously the same as found with the other a-PMMA samples. Calculation of the viscometric molecular weight in toluene and chloroform, based on the Staudinger-Mark-Houwink (SMH)-equation is in very good agreement with the M_w-value determined by GPC. Values for the constants K and a are taken from the literature.[5]

Solutions with a-7 PMMA revealed pseudoplastic behavior at concentrations above 0.1 g/dl, as shown in Figure 3. At concentrations above 1 g/dl the solutions are jelly. These elastic properties may be attributed to entanglements. The values of [η] indicate that at concentrations exceeding 0.05-0.1 g/dl^{-1} the polymer coils overlap. From these measurements there is no evidence for a specific association.

b) <u>Stereoregular PMMA</u>: Similar linear plots of reduced viscosity vs. concentration were determined in the case of i-1 and s-1 PMMA. In contradiction to the findings with a-PMMA, toluene seems to have a higher solvent power than DMF. Interesting deviations from this behavior were obtained using samples of higher molecular weight, greater polydispersity and higher stereoregularity. An intrinsic viscosity – molecular weight relationship for i-2 and s-2 PMMA could only be determined with reasonable results in chloroform. In this case the viscosity concentration curves still remained linear and pseudoplasticity could not be observed. The concentration dependence of i-2 PMMA

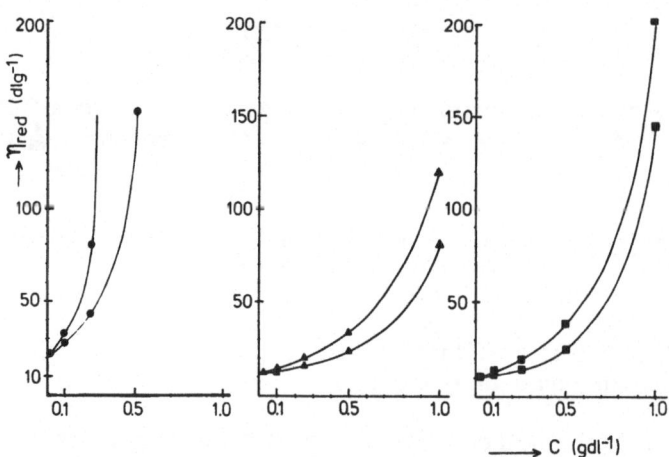

Figure 3.   Reduced viscosity η_{red} vs concentration c for a-7 PMMA at 25°C. Upper curves: shear rate D = 0.01 s^{-1}. Lower curves: shear rate D = 100 s^{-1}. ● chloroform; ▲ toluene; ■ DMF.

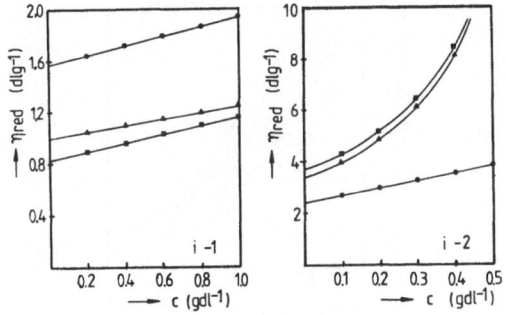

Figure 4. η_{red} vs c for the i-PMMA samples at 25°C; D = 0.01 s^{-1};
● chloroform; ▲toluene; ■ DMF.

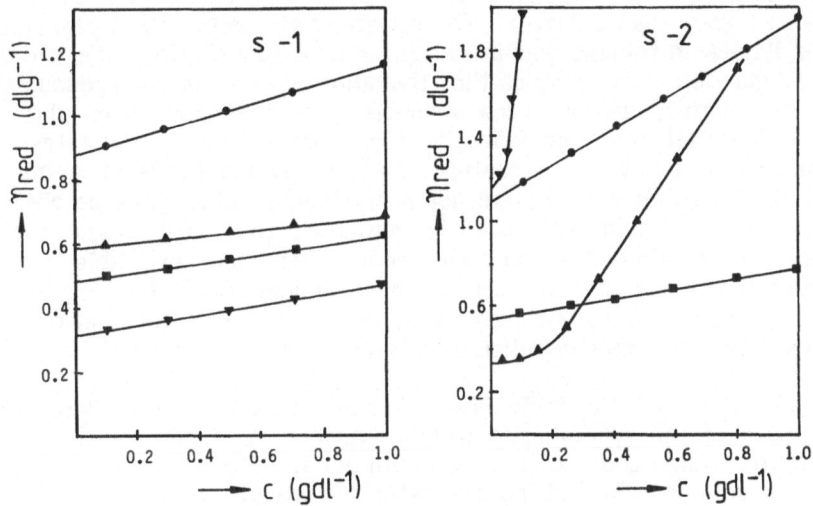

Figure 5. η_{red} vs c for the s-PMMA samples at 25° C; D = 0.01 s^{-1};
● chloroform; ▲ toluene; ▼ o-xylene; ■ DMF.

solutions in DMF and toluene is more complicated, and the solutions revealed a lightly pseudoplastics behavior, indicating the formation of a weak network, with gel-like properties at 0.5 g/dl.   This is probably caused by entanglements. With solutions of s-2 PMMA in toluene and o-xylene, but not in DMF, the reduced viscosity showed a remarkable concentration dependence far below the overlap concentration.   This makes us assume that associations of the syndiotactic sequences were formed from <u>intramolecular</u> associations at lower concentration to more <u>intermolecular</u> associations at higher concentrations.[6]   These are

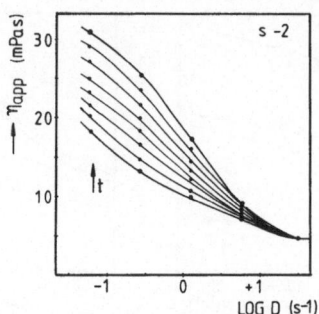

Figure 6.   Dependence of the apparent viscosity η_{app} upon shear rate
and time of s-2 PMMA in toluene at 25°C; c = 2 g/dl.
t: time elapse between each curve is 5 min.

responsible for the findings that the solutions are gel-like at concentrations higher than 2.0 g/dl. The apparent viscosity of a 2 g/dl solution
of s-2 PMMA in toluene was investigated as a function of time and shear
rate (Figure 6). It is obvious that the solution behavior is pseudoplastic,
but besides this, one observes a rheopexy which depends on the shear
rate.   The fact that the viscosity increases with time indicates   slow
formation of a network structure. After several hours at a constant
shear rate, neither a maximum nor a stationary value could be observed.
The result can be interpretated by assuming that the polymer chains
become more elongated through shear strain and are able to form
associations consisting of syndiotactic sequences.   In o-xylene the
viscosity increased so strongly, even at very low concentrations, that
further measurements could not be performed.

c) Mixed solutions: In Figure 7 the reduced viscosity is plotted as
a function of the syndiotactic weight fraction y_S in the solutions. With
dimethylformamide as solvent, a minimum is obtained at y_S = 0.67, in
a dilute solution; i.e. at an i/s ratio by weight of 1:2.   With rising
concentration, the minimum becomes smaller and eventually changes
into a maximum, indicating that a network structure is built up
gradually. The maximum broadens to an approximate value of y_S = 0.5.[7]
The mechanism of stereoassociation involves linking together i- and
s-units between different polymer chains which are assumed to be
helices.   At low concentrations, the molecules are contracted by
intermolecular stereoassociations without forming a network, reducing
only the hydrodynamic volume of the particles.   At higher concentrations, the isolated particles combine and lead to a more or less
perfect network formation. This leads to a high viscosity.   At still
higher concentrations the stereocomplexes between i- and s-sequences
become linked to microcrystals which then form crosslinking areas in the
gel.[8] If toluene is used as solvent and the concentration of the mixture
is < 0.2 g/dl, a minimum can also be observed, directly after mixing the
solutions.   The minimum changes into a maximum after several hours,

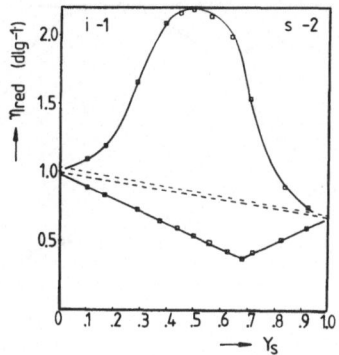

Figure 7.   Reduced viscosity η_{red} vs syndiotactic weight fraction y_s
for mixtures of i-1/s-2 PMMA in dimethylformamide at
25°C.   Lower curve: c = 0.02 g/dl.   Upper curve:
c = 0.6 g/dl. Shear rate D = 0.01 s^{-1}.

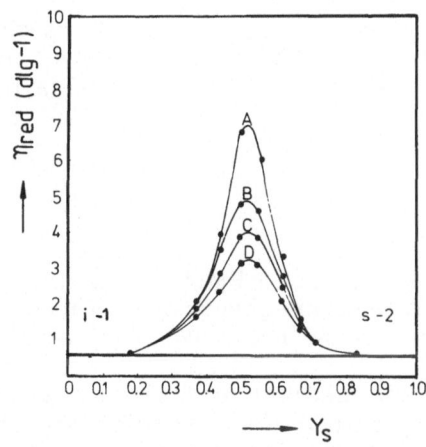

Figure 8.   Dependence of the reduced viscosity η_{red} upon shear rate
D for mixtures of i-1/s-2 PMMA in toluene at 25°C.
A: D = 0.05 s^{-1};   B: D = 1.285 s^{-1}; C: D = 5.59 s^{-1};
D: D = 27.7 s^{-1}. Total polymer concentration: 0.5 g/dl.

showing that the stereoassociation in toluene is a time-dependent
process.   The extent of the association is very high here.   From the
rheological point of view these solutions show properties of thixotropic
and pseudoplastic fluids (Figure 8).   When mixing solutions of a- with
i-PMMA, a broad maximum was observed in reduced viscosity, as shown
in Figure 9.   This can be attributed to the relatively high content of

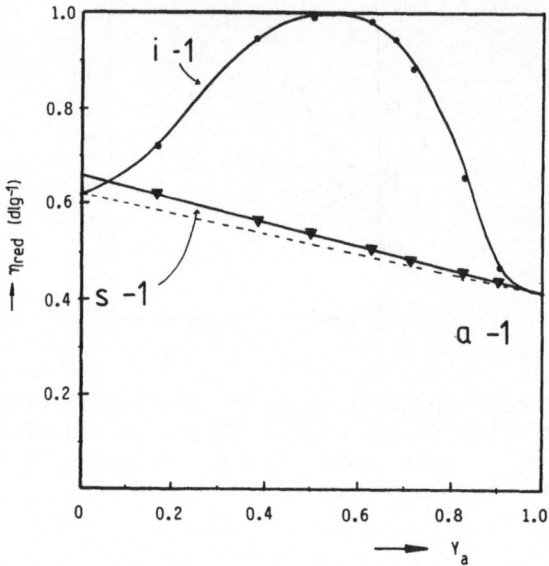

Figure 9.   Reduced viscosity η_{red} vs weight fraction Y_a for mixtures
of: ● i-l/a-1 PMMA; ▼s-1/a-1 PMMA; in toluene at 25°C;
D = 0.01 s^{-1}. Total polymer concentration c = 1.0 g/dl.

syndiotactic sequences in the a-PMMA.   Mixtures between a- and
s-PMMA indicate no association phenomena.   This is in accordance with
the result that s-1 PMMA did not show a self association.

Finally, η_{red} of the mixtures in chloroform yield a nearly additive
curve, pointing out the absence of any associations, probably because the
solvating power is too high to allow the formation of i/s-stereocomplexes
and self-associations (Figure 10).

d) <u>Stereoblock PMMA</u>: In Figure 11 the concentration dependence
of η_{red} of a stereoblock polymer of i- and s-PMMA sequences in
different solvents is shown. The results in DMF and toluene correspond
in a certain way with the mixtures, demonstrating the change from
intramolecular association at low concentrations to intermolecular
associations a higher concentrations.   These curves can be compared
with a cut through the plot in Figure 7 at constant y_s. It is also
remarkable that the reduced viscosity as a function of polymer
concentration in chloroform is linear; the value of the intrinsic viscosity
is much more higher than could be expected from the molecular
weight obtained by GPC measurements.

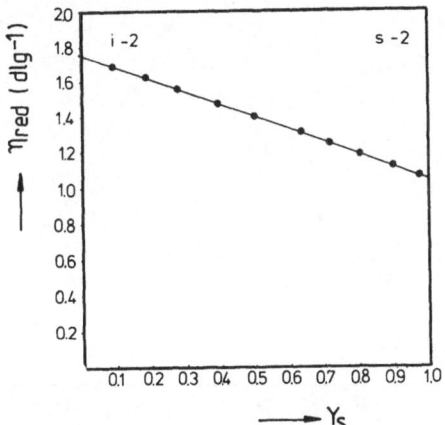

Figure 10. Reduced viscosity η_{red} vs weight fraction y_s for mixtures
            of i-2/s-2 PMMA in chloroform at 25°C; D = 0.01 s^{-1}; total
            polymer concentration c = 0.5 g/dl.

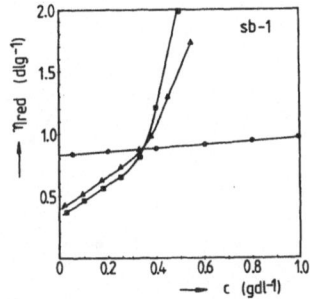

Figure 11. Reduced viscosity η_{red} vs c for the stereoblock PMMA sb-1
            at 25°C; ● chloroform; ■ DMF; ▲ toluene.

CALORIMETRIC MEASUREMENTS

The specific heat of the complex formation ($\Delta\tilde{H}$), which appears
when mixing solutions of PMMA of different tacticities in suitable
solvents, was measured in a Calvet calorimeter.[3]   It was always
negative.   The influence of solvent (o-xylene, chloroform, dimethylformamide), temperature, and polymer concentration on the value of ΔH,
was extensively studied.

According to these measurements the position of the maximum of
($-\Delta\tilde{H}$) was found near the composition of 2:1 (s:i) by mass fraction with
mixtures of i-1/s-1 PMMA.   The height of this maximum is nearly the
same using o-xylene or DMF (Figure 12).   No heat of the formation was
observed in chloroform.

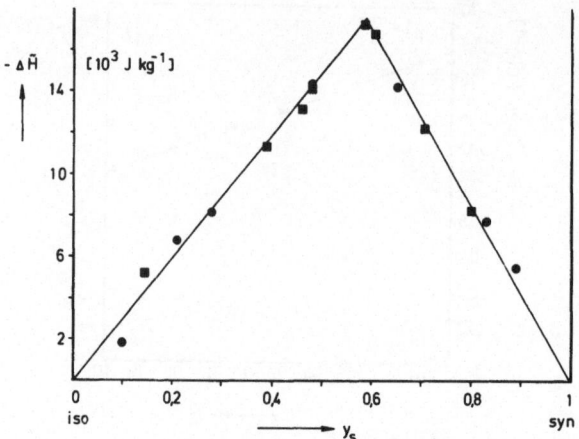

Figure 12.   Specific heat of complex formation $\Delta\widetilde{H}$ vs weight fraction y_s for mixtures of i-1/s-2 PMMA solutions at 25°C;   c = 1.0 g/dl; $\Delta\widetilde{H}$ = $\Delta H(m_i + m_s)$. m_i, m_s = masses of tactic components. ■ : o-xylene;   ● : DMF.

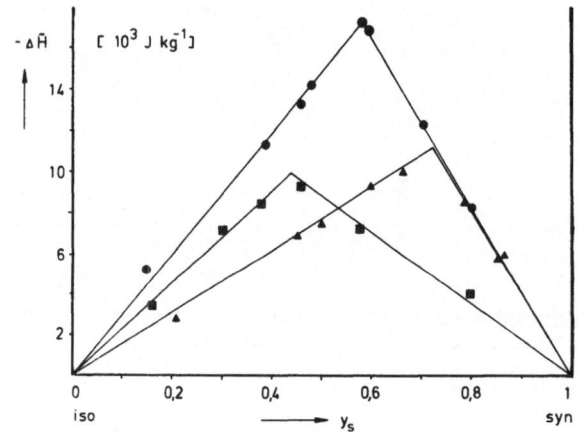

Figure 13. $\Delta\widetilde{H}$ vs y_s for different mixtures in o-xylene at 25°C; c = 1.0 g/dl. ● : i-1/s-2 PMMA;   ▲ : i-1/s-1 PMMA; ■ : sb-1/s-2 PMMA.

$\Delta\widetilde{H}$ is the total enthalpy of mixing, divided by the whole mass of the polymer: $\Delta\widetilde{H}$ = $H/m_i + m_s$). $\Delta\widetilde{H}$ should be proportional to the mass fraction of the syndiotactic or isotactic component.[3]

As shown in Figure 13 the position of this maximum is markedly influenced by the tacticity of the PMMA sample. If the trial content of the syndiotactic component is lower, the maximum shifts to higher values of y_s. If on the other hand the content of the isotactic component is lower, the maximum shifts to lower values of y_s.

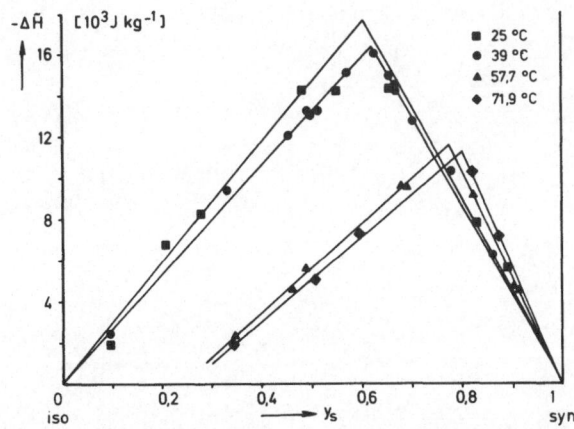

Figure 14. Temperature dependence of the heat of complex formation: $\Delta \tilde{H}$ vs y_s for mixtures of i-1/s-2 PMMA in DMF; c = 1.0 g/dl.

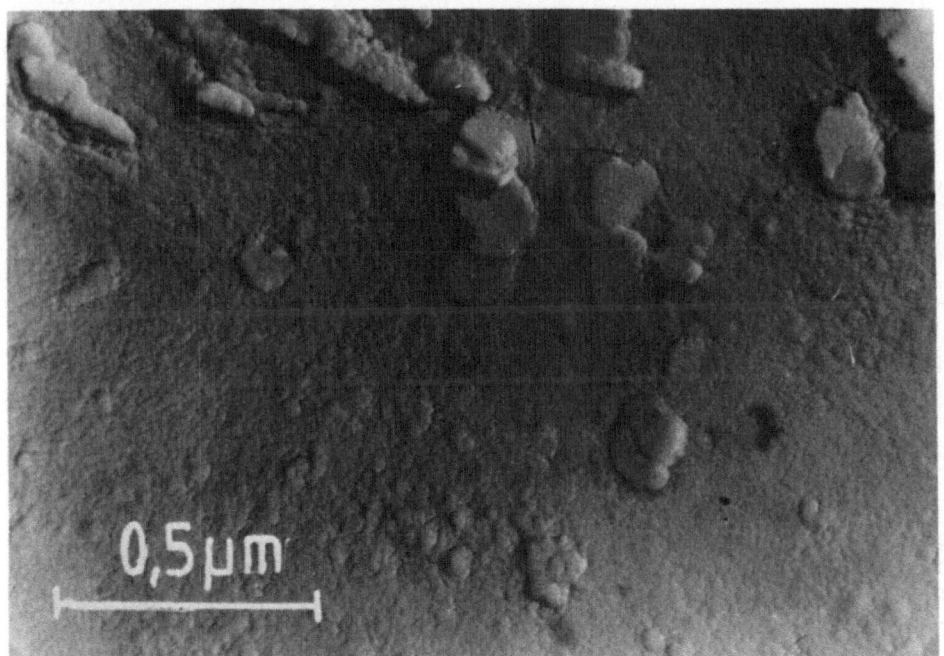

Figure 15. Electron micrograph of a 1:2 mixture of i-1/s-1 PMMA in acetone; c = 1.0 g/dl. (C/Pt-cast of the surface after breaking the frozen solution).

Figure 16. Electron micrograph of an atactic PMMA solution in chloroform; c = 10.0 g/dl. (preparation s.a.).

A similar effect was observed when measuring the heat of complex formation at higher temperatures (Figure 14). The sequence length distribution of the PMMA polymers was determined using results from NMR-measurements[9] and by means of Flory's theory of copolymer equilibrium crystallization.[10] The critical sequence length as a function of temperature was calculated. The sequence length required for a stable iso-syn microcrystal stereocomplex, respectively, becomes longer, when increasing the temperature. This was found more striking in the case of s-PMMA than with i-PMMA.[3,12] Therefore, a higher content of syndiotactic sequences is required for the formation of the stereo-complex at higher temperatures.

The maximum of $(-\Delta \tilde{H})$ becomes much lower upon increasing the total polymer concentration. At higher concentrations the stereo-associations or microcrystals, respectively, pervade the whole system like crosslinks, forming a network structure. It is obvious that this network formation decreases the segmental mobility and hinders a part of the isotactic and syndiotactic sequences to associate with each other.

ELECTRON MICROSCOPIC OBSERVATIONS

Within the scope of electron microscopic investigations[4] on structures in high molecular solutions, mixtures between i- and s-PMMA in acetone reveal morphological ordered structures, especially when the ratio s/i is 2:1 by weight (Figure 15).

With mixtures in chloroform no heterogenieties could be detected. A similar result was obtained with a concentrated solution of a-PMMA in chloroform (Figure 16).

CONCLUSIONS

These investigations show that the behavior of iso- or syndiotactic PMMA in solution is very complicated and multifarious. In dilute solutions, there may exist association equilibria as found by sedimentation velocity measurements.[2] At higher concentrations gelation occurs, caused by self-association and/or entanglements. Self association consists presumably of helical arrangements of the molecules. In mixtures of iso- and syndiotactic PMMA stereocomplex formation is predominant. The building up of the stereocomplexes depends strongly on the extent of the tacticity and the sequence length distribution of the pure iso- and syndiotactic components. At even higher concentrations stereoassociations combine and form microcrystals. These microcrystals are the junction areas in gels formed by mixtures of iso- and syndiotactic PMMA.[11,12] All these processes are influenced to a high degree by the nature of the solvent.

ACKNOWLEDGEMENTS

The authors wish to thank Dr. W. Siemens and Dr. B. Schriewer. The results of the calorimetric and electron microscopic investigations are taken from their Ph.D. theses. They thank Röhm GmbH, Darmstadt, for gifts and characterization of PMMA-samples. They are indebted to Deutsche Forschungsgemeinschaft, Verband der Chemischen Industrie—Fonds der Chemie and Dr. Otto Röhm-Gedächtnisstiftung for financial support of this work.

REFERENCES

1. W. Borchard, M. Pyrlik and G. Rehage, Die Makromol. Chem. 145, 169 (1971); M. Pyrlik, G. Rehage, Angew. Makromol. Chem. 29/30, 471 (1973).
2. B. Hermanns, Ph.D. Thesis, TU Clausthal, FRG (1979).
3. W. Siemens, Ph.D. Thesis, TU Clausthal, FRG (1976).
4. B. Schriewer, Ph.D. Thesis, TU Clausthal, FRG (1979).
5. S. Chinai et al., J. Am. Chem. Soc. V 77, 4763 (1955).

6. G. Meyerhoff, G. V. Schulz, Makromol. Chem. V 7, 294 (1952); J. Spěváček, B. Schneider, Polym. Bull. 2, 227 (1980).
7. G. Challa et al., Polymer 20, 59 (1979).
8. K. Könnecke, G. Rehage, Coll. Pol. Sci. 253, 1062 (1981).
9. U. Johnsen, Kolloid-Z. 178, 161 (1961); Kolloid-Z. Z. Polymere 210, 1 (1960).
10. P. J. Flory, Trans. Farad. Soc. 51, 848 (1955).
11. M. Pyrlik, G. Rehage, Rheol. Acta 14, 303 (1975); Coll. Polym. Sci. 254, 329 (1976).
12. M. Pyrlik, Ph.D. Thesis, TU Clausthal, FRG (1976).

DILUTE SOLUTIONS OF POLY(VINYLBUTYRAL): CHARACTERIZATION

OF AGGREGATED AND NON-AGGREGATED SOLUTIONS

P. Metzger Cotts and A. C. Ouano*
IBM Research Laboratory
San José, California, 95193

*Present address: General Products Division
                    IBM Corporation
                    San Jose, CA 95193

ABSTRACT

Poly(vinylbutyral) samples of two different degrees of butyralization and varying molecular weights were studied in several solvents using the techniques of low-angle light scattering, exclusion chromatography and viscometry. The polymers showed evidence of aggregation in certain solvents. The aggregation could be broken up by heating to moderate temperature (~70°C), but would reform at room temperature in a period of time dependent on concentration.   No dependence of apparent M_W on solvent was detected apart from that caused by aggregation. Chain dimensions calculated from intrinsic viscosities and molecular weights measured in non-aggregated solutions were compared with literature values of freely rotating dimensions calculated for a given degree of butyralization and tacticity.

INTRODUCTION

Poly(vinylbutyral), PVB, is widely used commercially as a laminate material for automobile windshields and also as a surface coating and adhesive. The polymer is prepared by condensation of poly(vinylalcohol) with butyraldehyde, producing a chain structure containing dioxane rings in the backbone. This presents the interesting possibility that the chain dimensions are affected by the chair structure of the dioxane ring. Isotactic dyads in the precursor poly(vinylalcohol) will place the chain ends in equatorial positions on the dioxane ring.   Similarly, Figure 1 shows that syndiotactic dyads place the chain ends in axial, equatorial positions. Thus the tacticity of the original poly(vinylacetate) will have

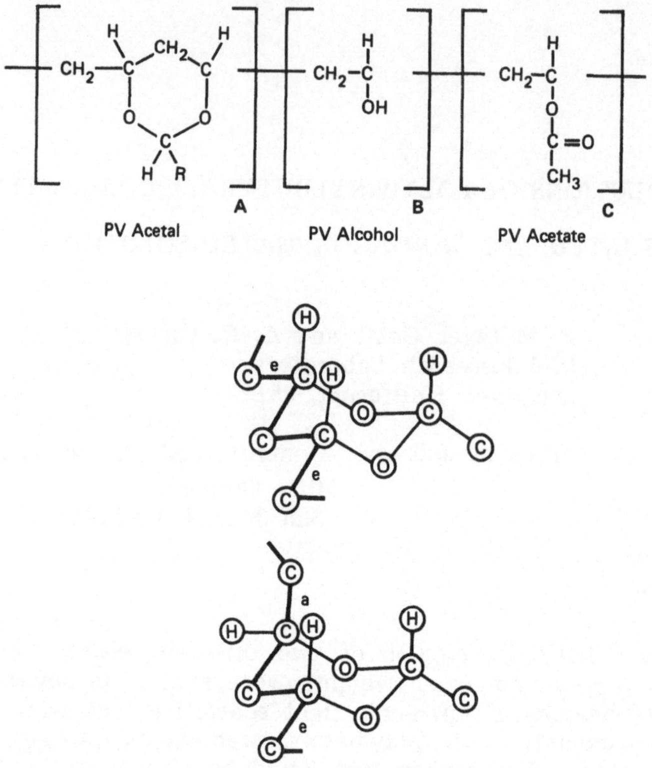

PV Acetal   PV Alcohol   PV Acetate

Figure 1.   Chemical and stereochemical structures of poly(vinylbutyral), showing residual alcohol and acetate groups, and chain positions on the dioxane ring. R = CH_2CH_3 for PVB.

an effect on the chain dimensions of the final poly(vinylbutyral). Residual amounts of both acetate and alcohol groups from the original poly(vinylacetate) and the precursor poly(vinylalcohol) are also present (Figure 1); the latter in substantial proportion (10-20% by weight). The presence of both the polar hydroxyl groups and the non-polar butyral groups in the chain presents the possibility for strong intermolecular or intramolecular interactions. The viscosity of concentrated solutions of poly(vinylbutyral) is known to vary substantially with the solvent composition indicating strong interactions between polymer and solvent.[1] Previous work indicated that molecular weights obtained by light scattering varied with solvent, and attributed this discrepancy to inhomogeneity in the distribution of functional groups on the chain.[2] A variation in functionality among chains can produce a variation in the differential refractive index increment (dn/dc) with the result that use of the average dn/dc measured can lead to molecular weights which are erroneously large.[3] The error becomes progressively larger as the average dn/dc becomes smaller. We have investigated the dilute solution behavior of PVB using the techniques of Low Angle Light Scattering (LALS), viscometry (η), and Size Exclusion Chromatography (SEC) coupled with both LALS and IR detectors in addition to the conventional refractive index (RI) detector with the aim of explaining some of these effects. The samples used were commercial materials available from Monsanto under the tradename Butvar. A summary of samples studied with their molecular weights and hydroxyl contents is shown in Table 1.

EXPERIMENTAL

Low Angle Light Scattering

Weight average molecular weights for the polymers studied were determined with a Chromatix KMX-6 Low Angle Light Scattering photometer. This instrument has several advantages over conventional light scattering photometers for the measurement of molecular weights. Both the sample volume (~150 $\mu\ell$) and the scattering volume (~5 $\mu\ell$) are extremely small, thus large particles are usually observed as spikes in scattering intensity as they move through the scattering volume and can easily be eliminated from the data. This reduces the problem of sample clarification. The scattering intensity is detected at very small angles, typically ~ 4°, permitting use of the equation:

$$Kc/R_\theta = 1/M + 2A_2c + \dots \qquad (1)$$

which is strictly valid only at zero angle. Here R_θ is the Rayleigh factor measured at the angle θ, c is the concentration of the polymer solution in g/mℓ, and K is the usual optical constant including the wavelength of light (632.8 nm), solvent refractive index n, and the differential refractive index increment dn/dc. The sample is contained between two 2 inch long silica windows, permitting the reduction of background to very low levels despite the low scattering angle. In addition, the Rayleigh factor can be calculated directly, without reference to any calibration standards, using geometric factors. The molecular weight M is a weight average molecular weight, M_w, for the polydisperse systems measured here, and the second virial coefficient, A_2, is a light scattering average $A_{2,LS}$ since A_2 shows a dependence on molecular weight.

Samples were dissolved at room temperature overnight and filtered directly into the light scattering cell through 0.2 μm Fluoropore filters (Millipore Company). The membrane filter would usually become clogged after 10-15 mℓ of solution had been filtered. High-capacity prefilters did not prolong the life of the membrane filter. Solutions were often difficult to filter and the higher molecular weight samples were more difficult to filter than lower molecular weights.

TABLE 1

Molecular Weights and OH Content of Samples Used in this Study

SAMPLE[a]	%OH[b]	M_w
A	18.6	58,000
B-1	--	110,000
C	19.3	84,000
C-1	17.0[c]	215,000
C-2	13.0[c]	76,000
C-3	12.0[c]	35,000
D	18.6	170,000
E	11.2	120,000
E-1	--	290,000
E-2	--	120,000
E-3	--	45,000

[a]Numbered samples are fractions of the whole polymer indicated by a letter, obtained by SEC using preparative columns.

[b]Weight percent poly(vinylalcohol), data supplied by Monsanto unless otherwise indicated.

[c]Determined by ^{13}C-NMR spectroscopy by J. Carothers in this laboratory.

The dn/dc values for each degree of butyralization in each solvent were measured at 25°C with 632.8 nm light using a Chromatix KMX-6 Laser Differential Refractometer.   The values obtained are shown in Table 2.   The measured dn/dc values were fit to the Dale-Gladstone relation:

$$dn/dc = v_{sp} (n_{polymer} - n_{solvent}) \qquad (2)$$

for each degree of butyralization, yielding polymer refractive indices of 1.495 and 1.499, and specific volumes of 0.951 and 0.860 mℓ/g, for the 20% and 10% hydroxyl contents respectively (Figure 2).   These values may be compared with reported values[1] obtained with ASTM tests of 1.490 and 1.485, and specific volumes 0.909 and 0.923 mℓ/g, for the 20% and 10% hydroxyl contents, respectively.

Size Exclusion Chromatography

The SEC instrument was composed of a Waters 6000 Solvent Delivery System and either a preparative column set (10^6 and 10^4 Å porosity, 22 mm ID) or analytical column set (10^6, 10^5, 10^4, and 10^3 Å porosity, 7.7 mm ID) of poly(styrene/divinylbenzene) (Polymer Laboratories).   Detectors used included a Waters Differential Refractometer R401, a Wilks Miran 1A IR detector and the Chromatix Low Angle Light Scattering Photometer described above.   When using the IR detector, the mobile phase chosen was methylene chloride (Burdick and Jackson, glass distilled) to provide windows in the IR regions of interest (2.9 μm and 9-10 μm).   The IR detector housed a flow-through sample cell of zinc selenide with a pathlength of 1.5 mm.   For all other SEC measurements, the mobile phase was tetrahydrofuran (THF, Burdick and Jackson, glass distilled) stabilized with 0.025% BHT (2,6-di-tert-butyl-4-methylphenol).

Intrinsic Viscosity

Viscosities were measured in capillary viscometers of the suspended level Ubbelohde type (Cannon Instrument Co.) using a Wescan Automatic Viscosity Timer.   Corrections for kinetic energy effects were not necessary since flow times were well over 100 seconds in all cases. Viscosities at 25°C of at least four concentrations were measured and extrapolated to infinite dilution using the usual relations:

$$\eta_{sp}/c = [\eta] + k[\eta]^2 c + \dots \qquad (3)$$

$$\ln(\eta_{rel})/c = [\eta]+(k-1/2)[\eta]^2 c + \dots \qquad (4)$$

All samples were filtered through 0.5μ m Fluoropore filters before being introduced into the viscometer.

TABLE 2

dn/dc Values in Various Solvents

SOLVENT	20% OH	10% OH
methanol	0.156	---
acetone	---	0.130
ethanol	0.127	---
acetic acid	0.116	0.112
isopropanol	0.111	---
n-butanol	---	0.085
THF	0.085	0.080
methylene chloride	0.074	0.066
chloroform	---	0.054
cyclohexanone	0.036	0.038

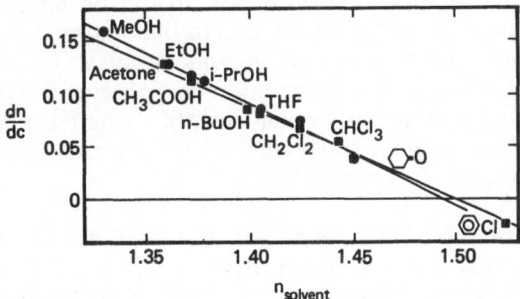

Figure 2. The differential refractive index increment, dn/dc, versus the refractive index of the solvent for the two different degrees of butyralization.

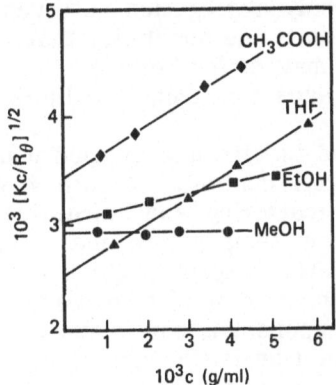

Figure 3. $[Kc/R_\theta]^{1/2}$ versus the concentration c for sample C in various solvents.

RESULTS AND DISCUSSION

Systems Exhibiting Aggregation

For samples containing ~20% by weight hydroxyl units (samples A–D), variation in apparent weight average molecular weight measured by LALS was observed among the solvents ethanol, methanol, THF, and acetic acid. Typical results obtained for one sample (sample C) are shown in Figure 3. Apparent weight average molecular weights obtained for this sample range from 84,000 in acetic acid to 160,000 in THF. This trend in apparent molecular weights was consistent for samples A–D. For sample E, a larger apparent molecular weight was only obtained in the poor solvent acetone, with several other solvents in agreement. This indicates less tendency toward aggregation for this sample with the lower hydroxyl content. These solutions were all filtered through 0.2 μm filters prior to the LALS measurement. As will be discussed below, this filtration appears to remove some but not all of the aggregated species. Matsuda and coworkers attributed a similar observed variation in apparent molecular weight to an inhomogeneity in the copolymer distribution.[2] Variations in dn/dc among the polymer chains can lead to erroneously large calculated molecular weights when the average dn/dc is used. However, we believe these variations are due to aggregation for reasons to be discussed further below.

The concentration dependence of the reciprocal scattering was linear in the relatively narrow range studied. Thus, for the concentration range studied here, (<10 mg/mℓ), the aggregation appears to be essentially independent of concentration. For samples A-D, the apparent weight average molecular weight obtained was not consistent with solvent quality, in the thermodynamic sense. The highest apparent M was obtained in the very good solvent THF for these samples, while a smaller apparent M was obtained in the poorer solvent methanol, indicating a higher degree of aggregation in the better solvent. For sample E, however, the only aggregation observed occured in the poor solvent actone. This presents the possibility that the mechanism for the formation of aggregates may differ between the two types of sample; however, we cannot conclude this from the limited data.

SEC results obtained in THF also showed unusual behavior for the samples of higher hydroxyl content. Figure 4 shows chromatograms obtained with the light scattering and refractive index detectors for sample B dissolved at room temperature and filtered through a 1.0 μm Fluoropore filter. The LALS response, which is proportional to cM, is very sensitive to small amounts of high molecular weight species. The high molecular weight component is barely visible with the RI detector, indicating that the weight fraction of aggregated species is very small. Filtration of the sample through a 0.2 μm filter (as was used in the LALS measurements) significantly reduced the high molecular weight peak visible in the LALS chromatogram. As mentioned above, filtration of the samples (especially those of higher molecular weight and/or hydroxyl content) through the 0.2 μm filters used for LALS and viscosity measurements was very difficult, with clogging occuring after a few mℓ had been filtered. These observations indicate that a significant portion of the aggregated species could be removed with the 0.2 μm filtration. However, the LALS results discussed above show that enough aggregated species remain to contribute to the apparent weight average molcular weight. Gravimetric analysis of unfiltered and 0.2 μm filtered solutions were in agreement, showing that the weight fraction of aggregated species removed by filtration, although easily detected with SEC/LALS, is in fact very small. This agrees with the chromatogram observed with the refractive index detector, which shows the high molecular weight component as a barely detectable shoulder on the main polymer peak.

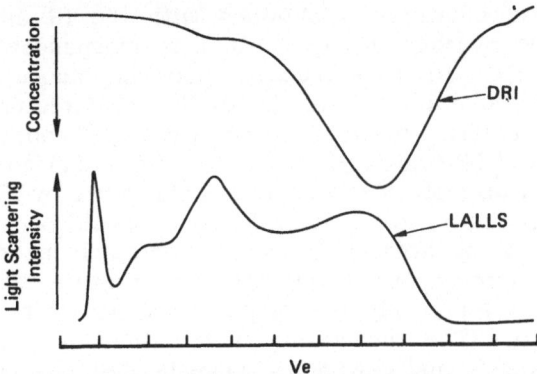

Figure 4. Chromatograms obtained for sample A in THF after filtration through a 1.0 µm fluoropore filter.

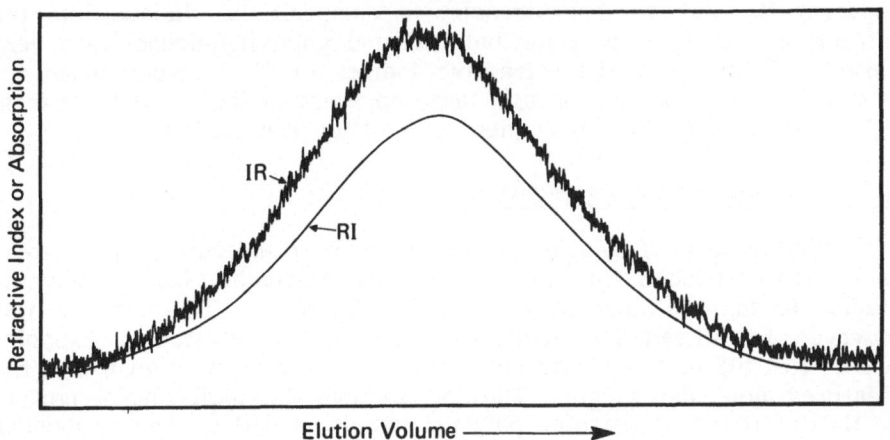

Figure 5. Chromatograms obtained for sample A in methylene chloride using the refractive index and IR detectors.

Heterogeneity in Functional Group Distribution

As mentioned above, heterogeneity in functional group distribution has been suggested in the literature as an explanation for the observed variation of apparent M_W obtained by light scattering in various solvents. The IR detector was used to assess the homogeneity of the hydroxyl content across the molecular weight distribution. Chromtograms were obtained in methylene chloride monitoring the OH stretch of the hydroxyl group (2.9 µm) and the butyral ring vibration (8.9 or 10.1 µm). In both cases, and for all samples studied, the normalized chromatograms obtained by IR were superposable with the refractive index chromatograms. Chromatograms obtained for sample A, monitoring the OH stretch, are shown in Figure 5.

The SEC chromatograms obtained with the IR and RI detectors showed that the hydroxyl group content is independent of molecular weight. Since the refractive index of poly(vinylalcohol) (1.50) is very close to that of poly(vinylbutyral) (1.49) the RI detector responds as a concentration detector even if there are significant differences in hydroxyl content. Matsuda and coworkers also found no dependence of hydroxyl content on molecular weight as determined by IR on fractions.[2] The fractional precipitation used by Matsuda should be a stringent test for heterogeneity in hydroxyl content since solvent quality can be expected to be more dependent on chemical structure than on molecular weight, especially for copolymers of polar and non-polar groups such as poly(vinylbutyral). The successful fractionation by molecular weight obtained by Matsuda and coworkers suggests that any heterogeneity in functional group distribution is small. We will show below that variations in the apparent M_W among different solvents can be removed with moderate heating. This is inconsistent with variations in the apparent M_W due to inhomogeneity in composition. In addition, the refractive indices of poly(vinylbutyral) and poly(vinylalcohol) are very close, and far from the refractive indices of the solvents used for samples A-D, so that erroneously large apparent molecular weights from light scattering due to copolymer composition are unlikely.

Effect of Heating on Aggregation

Mild heating of solutions to ~70°C prior to measurements at room temperature reduced the apparent molecular weights for all solvents studied to that obtained in acetic acid (Figure 6). This behavior was observed for all samples of higher hydroxyl content studied (samples A-D). Heating of the acetic acid solutions resulted in no change in the measured molecular weight. Thus, we believe this molecular weight to be that of the non-agregated polymer sample, and it is these molecular weights which are indicated for the samples in Table 1. For sample E, which had a lower hydroxyl content, molecular weights obtained in eight

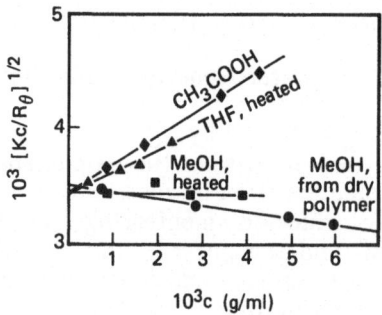

Figure 6. $[Kc/R_\theta]^{1/2}$ versus the concentration c for sample C in solvents after heating.

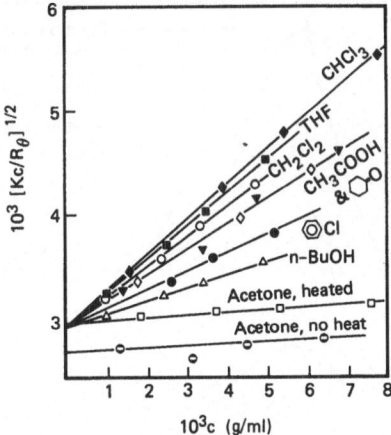

Figure 7. $[Kc/R_\theta]^{1/2}$ versus the concentration c for sample E in eight solvents.

different solvents which had not been heated were in excellent agreement as shown in Figure 7. These solvents spanned a large range in dn/dc (0.038 to 0.130) as well as a range of thermodynamic quality. Thus, these data provide additional evidence that functional group heterogeneity is not contributing to erroneously large apparent molecular weights since the measurement in cyclohexanone (dn/dc = 0.038) would be very sensitive to variations in dn/dc among chains. As mentioned above, the apparent molecular weight obtained in the poorest solvent, acetone, was too large, but decreased to agree with the other solvents when the solution was heated to ~70°C prior to the LALS measurement (Figure 7).

Heating the sample to ~70°C prior to injecting it onto the columns also resulted in a large reduction in the size of the high molecular weight peak in the LALS chromatogram. The amount of the high molecular weight component present depended on the time since heating. Given sufficient time, the LALS response would again show a substantial amount of large species, see Figure 8. The re-formation of the aggregated species at room temperature is slow, requiring time on the order of days to return to a similar state as existed prior to heating. This slow return to the aggregated state permitted LALS measurements to be made at room temperature without complications due to the formation of aggregated species during measurement. These changes would be undetectable with the RI detector alone, and serve to demonstrate the excellent sensitivity of the SEC/LALS technique to small amounts of high molecular weight material previously detectable only by measurement of mechanical properties. Light scattering alone, especially at low angles, can be dominated by the high molecular weight component, but in combination with SEC, the advantages of separation and sensitivity are both retained. The lower hydroxyl content sample, E,

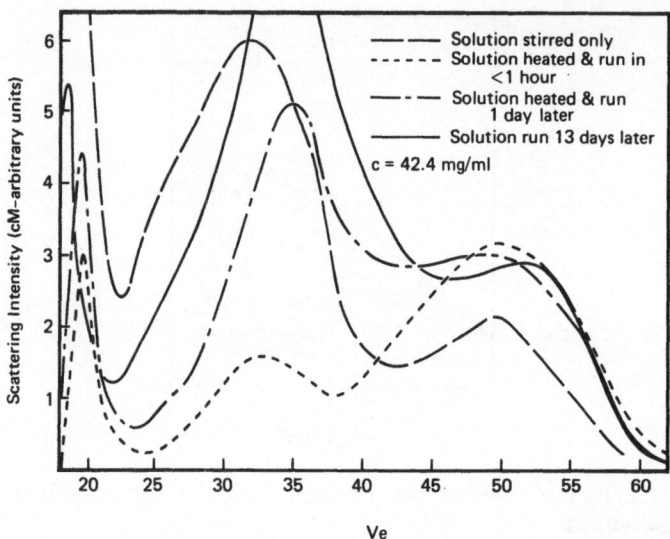

Figure 8. Chromatograms obtained with light scattering detector for sample A in THF as a function of time since heating.

showed none of these unusual effects in THF. Chromatograms obtained for sample E using the LALS detector showed only one peak and weight average molecular weights were in agreement with those from LALS alone, with no dependence on the filtration of the chromatography samples.

The intrinsic viscosity showed little or no change with prior heating of the sample. This indicates that this measurement is relatively insensitive to the presence of a small amount of aggregated species remaining in the solution after filtration with a 0.2 μm filter. An insensitivity of the intrinsic viscosity to low degrees of aggregation where the aggregates are expected to have a randomly branched configuration has been observed previously.[4] The branched configuration contributes less to the viscosity than a linear chain of the same molecular weight. For a weight average degree of aggregation of two, for a chain in a good solvent, it was shown that the intrinsic viscosity is only increased by ~25%. For the samples studied here, the degree of aggregation, given by the ratio of the apparent to the non-aggregated weight average molecular weight, is two or less. This model of branched aggregates is consistent with our observation that there is little change (less than 10%) in the measured intrinsic viscosity with the moderate heating which shows a dramatic change in LALS and SEC/LALS. The values for [η] reported in Table 3 are the minimum values measured for samples where measurements were made on heated and unheated solutions. In all cases the differences were less than 10%.

TABLE 3

Intrinsic Viscosities in Two Solvents

SAMPLE	SOLVENT	$[\eta]$ (ml/g)
A	THF	77
A	MeOH	52
B-1	THF	124
C	THF	101
C	MeOH	60
C-1	THF	198
C-2	THF	95
C-3	THF	55
D	THF	168
E	THF	116
E-1	THF	216
E-2	THF	113
E-3	THF	56

Molecular Parameters from Solutions with Minimal Aggregation

We have observed strong evidence of aggregate formation in this system through the techniques of LALS and SEC/LALS. The aggregation could be partially removed by filtration, and nearly eliminted with only moderate heating to ~70°C. The extent of the aggregate formation varied with molecular weight, hydroxyl content, solvent, and solution history. Aggregation phenomena have been observed frequently in polymers containing both polar and non-polar functional groups such as methyl methacrylate/methacrylic acid copolymers.[5-7] With poly(vinyl-butyral), the strongly polar solvent acetic acid appears capable of dissolving the aggregates into their molecular components. Sample E, containing a smaller percentage of hydroxyl groups, showed only slight evidence of this aggregation in the poor solvent acetone. Thus, it may be expected that the mechanism for formation of the aggregates involves hydrogen bonding of the hydroxyl group. The large variations in the extent of aggregation with history prevented a quantitative assessment of the aggregates' size or distribution, however significant scattering intensity was observed at the exclusion limit of the SEC columns corresponding to $M_W > 10.^7$

We have demonstrated above that both weight average molecular weights and intrinsic viscosities can be obtained which are only minimally affected by aggregation. This data, which we believe represents the non-aggregated chains, is reported in Tables 1 and 3, and is used for the calculation of chain dimensions below. In addition to obtaining agreement of M_W among several solvents using LALS alone, the validity of this M_W was also checked by SEC/LALS in THF on heated solutions, where LALS chromatograms showed little if any high molecular weight component. The SEC/LALS measurement yields an independent measure of M_W, as well as a molecular weight distribution, without reference to any calibration standard.[8,9] As shown above (Figure 8), the LALS chromatogram is by far the most sensitive technique to detect small amounts of aggregated material. These results then allow calculation of chain dimensions for the two degrees of butyralization for comparison with theoretical predictions such as those calculated by Matsuda and Inagaki.[10]

Analysis of Chain Dimensions

A log-log plot of the intrinsic viscosities, [η], in THF, versus the weight average molecular weight determined by light scattering on non-aggregated samples is shown in Figure 9. The Mark-Houwink relation for the higher hydroxyl content samples is given by:

$$[\eta] = 2.89 \times 10^{-4} M^{0.72} \qquad (5)$$

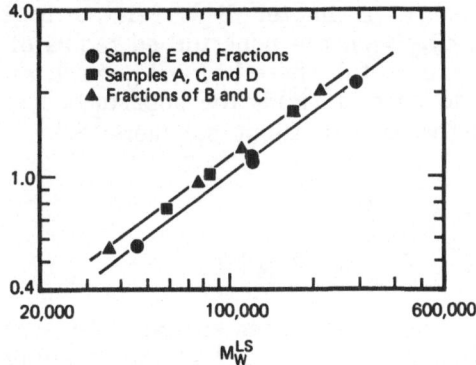

Figure 9. The Mark-Houwink relation (log [η] versus log M) for all
samples studied using M_W measured on solutions shown to be
nearly free of aggregation.

and for the lower hydroxyl content (sample E) by:

$$[\eta] = 2.52 \times 10^{-4} M^{0.72}. \tag{6}$$

The exponent of 0.72 is at the good solvent limit for flexible chains,
consistent with the large virial coefficients (8-10 x 10^{-4} mℓ/g dalton)
observed in THF.

Comparison of dimensions of the PVB chains with different
hydroxyl contents requires measurement of the chain dimensions under θ conditions where A_2 vanishes and the chain assumes its unperturbed
dimensions. The Matsuda and Inagaki calculation of freely rotating chain
dimensions[10] assumes that excluded volume effects are absent, and that
there is no hindrance to rotation about C-C bonds. The ratio of the
experimental unperturbed dimensions to the calculated freely rotating
dimensions is the steric factor σ, a characteristic parameter for a given
polymer which is dependent on the hindrance to rotation. Although we
have not directly measured the chain dimensions even in good solvents,
estimation of the unperturbed dimensions can be made with several
theories developed for use with intrinsic viscosities in good solvents.[11,12] These include the theory of Berry:[13]

$$[\eta]/M^{1/2} = 1.40 \ K_\theta + 0.30 \ K_\theta \ B \ (<R_G^2>_0/M)^{-3/2} \ M^{1/2} \tag{7}$$

and

$$([\eta]/M^{1/2})^{1/2} = K_\theta + 0.42 \ K_\theta^{3/2} \ B \ (<R_G^2>_0/M)^{-3/2} \ M/[\eta] \tag{8}$$

where the free energy parameter B is related to the second virial coefficient A_2, and $<R_G^2>_0$ is the unperturbed radius of gyration. These empirical relations are useful where $[\eta]/[\eta]_0 > 1.8$, as is the case here. Thus extrapolation to zero M yields the unperturbed dimensions through theories developed for random flight polymers:[14,15]

$$[\eta]_\theta = K_\theta \ M^{0.5} \tag{9}$$

$$[\eta]_\theta = (<r^2>_0/M)^{3/2} \ M^{0.5} \tag{10}$$

where $<r^2>_0$ is the unperturbed mean square end-to-end distance. Here, the viscosity function Φ is 2.68×10^{23} at the non-draining limit,[16] with $[\eta]$ in $m\ell/g$. Corrections for polydispersity effects were calculated from the LALS/SEC results obtained from samples with minimal aggregation, and corrections were only applied for whole polymer samples. Although the fractions were relatively broad ($M_w/M_n \sim 2$), the ratio $M_n : M_w$, given by:[17]

$$M_n : M_n : M_w = 1 : [(1+a) \Gamma (1+a)]^{1/a} : 2 \tag{11}$$

where a is the exponent in the Mark-Houwink relation, is ≤ 1.05 or less than 5% difference which is within the limitations of the data. The plots obtained for the two degrees of butyralization are shown in Figure 10. For the higher hydroxyl content polymers (samples A-D), the two data points obtained in methanol, where A_2 is nearly zero at room temperature, are in agreement with the extrapolated K_θ obtained from the data in the good solvent THF. Table 4 shows the results obtained for the two different degrees of butyralization, and the comparison with the freely rotating values calculated by Matsuda and Inagaki.[10] The range of values shown for $<r^2>_{FR}/N$ are the two limits of possible tacticity for the poly(vinylalcohol) precursor, with the lower limit corresponding to all syndiotactic (axial,equatorial), and the upper limit corresponding to all isotactic (equatorial,equatorial). These in turn lead to a range of values for the steric factor σ as shown. Since the tacticity of the commercial samples used in this study is unknown, the values for σ cannot be more precise.

The higher degree of butyralization would be expected to have the larger unperturbed dimensions, since more 6-membered rings are contained in the chain backbone, preventing rotation. The opposite trend is observed here, with the lower hydroxyl content sample yielding smaller unperturbed dimensions. Assuming that the precursor poly(vinylalcohol) is the same for both and that the steric factor σ is also identical, this discrepancy could be due to preferential formation of the butyral ring at dyads of poly(vinylalcohol) with a given stereochemistry. A study of this effect by Fujii and co-workers[18] using acetaldehyde showed that equilibrium acetalization occurred preferentially at isotactic portions of the chain. The same group also demonstrated that the rate of hydrolysis

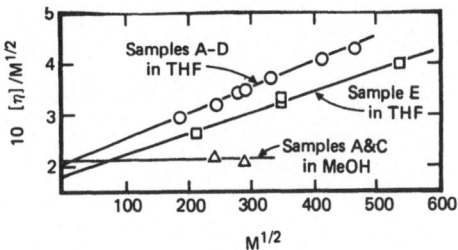

Figure 10. $[\eta]/M^{1/2}$ versus $M^{1/2}$ to estimate unperturbed chain dimensions through the empirical relation of Berry (see text).

of acetal rings formed from syndiotactic dyads of poly(vinylalcohol) was faster than that of rings formed from isotactic dyads.[19] The group used both stereoregular polymer and the cis and trans isomers of pentane-2,4-diol as models of the isotactic and syndiotactic portions, respectively. Later work by the group indicated that the rate of acetalization of the poly(vinylalcohol) dyads was nearly independent of the stereochemistry, using NMR and chemical techniques.[20,21] If in fact the isotactic dyads were preferentially acetalized to butyral rings, then an essentially atactic precursor poly(vinylalcohol) would progress from having a higher population of e,e butyral rings when only partly acetalized to a final equal population of e,e and a,e only when 100% acetalized. Since the e,e rings lead to larger chain dimensions, this effect might offset the differences in chain dimensions due only to degree of acetalization. Calculation of freely rotating chain dimensions assuming that the original poly(vinylalcohol) was atactic yield the results shown in Table 4 for the extreme cases of equal reactivity ($\sigma_{atactic}$), and 100% isotactic reaction prior to any syndiotactic reaction ($\sigma^*_{atactic}$). Although the dimensions of the higher hydroxyl content sample is increased more by the assumption of non-equal reactivity, the difference is only a few percent and would be indistinguishable with the experimental techniques used here. Direct measurement of unperturbed dimensions of samples of known tacticity and widely varying hydroxyl content would be required.

CONCLUSIONS

We have been able to measure weight average molecular weights of poly(vinylbutyral) samples in solutions shown to be nearly free of aggregation. Chain dimensions calculated from intrinsic viscosities using the method of Berry yielded a steric factor σ of ~1.7 in agreement with 1.8 obtained by Matsuda and coworkers.[2] These workers used number average molecular weights obtained from osmometry, and were unable to obtain consistent molecular weights by light scattering. The recent development of the technique of low angle light scattering as a detector for size exclusion chromatography has allowed us to show that inconsistencies in light scattering results, such as curving Zimm plots and variation in $M_{w,app}$ with solvent are due primarily to variations in the state of aggregation in these systems.

TABLE 4

Calculation of Unperturbed Chain Dimensions

SAMPLE	~20% OH	~10% OH
K_θ	0.152 ± 0.008	0.115 ± 0.008
$<r^2>_0/N$	$43 \pm 2 \text{ A}^2$	$38 \pm 2 \text{ A}^2$
$<r^2>_{FR}/N$	$11.8\text{--}15.8 \text{ A}^2$	$12.3\text{--}17.1 \text{ A}^2$
σ	$1.66\text{--}1.92$	$1.49\text{--}1.76$
$\sigma_{atactic}$	1.77	1.60
$\sigma^*_{atactic}$	1.73	1.59

σ^* is calculated assuming preferential reactivity of the isotactic portions of the chain.

REFERENCES

1. Monsanto Technical Bulletin No. 6070C, Monsanto Plastics and
    Resins Company, St. Louis, Missouri.
2. H. Matsuda, K. Yanano, and H. Inagaki, Kogyo Kagaku Zasshi (J.
    Ind. Chem., Japan) 73(2), 390-397 (1970).
3. W. Bushuk and H. Benoit, Canad. J. Chem. 36, 1616 (1958).
4. D. W. Tanner and G. C. Berry, J. Pol. Sci., Pol. Phys. Ed. 12, 941
    (1974).
5. H. Morawetz and R. H. Cobran, J. Pol. Sci. 12, 133-148 (1954).
6. H. Morawetz and R. H. Cobran, J. Pol. Sci. 18, 455-460 (1955).
7. S.-Y. Chang and H. Morawetz, J. Phys. Chem. 60, 782-799 (1956).
8. A. C. Ouano and W. Kaye, J. Pol. Sci. A-1 12, 1151 (1974).
9. A. C. Ouano, J. Chromatography 118, 303 (1976).
10. H. Matsuda and H. Inagaki, Macrom. Sci. Chem. A 2(1), 191-208
    (1968).
11. W. H. Stockmayer and M. Fixman, J. Pol. Sci. C 1, 137 (1963).
12. M. Kurata and W. H. Stockmayer, Fortschr. Hochpolymer. Forsch 3,
    196 (1963).
13. G. C. Berry, J. Pol. Sci. B 4, 161 (1966).
14. J. C. Kirkwood and J. Riseman, J. Chem. Phys. 16, 565 (1948).
15. P. Debye and A. M. Bueche, J. Chem. Phys. 16, 573 (1948).
16. G. W. Pyun and M. Fixman, J. Chem. Phys. 42, 3838 (1965).
17. J. R. Schaefgen and P. J. Flory, J. Am. Chem. Soc. 70, 2709 (1948).
18. K. Fujii, J. Ukida and M. Matsumoto, J. Pol. Sci., Polymer Letters
    1, 693 (1963).
19. K. Fujii, J. Ukida and M. Matsumoto, Makrom. Chem. 65, 86-90
    (1963).
20. M. Matsumoto and K. Fujii, Kogyo Kagaku Zasshi (J. Ind. Chem.)
    68, 843 (1965).
21. K. Fujii, K. Shibatani, Y. Fujiwara, Y. Ohyanagi, I. Ukida, and M.
    Matsumoto, Polymer Letters 4, 787 (1966).

GELATION ACCOMPANYING CRYSTALLIZATION FROM DILUTE SOLUTION: SOME GUIDING PRINCIPLES

L. Mandelkern, C. O. Edwards, R. C. Domszy, and
M. W. Davidson
Department of Chemistry and
Institute of Molecular Biophysics
Florida State University
Tallahassee, Florida 32306

ABSTRACT

Gelation accompanying the crystallization from dilute and moderately dilute solution, which has been known to be characteristic of copolymers, is demonstrated to take place in homopolymers as well. The guiding principles by which to study this phenomenon are established. Carefully controlled experiments with both model homopolymers and copolymers are described in terms of these principles. Distinction must be made between homopolymers and copolymers as well as the thermodynamic nature of the solutions from which crystallization is occurring. For homopolymers covering a very wide molecular weight range, as well as for copolymers containing a surprisingly large co-unit content, the gels are found to be comprised of an overlapping supermolecular structure. The often postulated fringed micelle model for gel structure is only applicable, if at all, to copolymers of high co-unit content.

INTRODUCTION

Gels are characterized by the existence of a continuous structure in the system. Within this definition there are, however, a large number of molecular mechanisms which are known to lead to gel formation in a diverse set of macromolecular systems. These gels outwardly display very similar appearances and macroscopic properties. A categorization of the different type gel systems involving macromolecules has been given by Flory.[1] The great diversity that is found has made difficult a molecular description of gel formation and gel properties. Therefore, the development of a set of general principles governing the phenomena are not as yet in hand.

121

There are two very general classes of gel structures. One of these involves amorphous chains as is found, for example, in cross linked swollen networks. The other class involves systems of macromolecules which are in ordered conformation in the gel. This category can be further subdivided. Polymer chains in statistical conformation can undergo a liquid to crystal phase transition in dilute or moderately dilute solutions, and form gels in the process. This mechanism involves a major conformational change for a significant portion of the chain. In another type of ordered system, the molecules are highly asymmetric, or rodlike in character. Such molecules can also form gels in dilute solution. In this case, however, they do so without undergoing any structural or conformational change. The molecular basis for gelation in this case would appear to be quite different from the process involving statistically conformed chains. Examples of gel forming rod-like molecules are collagen and the alpha-helical polypeptides.[2-5] In this present report, we shall restrict our discussion to a consideration of gelation of statistically coiled chains in dilute and moderately dilute solutions which accompanies crystallization.

When gels are formed as a consequence of crystallization the polymer system pervades the complete volume. Hence a very fluid dilute solution is converted to a rigid medium of high viscosity upon gelation. Concomitantly, major changes are observed in mechanical properties. Gels formed by this process are almost invariably thermally reversible. Gelation as a consequence of the crystallization of copolymers from dilute and moderately dilute solutions has been a well known and widely studied phenomenon.[6-9] It has recently been receiving renewed attention.[10-14] The presence of randomly distributed non-crystallizing units in many examples has led to the concept that the gel structure is micellar,[1,11,12,15] as compared to the more conventional lamellar-like crystallites that are observed to form from dilute solution. The logic of the arguments leading to the fringed micelle structure under these circumstances has appeared to be strong and very compelling. However, we must bear in mind that there has not as yet been made available any substantive, direct evidence for the existence of such structures.

In contrast to copolymers, the observation that homopolymers also form gels has been cursorily mentioned in earlier literature. It is only more recently that it has received serious study. Pennings and co-workers demonstrated about ten years ago that gelation occurred during the crystallization of stirred dilute polyethylene solutions.[16,17] More recently, it has been suggested that gelation will take place in the unperturbed or quiescent state if the solution was pre-stirred.[18] It had been a widely held view that stirring or pre-stirring is a requirement for gelation.[18,19] However, the gelation of a very high molecular weight sample of linear polyethylene under quiescent conditions has recently been described by Smith and co-workers.[20,21] We have also recently reported that gels can be observed for a very wide range of linear polyethylene fractions ($\bar{M} = 1\times10^4 - 1.6\times10^6$) without the necessity of any stirring or pre-stirring.[22,23]

The observation that homopolymers can form gels under quiescent conditions raises again the question as to the gelation mechanism and the general validity of the micellar concept. The concept that copolymers are required for gelation and that they lead to fringed micelle type structures, is obviously a very restricted one. The general principles involved clearly need to be re-examined although the fringed micelle structure may be valid in particular circumstances. It remains, however, a restrictive concept in terms of the general phenomenon of gel formation accompanying crystallization.

STATISTICALLY CONFORMED CHAINS

| HOMOPOLYMER | | COPOLYMER |

HOMOGENEOUS SOLUTION

LIQUID-LIQUID PHASE SEPARATION

$$\left\{ \begin{array}{l} \text{METASTABLE} \\ \text{UNSTABLE-SPINODAL DEC.} \end{array} \right\}$$

ORDERED ROD-LIKE CHAINS

Figure 1.   Schematic diagram of categories involving polymer structure and solvent thermodynamics by which to study
crystallization/gelation.

Based on experience it is very convenient to consider certain specific classifications within the major category of gelation accompanying crystallization.   We list these in Figure 1 for   statistically conformed chains crystallizing (geling) under quiescent, non-stirring,conditions. Similar considerations should also be applicable to stirred solutions as well as dilute solutions of ordered rod-like molecules.   As we examine the classification listed in the chart of Figure 1, it becomes very important to distinguish between the crystallization of homopolymers and copolymers.   From the point of view of crystallization behavior any chain irregularity can act as a copolymeric unit.[25]   Besides chemically different units, these include stereo-irregularities, geometric isomers, branch points and head-to-head structures, to cite a few examples. Thus, in the present context of crystallization and gelation isotactic polystyrene, syndiotactic polyvinyl chloride, and the so-called linear low-density polyethylene, for example, are properly to be considered copolymers, in addition to chains containing obviously chemically distinct co-units.

There are two types of solution from which crystallization can occur. One of these is a homogeneous solution, that is, one liquid phase. The other, a heterogeneous system, or a liquid-liquid phase separated system.[26] We also have to distinguish in the latter case as to whether the crystallization (gelation) is occurring from the metastable region, i.e. between the binodal and the spinodal or in the unstable region, within the spinodal. In the metastable region large fluctuations are required for the liquid-like phase separation to take place. In the unstable region infinitesimal fluctuations will continue to grow and pervade the whole phase and crystallization will occur by the so-called spinodal decomposition mechanism.[27] Crystallization, and consequently the accompanying gel structure, should be affected accordingly. Kinetic studies have shown that dilute solution crystallization from a homogeneous phase is a nucleation and growth process.[28,29]

With these guides, it becomes necessary to study and analyze the crystallization and gelation of several different polymer types, in each of the categories, in order to determine the basic principles involved in gel mechanism and in establishing gel properties. The classification in Figure 1 serves as a basis for which diverse experiments that have been reported can be analyzed in a unified manner and as a direction for new experiments with specifically chosen model systems. It makes clear the extreme importance of determining the microstructure of the chain, particularly whether it possesses any copolymeric character, and also establishing the thermodynamic nature of the polymer-solvent system.

In the study of crystallization from solution quantitative data of general validity can only be obtained when molecular weight and compositional fractions are used. Studies with unfractionated systems must be undertaken with prudence for only qualitative indications of the phenomena can be discerned. Fractionation, upon crystallization from solution, because of molecular weight or co-unit concentration, will most often confuse and complicate the experiment.

RESULTS AND DISCUSSION

In the first category to be examined, the gelation of homopolymers from an initial homogeneous solution, we consider molecular weight fractions of linear polyethylene crystallized from xylene.[22,23] This system forms gels when crystallized quiescently, without any stirring or pre-stirring, over the complete molecular weight range. The minimum polymer concentration required for gelation depends on the crystallization temperature and molecular weight. When the homogeneous solutions are rapidly cooled, that is, quenched to 22-25°C, the minimum concentration ranges from 2 wt percent polymer at the lowest molecular weights, to a concentration as low as 0.3 wt percent for the highest molecular weight fraction studied (1.6×10^6). For isothermal crystallization (gelation), at an elevated temperature, such as 86°C, the concentration required for gelation increases. However, the molecular weight dependence remains qualitatively similar. Photographs of typical gels have already been reported.[22] Gels formed with linear polyethylene are turbid and thermally reversible.

In contrast to the gel formation at these concentrations, it is well known,[30] that from more dilute solution, the polymer will precipitate, or crystallize, in the form of isolated lamella-like crystallites. It is theoretically possible to prepare both the lamella-like crystallites and the gels, of the same molecular weight fraction, at the same undercooling. This possibility exists since there is only an imperceptible change between the equilibrium melting temperature for a concentration of about 0.1%, where the platelets form, and the higher concentrations typical of gel formation.[24] Hence, a rationale or natural comparison can be made between the thermodynamic properties of the well-known lamella-like crystallites typical of dilute solution crystallization and those of the crystallites involved in the gel formation when both are crystallized at the same temperature. In Table I, a typical set of such data is given.

In this Table, the dissolution temperature, T_S, i.e. the melting temperature at 0.08% of the undried gel, the small-angle X-ray spacing (SAXS), and the enthalpy of fusion of the dried samples (ΔH), for a gel formed at 86°C from a molecular weight fraction M = 4.45×10^5 are given along with the comparable data for crystallites formed from dilute solution. Extensive data in the literature for dilute solution crystals show that these properties are independent of molecular weight in this range.[31,32] The data in Table I make abundantly clear that the crystallite thicknesses and the accompanying thermodynamic properties

TABLE I[a]

Properties of Dilute Solution Crystals and Gels
Formed at 86°C, for Polyethylene Fraction $\overline{M}_W$ = 4.5 x 10[5]

	Dilute solution[b] crystal 0.08%	Gel formed from xylene at: 2%	5%
SAXS A	133 ± 4	141 ± 4	146.5 ± 4
T_s °C	95.5 ± 1		95.8 ± 1
ΔH(cal/g)	53.3 ± 1		50.6 ± 1

(a) From ref. 22
(b) Literature values 31, 32, 33

for the dilute solution crystals and those of the gel are identical. In addition, we have also found that the enthalpies of fusion of "the as formed gel" and that for the dilute solution crystals, at a comparable concentration, are also very close to one another.

We must conclude from this set of typical data that the crystallite properties of the gel and that of the lamellae formed in dilute solution are the same. Hence, barring an unusual set of coincidences the basic crystallite characteristic of the gel should be of lamellar form for the linear polyethylenes when crystallized from homogeneous solutions. Thus we have at least one system where the crystallization-gelation phenomenon does not involve a fringed micellar structure.

This conclusion with regard to lamellar habit of these gels is supported by direct scanning electron microscopic studies. The results for a very high molecular weight sample was recently published by Smith and co-workers[15] and intimated the presence of a lamellar structure. This type of observation is not limited to the very high molecular weights. From our own work, we show in Figure 2 scanning electron micrographs for linear polyethylene gels of different molecular weights. The character of these micrographs is similar to the one reported by Smith[15] and is not dependent on molecular weight. The lamellar nature of the crystal habit characteristic of the gels, that has been deduced from the physical-chemical studies, thus receives support from the gross morphological features as revealed by the electron microscope studies.

The question that must be addressed next is why thermally reversible gels, with these crystallite properties, are formed under these circumstances. The thermal reversibility is clearly a matter of melting-recrystallization. The reason that gels are formed can be found by examining polarized light micrographs of the undried gels. It is found

Figure 2. Scanning electron micrographs of gels of linear polyethylene fractions formed in xylene at 23°C. (a) M = 4 x 10^3; (b) M = 1.54 x 10^4; (c) M = 5.31 x 10^4; (d) M = 1.13 x 10^5; (e) M = 2.25 x 10^5.

that such gel systems possess well organized supermolecular structures at levels above that of the elementary lamellar crystallite.[23] A typical set of polarized light micrographs is given in Figures 3a and 3b for a high molecular weight linear polyetheylene fraction crystallized at concentrations above and below the critical concentration required for gel formation. Figure 3a represents a case where gel has formed, while in Figure 3b, the concentration is such that gel has not formed. We can see from these examples that the lamellae are organized relative to one another, even at these low polymer concentrations, and a well defined supermolecular structure develops in both cases. These structures have the same character at concentrations below and above the minimum that is required for gelation. Moreover, the size of these structures is about the same in both cases. The difference between the two situations is that gels are formed when the superstructures are significantly over-lapped. When they are relatively isolated from one another gels do not form. At sufficiently low concentrations the isolated lamellar-like crystallites precipitate.

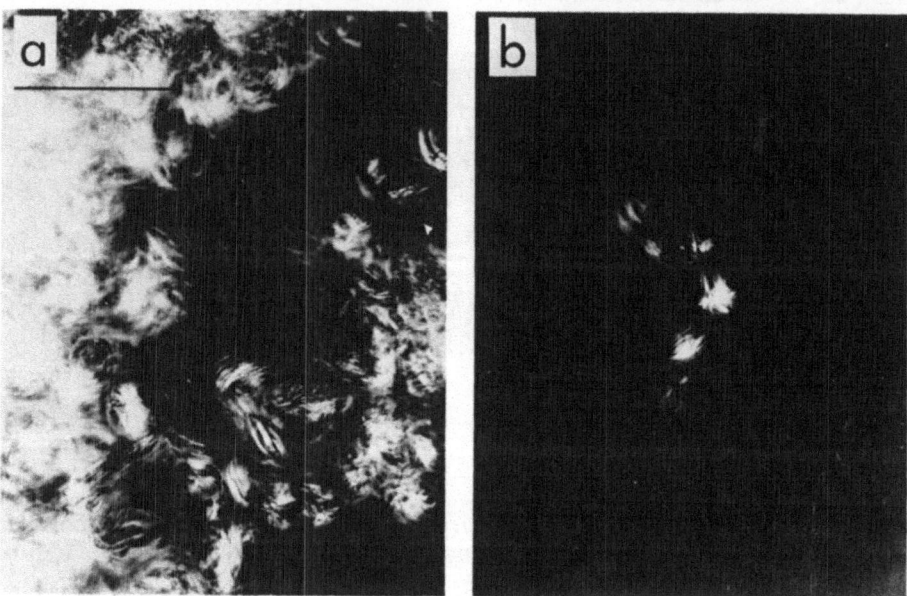

Figure 3. Polarized light micrographs of a linear polyethylene fraction, M = 1.62 x 10^6, crystallized in xylene at 86°C. (a) 1.3 wt. % polymer, forms gel; (b) 0.47 wt. % polymer, does not form gel.

Gelation by this type of crystallization, which yields organized superstructures is not limited to linear polyethylene. A similar phenomenon is also observed in other homopolymers which crystallize from homogeneous solution. For example, for molecular weight fractions of poly(ethylene oxide), crystallized from xylene solutions, gelation is observed over the molecular weight range from $2x10^4$ to $2x10^6$. This molecular weight range could most probably be extended with further study. Typical polarized light micrographs and accompanying small-angle light scattering patterns (SALS), for the undried poly(ethylene oxide) gels, are given in Figures 4, 5 and 6. The light micrographs for each of the molecular weight fractions indicate that the gels are comprised of well-defined overlapping superstructures. From the corresponding SALS patterns we conclude that for the lowest molecular weight sample, M = $2.0x10^4$, poorly developed spherulites are formed. For the two higher molecular weight fractions M = $1.16x10^5$ and $1.77x10^6$ rod-like structures are deduced.[34,35]

This relatively simple gelation mechanism which results from the development of overlapping organized superstructures, with the basic lamellar-like crystallite being maintained, appears to be quite general for homopolymers crystallizing from homogeneous solutions of appropriate concentrations above and below that required for gelation. The major criterion for gelation to occur by this process is that there be sufficient concentration of such superstructures so that they can extensively overlap.

The next classification which we will consider, according to Figure 1, is that of copolymers crystallizing from homogeneous solution. We shall treat in detail chemically distinct copolymers and focus attention on those based on ethylene. The first example that we shall discuss are ethylene-butene copolymers, which are in fact hydrogenated poly-butadienes. This type copolymer has a very narrow molecular weight and composition distribution and the ethyl side-groups are randomly distributed.[36,37] The particular copolymers studied in this group contain 2 mole percent ethyl branches. Thermally reversible gels could be formed over the available molecular weight range of $1.6x10^3$ to $4.2x10^5$. The minimum polymer concentration required for gelation is increased substantially over that required for homopolymers at comparable molecular weights.

A comparison of the properties of the gels formed from this copolymer, with the lamellar crystallites formed from dilute solution at the same temperature and undercooling is again of interest. A typical set of data are given in Table II. The data presented in this Table demonstrate, as was found for homopolymers, that the properties of the crystallites formed from dilute solution and those involved in the gel structure are identical. Thus, for these chemically distinct copolymers the gels do not display a micellar character. This conclusion is supported by the scanning electron micrograph given in Figure 7a. For

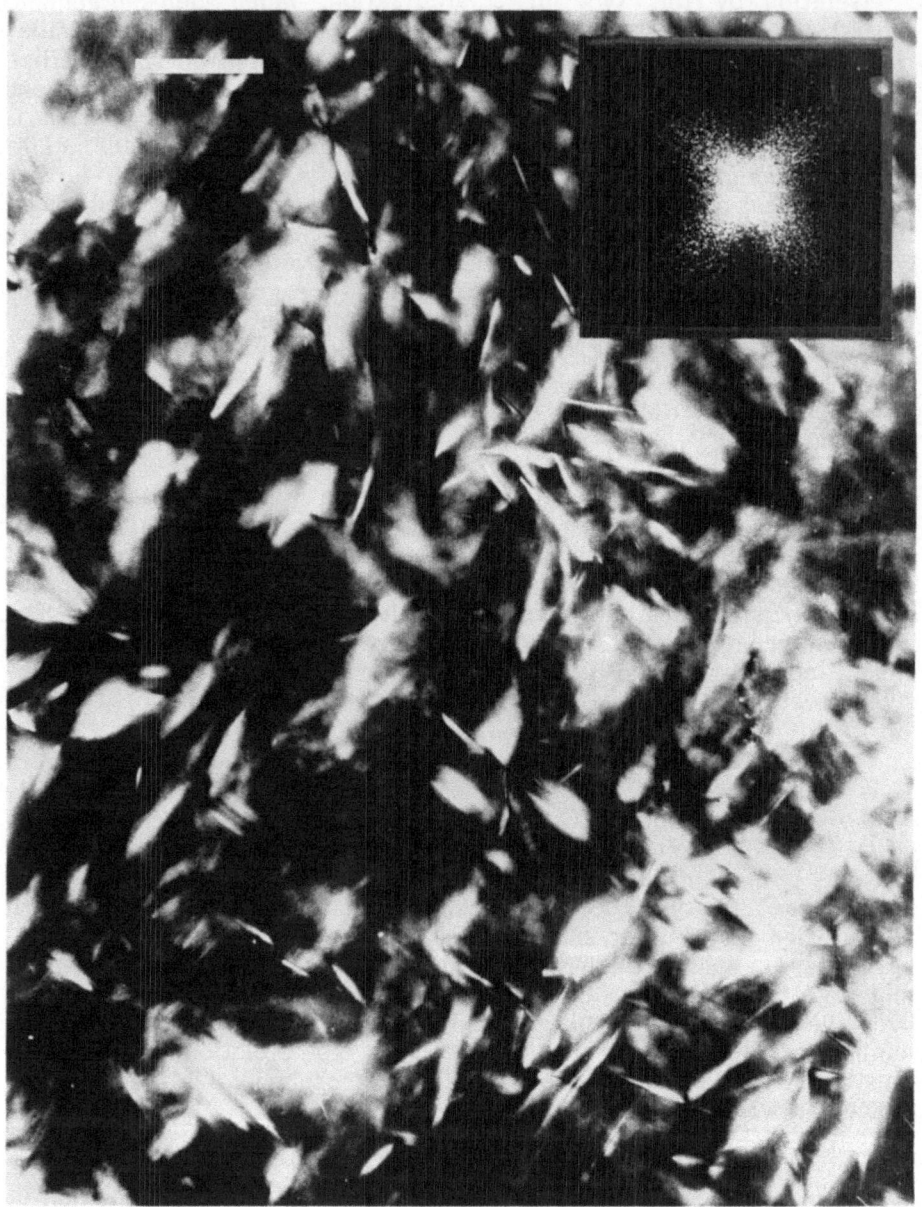

Figure 4.   Polarized light micrograph and SALS pattern of undried gel
             of poly(ethylene oxide) fraction. $M = 2.0 \times 10^4$, 1.8 wt. %
             polymer, formed at 23°C in xylene.

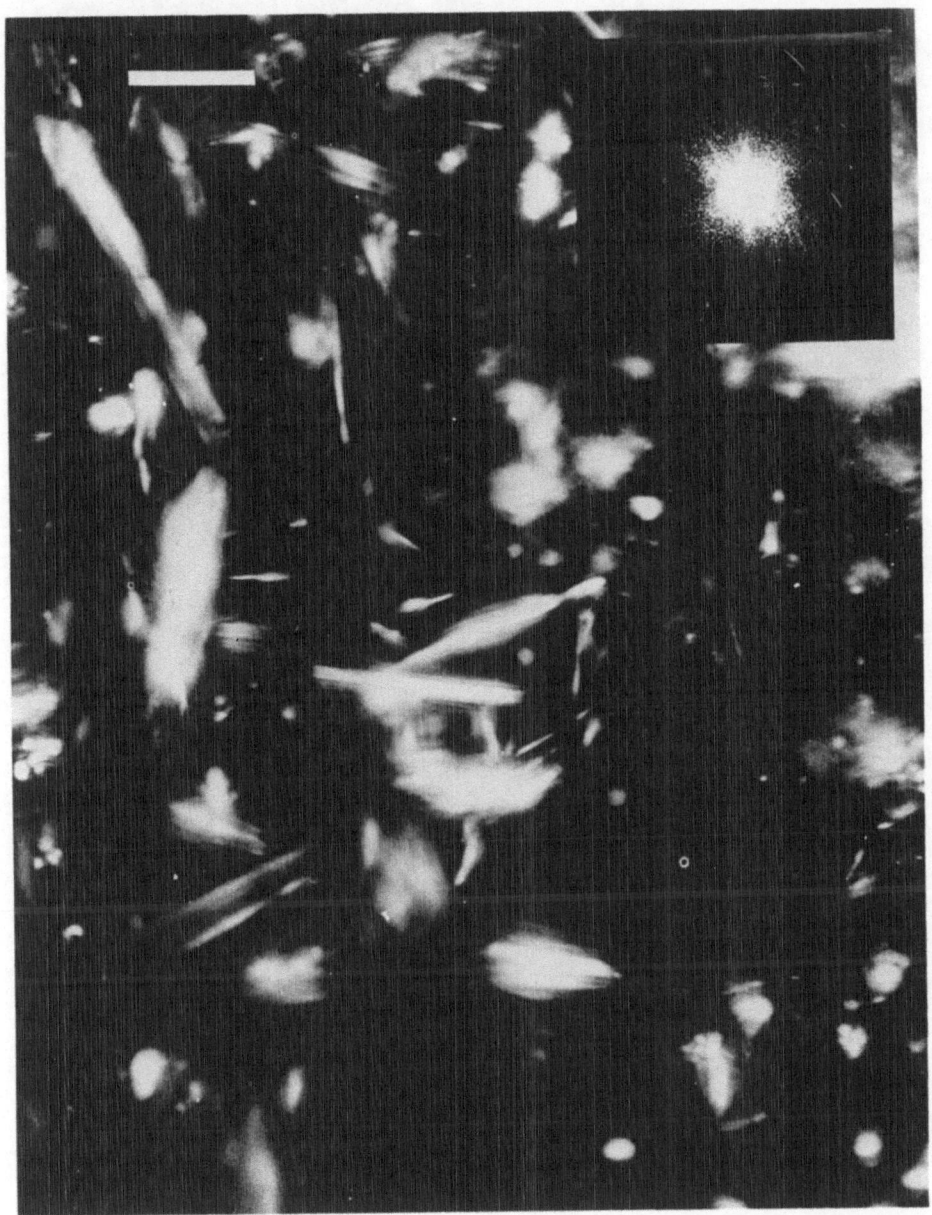

Figure 5. Polarized light micrograph and SALS pattern of undried gel of poly(ethylene oxide) fraction $M = 1.16 \times 10^5$, 1.8 wt. % polymer, formed at 23°C in xylene.

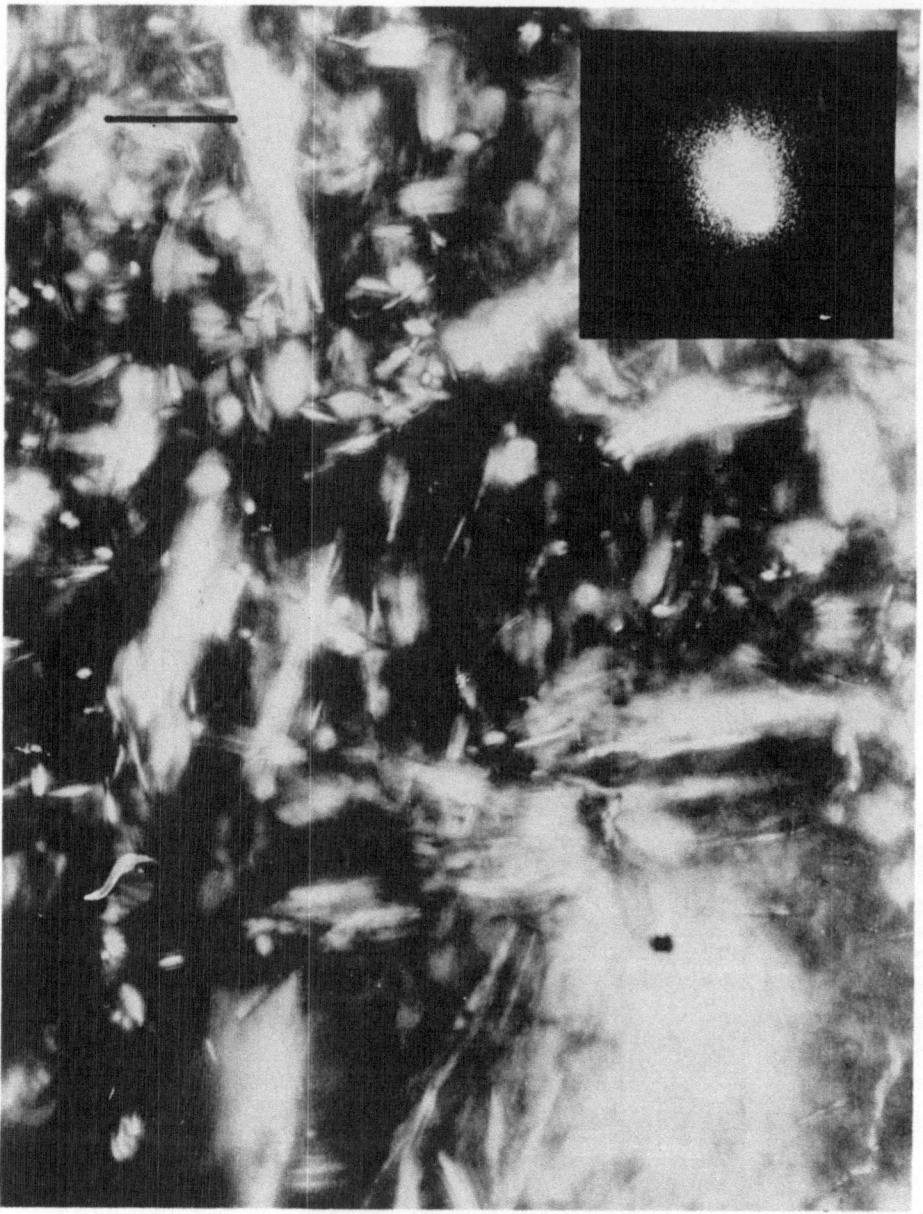

Figure 6.   Polarized light micrograph and SALS pattern of undried gel
             of poly(ethylene oxide) fraction. $M = 1.77 \times 10^6$, 1.48 wt.
             % polymer, formed at 23° in xylene.

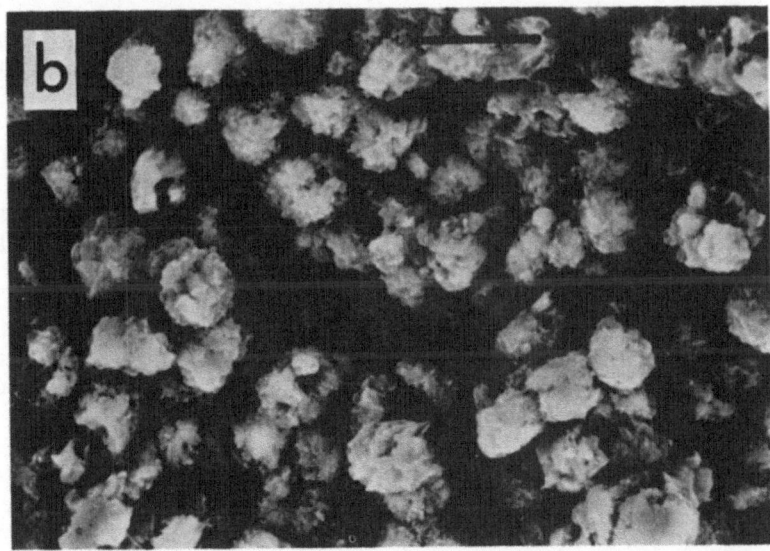

Figure 7.   Scanning electron micrographs of polyethylene copolymers.
(a) Hydrogenated polybutadiene (ethylene–butene)
$M = 108 \times 10^5$ from xylene at $23^{\circ}C$;   (b) Branched
polyethylene $M_W = 2.6 \times 10^5$ from xylene at $23^{\circ}C$.

this case an organized structure is again evidenced although the lamellar character is not as well defined as with the homopolymers illustrated in Figure 2. The lamellae in this case are undoubtedly thinner and restricted in their lateral dimensions and appear to be curled up and packed into a spherical array.[38]

TABLE II

Properties of Dilute Solution Crystals and Gels Formed at 52.8°C in Xylene for an Ethylene-Butene Copolymer Fraction $\overline{M}_w = 4.06 \times 10^5$ [55(a)]

	Dilute Solutione	Gel
SAXS A	109.6 $\pm$ 2	113.9 $\pm$ 2
T_s °C	65.2 $\pm$ 1	66.4 $\pm$ 1
$\Delta H(cal/g)$	36.4 $\pm$ 4	38.3 $\pm$ 4

(a)   Date from ref. 22

Polarized light micrographs of the undried gels formed by these ethylene-butene copolymers show well developed supermolecular structures having a great deal of spherulitic character. Typical examples of the superstructures that are found in these gels are given by the polarized light micrographs of Figure 8a and b. In both these micrographs the spherulitic character of the supermolecular structure is evident. For the sample in Figure 8a (M = 4.1×10^5, 1.6 wt. % polymer) gel is not formed. However, for the sample in Figure 8b (M = 1.94×10^5, 2.3 wt. % polymer) gel is formed. The character of the superstructure has not changed from Figure 8a. However, the concentration of these structures has changed considerably and there is an extensive overlapping. The basis of gel formation for this copolymer system is thus the same as for homopolymers, namely the developing of an overlapping supermolecular structure.

The more conventional type branched polyethylenes that are formed by high pressure free radical polymerization, and are copolymers from a crystallization point of view, are also known to form gels.[39,40] We have confirmed this phenomenon and a scanning electron micrograph of such a gel is given in Figure 7b.*

---

*The characterization of this polymer is as follows: $M_w = 2.6 \times 10^5$; $M_w/M_n = 20$; short chain branches 15.8 per 1000 C atoms; long chain branches 1.2 per 1000 C.[41]

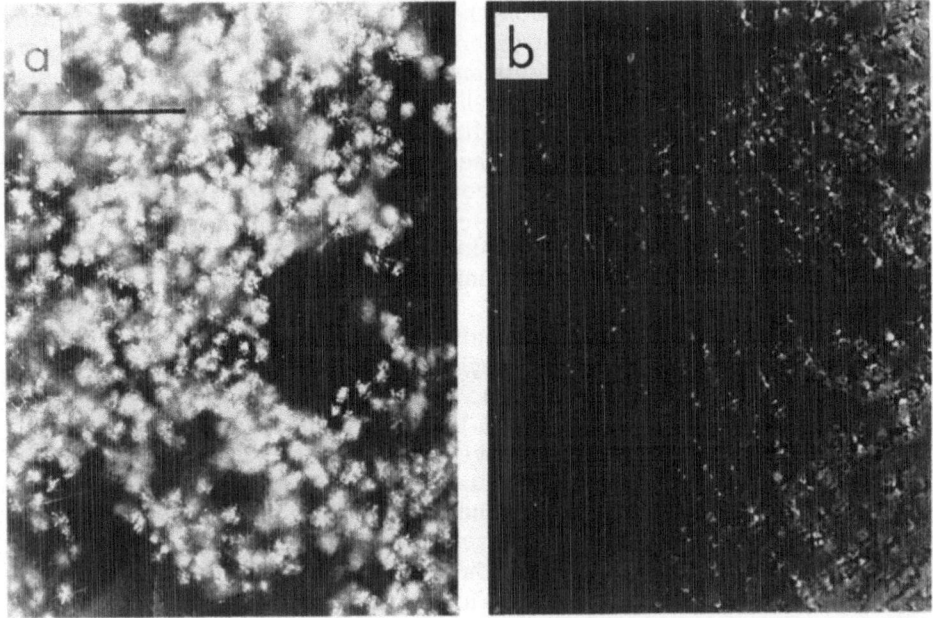

Figure 8. Polarized light micrographs of undried hydrogenated ethylene-butene fractions in xylene. (a) $M = 4.1 \times 10^5$, 1.6 wt. % polymer, gel not formed; (b) $M = 1.9 \times 10^5$, 2.3 wt. % polymer, gel forms.

The structure found here is very similar to that characteristic of the ethylene-butene copolymers (Figure 7a). We conclude that we have a very similar mechanism for gel formation. This conclusion is supported by the well-developed, overlapping spherulitic structures that can be observed in the polarized light micrographs of the undried gels formed by this polymer. Theories that are based on other type structures and mechanisms will need to be reexamined.

Despite the inherent initial attractiveness of the fringed micelle concept, these type structures cannot be the general characteristic for gelation accompanying crystallization of either homopolymers or co-polymers. These structures certainly are not required for the gelation of copolymers despite the attractiveness of the concepts that led to this postulate.[6,11,12] The essential fact that has been established here is that gelation can be associated with lamellar crystallites, and the associated supermolecular structure despite a branching or co-unit content of as much as 2 mole percent.

The very important question then arises, as to what happens when the co-unit concentration is increased further. It is well known that for crystallization from the pure melt an increase in co-unit content lowers the melting temperature and level of crystallinity of random type copolymers. The level of crystallinity can become very small and eventually vanishes.[24,25] We are then concerned with the question as to whether gels can form at high co-unit content and if so, are the crystallites still lamellar in character. The companion question is whether the gel mechanism remains that of an overlapping super-molecular structure for higher co-unit copolymers.

We can get our first insight into this problem from the results of a preliminary study of ethylene-vinyl acetate copolymers. For explora-tory purposes we have used an unfractionated set of samples.* The macroscopic characteristics of these gels and the critical concen-tration required for gelation are given in Table III. For the lower co-unit contents, up to about a four mole percent total, turbid gels are formed and an overlapping type spherulitic supermolecular structure is observed by polarized light microscopy. However, for the last two copolymers listed in the Table, where the total co-unit content has increased to the order of five and six mole percent, although gels are still formed they are now clear. The polarized light micrographs no longer yield such well-defined supermolecular structures. In fact a decrease in the turbidity of the gels can be noted when the acetate branches become greater than the order of one mole percent. The turbidity thus decreases with increasing co-unit concentration. Except for the turbidity, all the gels described display the same outward or macroscopic appearance. There is, therefore, a co-unit content above which there is a change in the structure and presumably in the mechanism of gel formation. This qualitative conclusion can be reached from studies with these unfractionated systems. For the reasons cited previously a more quantitative analysis must await further studies using compositional and molecular weight fractions.

The structural changes that occur with increasing co-unit content are supported by the scanning electron micrographs for this system that are given in Figure 9. For the lower co-unit content copolymers, spherical structures encompassing a collection of lamellae are observed. These are similar to the structures shown in Figure 7 for other copoly-mers. However, as the co-unit content increases these well-defined spherical structures become smaller and begin to deteriorate. For the

---

*For these copolymers the co-units consist of both acetate groups and butyl branches. The composition of each of the copolymers studied is listed in Table III. Since an unfractionated set of copolymers is being used the results should only be used for qualitative guidance.

highest co-unit copolymers studied no structure can be seen although a gel has definitely formed. The suggestion could be put forth that at the very highest co-unit content structures similar to that envisaged for the fringed micelle might be manifesting themselves.

We consider next the gelation/crystallization process from heterogeneous solutions, i.e. at temperatures and concentrations which are below the liquid-liquid phase separation boundary. Here several additional factors need to be established and considered. These include whether the process is being conducted from the metastable or unstable region and the relative rates of phase separation and crystallization. Extensive studies in this category have not as yet been completed to yield definitive quantitative results. However, we have found that crystallization from amyl acetate, a typical liquid-liquid system for the polyethylenes, does not yield gels for either linear polyethylene fractions or the ethylene-butene copolymer. Instead, at concentrations at which gels would normally be observed white precipitates are formed from this solvent. The basis for this result and the question as to whether it is a general phenomenon for crystallization from liquid-liquid systems is being investigated at present.

TABLE III

Characteristics of Gels Formed from Ethylene
Vinyl Acetate Copolymers in Xylene at 24°C

Acetate Branches[a]	Butyl Branches[a]	Appearance	Critical concentration for gelation[b]
0.33	1.10	Turbid	2-3
0.91	1.90	Turbid	4-5
0.90	1.30	Turbid	4-5
1.7	0.88	Turbid	
2.6	1.30	Turbid	
4.3	1.0	Clear	4-5
5.1	0.78	Clear	8.5-11

(a) mole percent
(b) weight percent

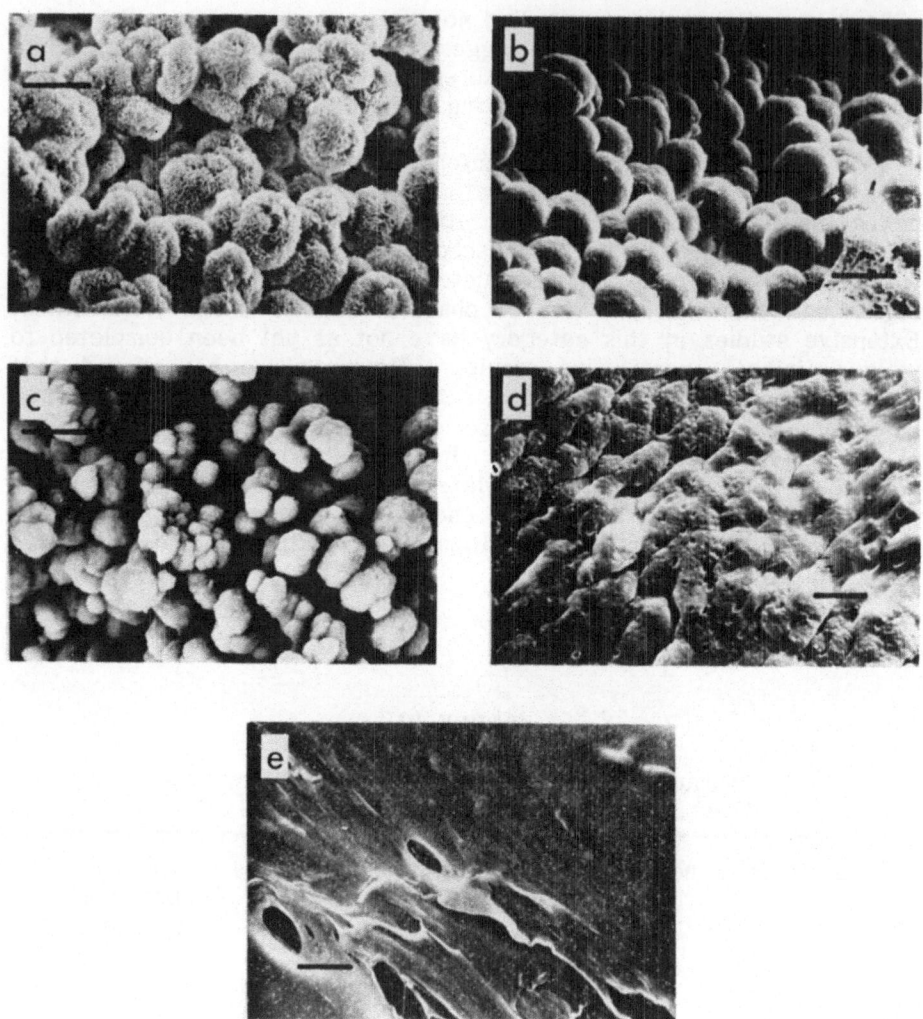

Figure 9. Scanning electron micrograph of gels of ethylene–vinyl acetate copolymers formed in xylene at 23°C. (a) 0.33 mole % acetate; 5 wt. % polymer; (b) 0.91 mole % acetate; 7 wt. % polymer; (c) 0.90 mole % acetate; 6.5 wt. % polymer; (d) 4.3 mole % acetate; 6 wt. % polymer; (e) 5.1 mole % acetate 11.4 wt. % polymer.

The gelation of stereo-irregular, vinyl type polymers such as iso-tactic poly(styrene) and syndiotactic poly(vinyl chloride) has been receiving a great deal of interest.[11,13,14] Following our guiding principles, we note that such polymers are properly to be considered copolymers, because of stereoirregularities and head-to-head placements.[12] Consequently sequence distributions, rather than overall co-unit content will play a crucial role in governing crystallization and the accompanying gelation process. Supermolecular structures have not as yet been observed for gels formed from these type systems. The absence of such structures is of course consistent with their copolymeric character.

It has been thought that the distinction between platelet crystal-lization and gelation in the polystyrene decalin system was caused by fractionation.[12] However, recent work has shown that a similar pheno-menon is found with fractionated polymers.[42] More quantitative results for these type polymers are clearly in order and will be forthcoming as the thermodynamic nature of the solvent systems are established and the microstructure of fractionated polymers delineated.[42]

In summary, we find that at present the classification outlined in Figure 1 provides a logical and useful framework within which to examine the problem of gelation accompanying crystallization from dilute and moderately dilute polymer solutions. Gels, without micellar structure, are easy to obtain and are quite commonly observed. These findings, although contrary to our initial expectations follow naturally from a consideration of the morphological factors which accompany the crystallization of polymers. The micellar, or closely related, type structure appears to be associated with gels formed from chains con-taining a significant amount of structural irregularities.

We have not considered here gel formation involving highly ordered, geometrically asymmetric macromolecules such as collagen[2] and the α-helical poly-α amino acids.[3-5] For these systems gelation by means of a spinodal decomposition mechanism has been proposed.[3] More quantitative studies are needed to establish the validity of this principle.

ACKNOWLEDGEMENT

Acknowledgement is made to the Donors of the Petroleum Research Fund, administered by the American Chemical Society for the support of this research.

REFERENCES

1. P. J. Flory, Disc. Farad. Soc. 57 7 (1974).
2. J. M. Cassel, L. Mandelkern and D. E. Roberts, J. Amer. Leather Chem. Assoc. 57 556 (1962).
3. W. G. Miller, Ann. Rev. Phys. Chem. 29 519 (1978); K. Tohyama and W. G. Miller, Nature 289 813 (1981).
4. R. J. Oteri, Masters Thesis, Florida State U., December 1980.
5. S. Sasani, M. Hikato, C. Shiraki and I. Uematso, Polymer J. 14 205 (1982).
6. L. Mandelkern, Crystallization of Polymers, McGraw-Hill, New York 1964, pp. 113 and 308.
7. A. Takahashi, T. Nakamura and I. Kagawa, Polym. J. 3 207 (1972).
8. A. Takahashi, Polym. J., 4 379 (1973).
9. A. Takahashi and S. Hiramitsu, Polym. J. 6 103 (1974).
10. J. Lemstra and G. Challa, J. Polym. Sci. Polym. Phys. Ed. 13 1809 (1975).
11. M. Girolamo, A. Keller, K. Miyasaka and N. Overbergh, J. Polym. Sci., Polym. Phys. Ed. 14 39 (1976).
12. R. Benson, J. Maxfield, D. E. Axelson and L. Mandelkern, J. Polym. Sci., Polym. Phys. Ed. 16 1583 (1978).
13. P. J. Lemstra, A. Keller and M. Cudby, J. Polym. Sci., Polym. Phys. Ed. 16 1507 (1978); S. J. Guerroro and A. Keller, J. Macrom. Sci. Phys. B20 167 (1981).
14. S. M. Aharoni, G. Chailet and G. Delmar, Macromolecules 14 1394 (1981).
15. A. Keller, Farad. Soc. Disc. 68 145 (1979).
16. A. J. Pennings, I. M. A. A. van der Mark and H. C. Booij, Koll. Z. Polym. 236w 99 (1970).
17. A. J. Pennings, J. Polym. Sci., Polym. Symp. 59 55 (1977).
18. P. J. Barham, M. J. Hill and A. Keller, Coll. and Polym. Sci. 258 899 (1980).
19. D. C. Bassett, Principles of Polymer Morphology, Cambridge U. Press (1981).
20. P. Smith, P. J. Lemstra and H. C. Booij, J. Polym. Sci., Polym. Phys. Ed. 19 877 (1981).
21. P. J. Lemstra and P. Smith, British Poly. J. pg. 212 (1980).
22. C. O. Edwards and L. Mandelkern, J. Polym. Sci., Polym. Lett. Ed. 20 355 (1982).
23. C. O. Edwards, Ph.D. Dissertation, Florida State U., December 1981.
24. P. J. Flory, J. Chem. Phys. 17 223 (1949).
25. Ref. (6) pp. 74 ff.
26. P. J. Flory, Principles of Polymer Chemistry, Cornell U. Press (1953) pg. 495 ff.
27. J. W. Cahn, J. Chem. Phys. 42 93 (1965).
28. C. Devoy, L. Mandelkern and L. Bourland, J. Polym. Sci. A2 8 869 (1970).
29. L. Mandelkern, Polymer 5 637 (1964).

30. Ref. (6) pg. 307 ff.

31. J. F. Jackson and L. Mandelkern, Macromolecules 1 546 (1968).

32. R. K. Sharma and L. Mandelkern, Macromolecules 3 758 (1970).

33. L. Mandelkern, J. Phys. Chem. 75 3909 (1971).

34. J. Maxfield and L. Mandelkern, Macromolecules 10 1141 (1977).

35. L. Mandelkern, Disc. Farad. Soc. 68 310 (1979).

36. L. Mandelkern, M. Glotin and R. A. Benson, Macromolecules 14 22 (1981).

37. M. Glotin and L. Mandelkern, Macromolecules 14 1394 (1981).

38. I. G. Voigt-Martin, E. W. Fischer and L. Mandelkern, J. Polym. Sci., Polym. Phys. Ed. 18 2347 (1980).

39. A. Takahashi, M. Sakai and T. Kato, Polymer J. 12 335 (1980).

40. H. Matsuda, H. Fujimatsu, M. Imaizumi and S. Kuroiwa, Polym. J. 13 807 (1981).

41. D. E. Axelson, G. C. Levy and L. Mandelkern, Macromolecules 12 41 (1979).

42. R. Domszy and L. Mandelkern, to be published.

AGGREGATION, PHASE BEHAVIOR AND THE NATURE OF NETWORKS

FORMED BY SOME ROD-LIKE POLYMERS

Wilmer G. Miller, Sumana Chakrabarti and Kathleen M. Seibel
Department of Chemistry
University of Minnesota
Minneapolis, Minnesota 55455

ABSTRACT

The rheological properties of four polymer-solvent systems involving rod-like polymers which form gels were investigated. The systems studied were rod-like polybenzylglutamate (PBLG) in dimethylformamide (DMF), a solvent in which it does not aggregate, and in toluene, a solvent in which it end-to-end aggregates; polymerized deoxy-hemoglobin S (HbS) in aqueous solution; and bronchial mucus, having as a main component a stiff chain glycoprotein. The storage modulus of PBLG was found to be remarkably independent of concentration, molecular weight, temperature and solvent, and did not correspond to the "brush-pile" (rod polymer entanglements) theory of Doi. Sickle cell hemoglobin gels were found to have rheological properties very similar to PBLG gels. We propose that HbS does not show gel properties due to "brushpile" behavior, but indeed forms a three dimensional network. The rheological properties of the bronchial glycoprotein were found to be quite distinct from PBLG or HbS. A strong concentration dependence of the storage modulus was observed. At concentrations delivered from the lungs bronchial mucus is a non-integral, weak gel. Only upon concentrating do the rheological properties indicate an integral network.

INTRODUCTION

There are numerous examples of random coil polymers forming networks (gels) not involving covalent bonds, e.g., poly(vinyl chloride) in di-2-ethylhexyl phthalate,[1] polyacrylonitrile in dimethylformamide,[2] poly-acrylglycinamide in water,[3] celluloseacetate in benzyl alcohol,[4] and polyethylene in xylene.[5] In some systems the network branch points are clearly a result of specific interchain aggregation as, for example,

143

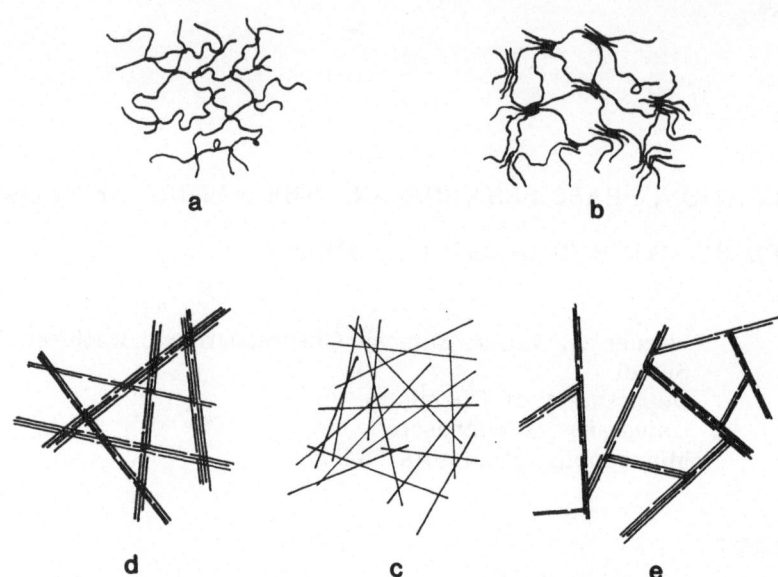

Figure 1.   Representation of a noncovalent network of random coils
             crosslinked by interchain association (a) or by micro-
             crystalline domains (b); non-covalent network of rod-like
             polymers or polymer aggregates whose "crosslinking" is a
             manifestation of a brush-pile of rods (c) or fibers
             (aggregates of aligned rods) (d), or a result of non-
             nucleated phase separation kinetics (e).

interchain hydrogen bonding. In other systems there is substantial
evidence that the branch points involve microcrystallites, although their
nature is not always clearly established.[6,7] The existence of crystalline
order indicates that the network was formed at a temperature below the
crystalline melting point of the polymer. In this case a relationship exists
between network formation and phase behavior. Crystallinity may also be
induced after covalent crosslinking above the melting point.[7,8] Irrespective of the nature of the branching in these noncovalent networks,
their rheological properties are consistent with that expected for random
coils spanning branch points, as depicted in Figure 1a,b with energy
storage upon deformation residing in the change in conformational
entropy.

Numerous polymers of both synthetic and biological origin are rod-
like or at least locally stiff due to their primary or secondary structure,
or form specific aggregates which are rod-like.   Polyphenylene, poly-
isocyanate, para-aromatic polyamides and poly(bisbenzoxazole) are rod-
like as a consequence of their primary structure, and the intramolecularly
hydrogen-bonded α-helical polyamino acids are rod-like as a consequence
of their secondary structure. The nature of rod-like aggregates is quite
varied. They are generally of biological origin. One type consists of

multichain helices, e.g., double stranded DNA, native collagen and the polysaccharide schizophyllan.[9] A far more numerous type is composed of a globular protein, which under suitable conditions reversibly aggregates through specific interactions to form rod-like (generally helical) structure.[10] These include F-actin, flagellin, sickle cell hemoglobin and tubulin. A still more complex type involves specific interaction of globular protein with a nonglobular polymer as, for example, the tobacco mosaic virus particle.

Many of the rod-like systems are known to form reversible gels (networks) under appropriate conditions through some type of interaction among the rod-like particles.[11-15] Regardless of the nature of the branchpoints, the polymer spanning from branchpoint to branchpoint is not a random coil, i.e., the system cannot be thought of as the traditional Gaussian network. What, then is the structure of the network, and what are its properties? We summarize here our work on three polymers - poly-γ-benzyl-α, L-glutamate (PBLG), sickle cell hemoglobin (HbS) and bronchial mucus, whose main component is a glycoprotein.

SYSTEMS STUDIED

PBLG - Dimethylformamide (DMF)

Dilute solutions of PBLG in DMF at 25°C disperse without aggregation[12] in the α-helical conformation with a persistence length of about 1200Å.[16] Molecular weights below 10^6 have viscosities which differ only slightly from that expected for rigid rods.[12] At higher concentrations the liquid crystalline phase becomes the stable phase.[17] All of the features of the phase behavior of rigid rod polymers predicted by Flory,[18] Figure 2, have been observed in the PBLG-DMF system; this includes the narrow biphasic region,[19,20] the wide biphasic region[21] and the LC-LC cap.[22] In dry DMF the wide biphasic region is entered slightly below room temperature, dependent upon the concentration.[20] The impetus for all of the work reported herein was the observation that a three dimensional network was formed upon lowering the temperature and crossing into the wide biphasic region.[20] Why was network formation dependent on the phase boundary, and what are the properties and structures of this network?

PBLG - Toluene

In solvents other than DMF or benzyl alcohol PBLG is generally observed to associate,[12,23-26] although there is not always agreement as to whether the association is specific end-to-end (utilizing the unmade hydrogen bonds at each end of the helix) or nonspecific side-by-side interaction.[24,25,27,28] By a study of the temperature, concentration and molecular weight dependence, we have shown that PBLG in toluene can

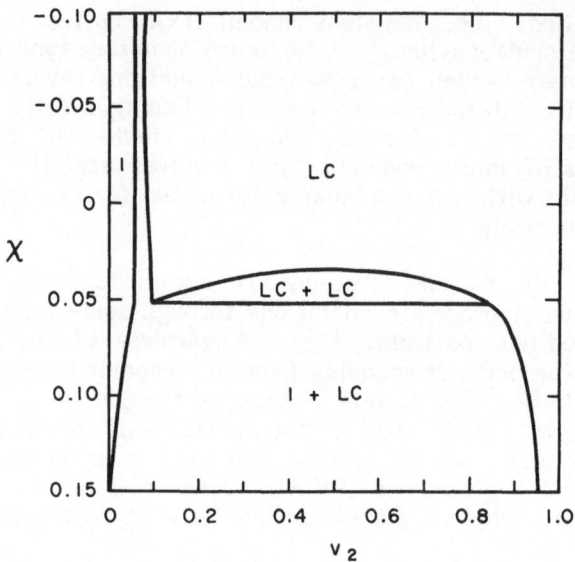

Figure 2. Flory phase diagram[18] for a rigid rod-diluent. Phases present are isotropic(I) or liquid crystalline (LC).

be quantitatively modeled as an open, reversible, end-to-end aggregation.[29] Shown in Figure 3 is the temperature dependence of the aggregation of mixed molecular weights, using the association constant determined for aggregation in the 138000 molecular weight sample to predict the values for the mixed molecular weights. There can be little doubt that the dominant mode of aggregation is end-to-end. The enthalpy of association, -4.3 kcal per mole of PBLG, is sufficient to lead to very viscous solutions at rather low concentrations, thus making the effect of aggregation on the phase boundaries difficult to assess.

Sickle Cell Hemoglobin (HbS)

In the absence of oxygen HbS, a globular protein which is a mutant of normal hemoglobin, is well-known to gel.[15] Long rodlike aggregates are formed through specific non-covalent interaction.[30-32] The enthalpy of aggregation is positive.[33,34] Thus aggregation is enhanced by an increase in temperature, which is opposite to that of PBLG in toluene.[29] Of the several types of processes which one might identify (Figure 4) only the first two are deemed necessary to form a gel, although it is not made clear as to why aligned rods should behave as a gel.[35,36] The kinetics of the gelation process is considered to be rate limited by the nucleation of rod formation. The various processes identified in Figure 4 cannot easily be studied independently.

g1/g2
∞
1.41
0.625
0
ηr
T(°C)
1
2
3
4
5
6
30
40
50
60
70
80
90

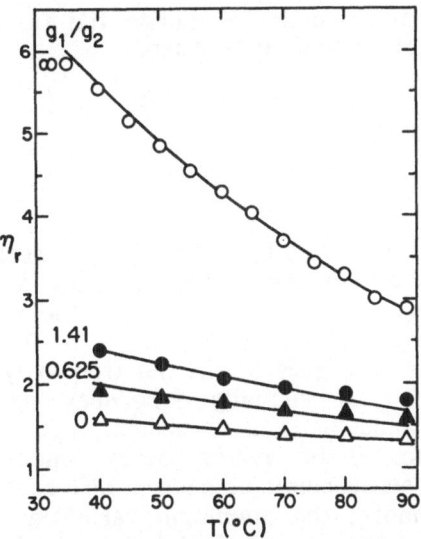

Figure 3. Temperature dependence of the relative viscosity of 0.1 gm dl^{-1}PBLG in toluene, using mixed molecular weights as indicated (g_1/g_2; gm(138000)/gm(28000)). The solid line was calculated using the association constant for open end-to-end aggregation determined with the 138,000 molecular weight data.[29]

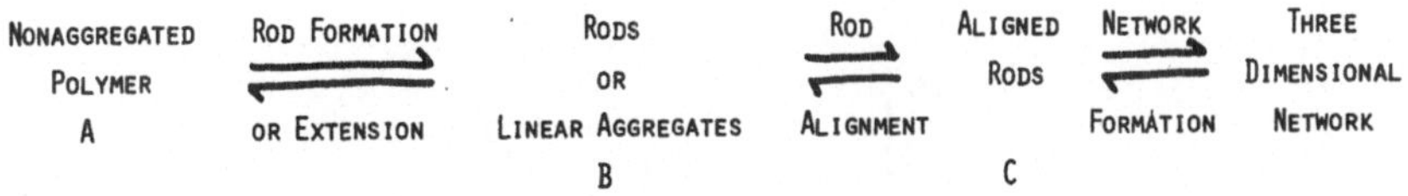

NONAGGREGATED POLYMER A
ROD FORMATION OR EXTENSION
RODS OR LINEAR AGGREGATES B
ROD ALIGNMENT
ALIGNED RODS C
NETWORK FORMATION
THREE DIMENSIONAL NETWORK

Figure 4. Steps which may lead to rod "gels."

Bronchial Mucus

Several organs, e.g., the lungs, the mouth, the stomach, the intestines and the vagina produce mucus, whose principal component is a glycoprotein.[37,38] Although the carbohydrate as well as the amino acid sequence may differ among the various glycoproteins produced by these organs, there appear to be many features in common. The hydrodynamic properties of the purified glycoprotein suggest a molecule that is rather rod-like. On a weight basis the carbohydrate component is considerably larger than the protein component. In an idealized view the glycoprotein may be considered to look like a bottle brush - a protein core, attached carbohydrate (the bristles), and an exposed section of protein not containing carbohydrate (the handle). Although aggregation as well as network formation may involve covalent disulfide bonds inasmuch as sulfhydryl reducing agents appear to change the fluidity, the role of the disulfide bonds is not clearly understood.

RHEOLOGICAL BEHAVIOR

PBLG Networks

Solutions not in the wide biphasic region, whether isotropic or liquid crystalline, flow and do not exhibit gel behavior. If a solution is brought into the wide biphasic region by lowering the temperature, a gel is formed. The dynamic mechanical behavior of a 10.6 wt.% PBLG in DMF gel at 10°C is shown in Figure 5. In the frequency range measured, the storage modulus (G') is virtually frequency independent. At low frequencies the loss modulus is so small compared to the storage modulus that we cannot obtain reliable values. Measurement on a given sample is very reproducible. Measurements on several different samples of the same composition differ more, the maximum variation in G' was at most a factor of 3. The dynamic mechanical behavior of these gels has been studied as a function of concentration, temperature, temperature of formation, aging, molecular weight, strain and solvent.[39] The effect of concentration, molecular weight, solvent and temperature of formation on the storage modulus is shown in Figure 6. The temperature dependence of a network formed at one temperature and then aged at other temperatures is shown in Figure 7.

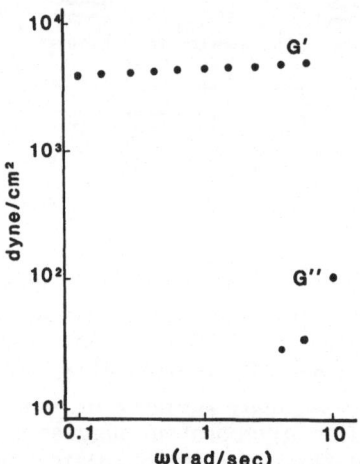

Figure 5. The storage (G') and loss (G") modulus as a function of frequency (2% strain) of 10.6 wt.% PBLG (MW 138000) in DMF at 10°C, measured on a Rheometrics fluids rheometer.

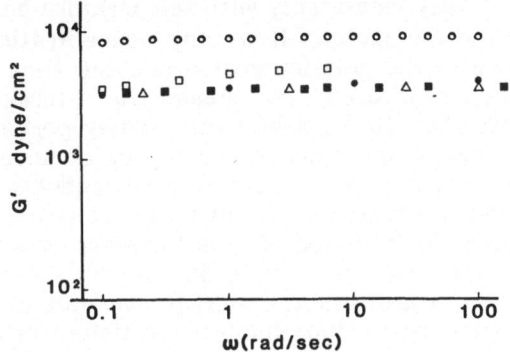

Figure 6. Representative plots of the frequency dependence of the storage modulus as a function of concentration, molecular weight, solvent and temperature of gelation.

Symbol	Wt.%	Mol. Wt.	Solvent	Gelation Temp.
(0)	75	310,000	DMF	2.8°C
(●)	60	310,000	DMF	1.5
(□)	3	138,000	Toluene	23
(△)	1	8,000	Toluene	24
(■)	.25	138,000	Toluene	24

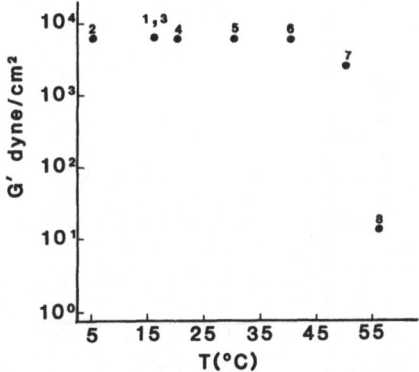

Figure 7. The storage modulus at $\omega=0.1$ rad sec^{-1} of a 1 wt.% PBLG in toluene sample gelled at 16°C and measured as a function of temperature in the order as indicated.

The modulus does not vary measurably with temperature until close to the temperature where the gel melts. If at any concentration a sample is temperature jumped into the gelation region and the time dependence of the dynamic moduli monitored, no meaningful storage modulus is measured (torque less than 10^{-3} gm-cm) for a delay period ranging from minutes to hours, depending inversely on the degree of undercooling below the phase boundary. Once a meaningful storage modulus is observed, it increases rapidly and approaches an ultimate value typically within minutes. Further aging for a period of days to weeks does not change the modulus. However, placing the sample in steady shear, followed by dynamic measurements, results in the storage modulus dropping to very low and frequently meaningless (unreliably measurable) values. Aging of the gel for hours to days does not restore the storage modulus of the sheared gel.

HbS Gels

The rheological properties of HbS gels have been studied as a function of concentration, temperature of gelation, temperature, mixtures with HbA and met-HbS, pH and electrolyte concentration.[40,41] Typical results are shown in Figure 8 for the concentration dependence. The storage modulus was independent of ionic strength (up to 2), pH (4.5-9.5) and nature of oxygen depletion (dithionite or N_2). The storage modulus of gelled samples was not affected by the presence of met-HbS in the sample (up to at least 20%) or of HbA (up to 50%). In aged samples the modulus was little dependent on the temperature of gelation or the temperature of measurement after gelation above a minimum tempera-

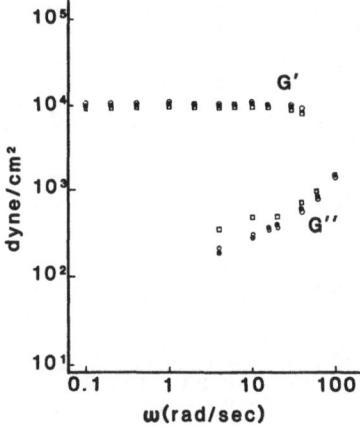

Figure 8. The storage and loss modulus for HbS gelled at 37°C at pH 6.86 by sodium dithionite addition at concentration 15 gm dl^{-1}(□), 20 gm dl^{-1}(0) and 30 gm dl^{-1} (●).

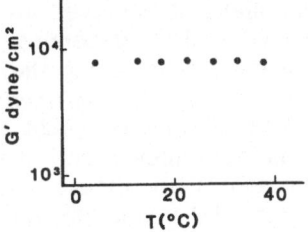

Figure 9.   The storage modulus at $\omega = 0.1$ rad sec^{-1} of an aged 30 gm dl^{-1} HbS gel as a function of temperature after being gelled at 37°C.

ture (depending on the concentration), as shown in Figure 9.   Analogous to the PBLG results, the kinetics of the magnitude of the storage modulus after a sample was temperature jumped into the gelation region depended on how close one was to the melting temperature.   Lowering the temperature below about 5°C resulted in loss of a measurable G', i.e., the sample was fluid.   As with PBLG gels, steady shear resulted in irreversibly lowering G', which could only be brought back to its original value by going through a temperature cycle through the "melting" temperature.

Bronchial Mucus

The rheological properties of bronchial mucus was investigated from patients with cystic fibrosis (CF) as well as from normal patients.

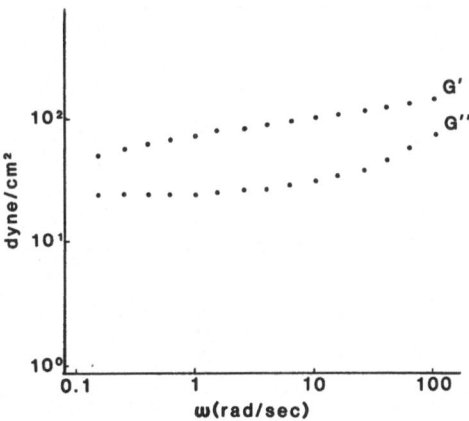

Figure 10.   The storage and loss modulus of bronchial mucus as a function of frequency at 25°C.

Typical results[42] from specimens as received are shown in Figures 10 and 11. The absolute values of G' and G" depended upon the specimen. The frequency dependence, however, was nearly the same in all samples, CF or normal. The dynamic viscosity, $_\eta$*, was linear with frequency and had a slope that was independent of sample (-0.84 $\pm$ 0.03 averaged over 250 samples). Concentrating the specimen resulted in a dramatic change in rheological properties, as shown in Figure 12. It is clear that the rheological properties of bronchial mucus are quite concentration dependent.

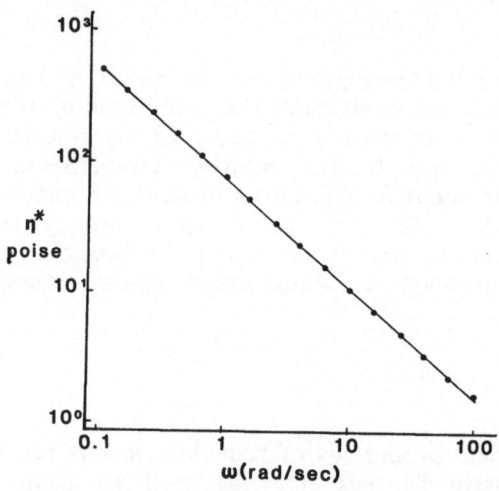

Figure 11.   The dynamic viscosity of bronchial mucus. or normal.

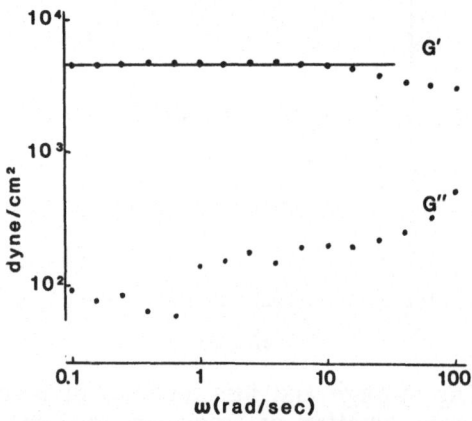

Figure 12.   Effect of concentrating the bronchial mucus.

DISCUSSION

The rheological properties of PBLG are remarkably independent of concentration, molecular weight, temperature and presence or absence of end-to-end aggregation. Below 0.25 wt.% in toluene, G' starts to drop off with concentration, presumably because the concentration is so low that an integral network can no longer be formed.[39] However, in the concentration range from 0.25 wt.% in toluene to 75 wt.% in DMF the maximum variation, taken as the extreme values measured but keeping away from the phase boundary, is $G' \propto c^{<|0.2|}$. The variation in G' in different samples of the same composition is as large as the variation among samples of different concentrations. The stated value of $|0.2|$ is simply a reflection of a factor of 3 variation over all samples spanning a 300 fold change in concentration. With a given sample the temperature dependece of G' shows no consistent trend and appears to vary less than T^{+1}. To our knowledge the concentration dependence is unprecedented. Gelatin is reported[43] to have a c^2 dependence, poly(vinyl chloride)[1] about c^4, and polyacrylonitrile[2] is reported as about c^{22}. We have proposed previously[44] that gelation in PBLG resulting from crossing the phase boundary is controlled by the kinetic mechanism of the phase separation, namely that the polymer network is formed by spindodal decomposition. Visualization of the network[45] shows strong evidence that the gel behavior is not the result of a "brush-pile," the stiff chain polymer equivalent of chain entanglements. A study of the gelation kinetics by light scattering strongly supports a non-nucleation mechanism.[46] A "brush-pile," whether of individual rods or of bundles (fibers) of rods as depicted in Figure 1c,d, would be expected to have rheological properties which are strongly concentration dependent (c^5).[47,48] Thus the morphology as depicted in Figure 1e, a result of spinodal decomposition coupled with a low driving force for ripening, would appear to be the structure of the gel. If one postulates that the wet bundles of aligned rods have imperfections at the branch points, the branch points may be the weak points in the network. Since the number of branch points is controlled by the phase separation mechanism, the molecular weight, concentration and temperature dependence can be at least approximately explained. What one must postulate, then, is that the amount of material in the branch points is not significantly concentration dependent, and that upon deformation, energy is stored in the branch points.

Turning to the HbS gels, the similarity in rheological behavior to the PBLG gel is remarkable. From the nature of the concentration dependence it seemed to us that the "brush-pile" effect, rod entanglement, does not explain our results. In addition, the equilibrium isotropic phase contains over 10% non-aggregated hemoglobin,[34] which should serve to rearrange the gel into its thermodynamically more stable crystal state,[49] which does not happen over a period of days. Thus we believe that steps A B C, Figure 4, are not sufficient to explain the results. Inasmuch as aggregation to give reversibly extended rods did not effect the properties of the PBLG gel (PBLG in toluene versus PBLG in DMF), we see no reason not to postulate a mechanism analogous to that discussed for PBLG gelation.   That is, when HbS aggregates to a rod length where the isotropic–liquid crystal phase boundary is crossed, the kinetics of the phase separation controls the formation of a three dimensional network. This is an entirely different phase formation model from that previously proposed.[33-35]

   Previous work done on mucus glycoproteins has shown the similarity among mucus from different organs.[50-54] The rheological properties of these stiff chain polymers is obviously different from that of PBLG or HbS gels.   The modulus is highly concentration dependent.   At the concentration normally excreted an integral network is not achieved. In fact, the rheological properties are very similar to gelatin gels at low concentrations where integral networks are also absent.[42]   At higher concentrations the frequency dependence approaches that expected for a classical Gaussian network. Although little is known about phase behavior in these glycoproteins, it is clear that the gel structure will be different from PBLG or HbS gels.

SUMMARY

   Rod-like polymers may aggregate, or their rod-like nature may be a result of reversible aggregation of a globular polymer. The rods may align to form fibers, or an ordered phase. Under appropriate conditions a phase boundary may be crossed leading to a biphasic (isotropic–ordered) system. Network-like behavior may result from single rod entanglements (Figure 4, state B; Figure 1c), from fiber entanglements (Figure 4, state C; Figure 1d), or from true three dimensional network formation (Figure 4, state D; Figure 1e). Our studies on PBLG under aggregating or non-aggregating conditions and on HbS gels indicate that true networks are formed in these systems. Bronchial glycoproteins behave differently, perhaps involving interactions of the exposed protein part of the molecule, as has been suggested.[55]

ACKNOWLEDGMENT

This work was supported by the National Institutes of Health.

REFERENCES

1. A. T. Walter, J. Polym. Sci., 13, 207 (1954).
2. J. Bisschops, J. Polym. Sci., 17, 89 (1955).
3. H. C. Haas, C. K. Chiklis and R. D. Moreau, J. Polym. Sci. A-1 8, 1131 (1970).
4. K. D. Goebel and G. C. Berry, J. Polym. Sci., Polym. Phys. Ed., 15, 555 (1977).

5. P. J. Barham, M. J. Hill and A. Keller, Colloid and Polymer Science, 258, 899 (1980).

6. L. Mandelkern, Polymer Preprints, 23, 25 (1982).

7. L. Z. Rogovina and G. L. Slonimskii, Russian Chemical Reviews, 43, 503 (1974).

8. T. Tanaka, D. J. Fillmore, I. Nislico, S. T. Sun, G. Swislow and A. Shah, Phys. Rev. Letters, 45, 1636 (1980).

9. K. Van, T. Norisuye and A. Teramoto, Mol. Cryst. Liq. Cryst., 78, 1233 (1981).

10. F. Oosawa and S. Asakura, "Thermodynamics of the Polymerization of Protein," Academic Press, New York, 1975.

11. F. C. Bawden and N. W. Pirie, Proc. Royal Soc. London, Ser. B, 123, 274 (1937).

12. P. Doty, J. H. Bradbury and A. M. Holtzer, J. Am. Chem. Soc., 78, 947 (1956).

13. T. I. Bair, P. W. Morgan and F. L. Killian, Macromolecules, 10, 1396 (1977).

14. A. J. Bur and L. J. Fetters, Chem. Rev., 76, 727 (1976).

15. V. M. Ingram, Nature, 178, 792 (1956).

16. N. Ookubo, M.Komatsubara, H. Nakajima and Y. Wada, Biopolymers, 15, 929 (1976).

17. C. Robinson, Trans. Faraday Soc., 52, 571 (1956).

18. P. J. Flory, Proc. Royal Soc. London, Ser. A., 234, 73 (1956).

19. C. Robinson, J. C. Ward and R. B. Beevers, Discuss. Faraday Soc., 25, 29 (1958).

20. E. L. Wee and W. G. Miller, J. Phys. Chem., 75, 1446 (1971).

21. W. G. Miller, J. H. Rai and E. L. Wee, in "Liquid Crystals and Ordered Fluids," eds. J. F. Johnson and R. S. Porter, Vol. 2, Plenum, New York, 1974, p. 243.

22. P. S. Russo and W. G. Miller, Macromolecules, in press.

23. H. G. Elias and J. Gerber, Makromol. Chemie, 112, 142 (1968).

24. A. K. Gupta, C. Dufour and E. Marchal, Biopolymers, 13, 1293 (1974).

25. H. Kihara, Polymer J., 9, 443 (1977).

26. C. M. Balik and A. J. Hopfinger, J. Colloid Interface Sci., 67, 118 (1978).

27. A. Wada, J. Polym. Sci., 45, 145 (1960).

28. A. K. Gupta, C. Strazielle, E. Marchal and H. Benoit, Biopolymers, 16, 1159 (1977).

29. S. Chakrabarti and W. G. Miller, Biopolymers, in press.

30. J. T. Finch, M. F. Perutz, J. F. Bertles and J. Dobler, Proc. Natl. Acad. Sci. USA, 70, 718 (1973).

31. G. Dykes, R. H. Crepeau and S. J. Edelstein, Nature, 272, 506 (1978).

32. T. E. Willems and R. Josephs, J. Mol. Biol., 135, 651 (1979).

33. P. D. Ross, J. Hofrichter and W. A. Eaton, J. Mol. Biol., 96, 239 (1975).

34. P. D. Ross, J. Hofrichter and W. A. Eaton, J. Mol. Biol., 115, 111 (1977).

35. J. Hofrichter, P. D. Ross and W. A. Eaton, Proc. Nat. Acad. Sci. USA, 71, 4864 (1974).
36. C. T. Noguchi and A. N. Schechter, Blood, 58, 1057 (1981).
37. "Mucus," ed. J. R. Clamp, Brit. Med. Bull., 34, No. 1, 1978.
38. "Modern Problems in Pediatrics," 19, S. Kagel AG, 1977.
39. S. Chakrabarti and W. G. Miller, Biorheology, to be submitted.
40. S. Chakrabarti and W. G. Miller, J. Mol. Biol., to be submitted.
41. S. Chakrabarti, J. Anderson and W. G. Miller, Biorheology, to be submitted.
42. K. M. Seibel and W. G. Miller, Biorheology, to be submitted.
43. P. R. Saunders and A. G. Ward, Proc. 2nd Intern. Congr. Rheol., Oxford (1953), p. 284, as given in A. Veis, "The Macromolecular Chemistry of Gelatin," Academic Press, 1964.
44. W. G. Miller, L. Kou, K. Tohyama and V. Voltaggio, J. Polym. Sci., Polym. Symp., 65, 91 (1978).
45. K. Tohyama and W. G. Miller, Nature, 289, 813 (1981).
46. P. Russo and W. G. Miller, Macromolecules, to be submitted.
47. M. Doi and N. Y. Kuguu, J. Polym. Sci., Polym. Phys. Ed., 18, 409 (1980).
48. M. Doi and S. F. Edwards, J. Chem. Soc., Faraday Trans. II, 74, 560, 918 (1978).
49. J. G. Pumphrey and J. Steinhardt, J. Mol. Biol., 112, 359 (1977).
50. M. T. Lopez-Vidriero and L. Reid, Mod. Probl. Paediat., 19, 1205 (1977).
51. M. Litt, Mod. Probl. Paediat., 19, 195 (1977).
52. M. King, Pediatr. Rec., 15, 120 (1981).
53. N. Eliezer, Biorheology, 11, 61 (1974).
54. M. Litt, M. A. Khan and D. P. Wolf, Biorheology, 13, 37 (1976).
55. F. A. Meyer and A. Silberberg, Respiratory Tract Mucus, CIBA Symposium 54, Elsevier, 1978, p. 203.

PART III

ORDERING IN POLYELECTROLYTE SOLUTIONS

PART III.

ORDERING IN POLYELECTROLYTE SOLUTIONS

POLYELECTROLYTIC ASPECTS OF CONFORMATIONAL TRANSITIONS AND INTERCHAIN INTERACTIONS IN IONIC POLYSACCHARIDE SOLUTIONS: COMPARISON OF THEORY AND MICROCALORIMETRIC DATA

S. Paoletti
A. Cesàro
F. Delben
Istituto di Chimica
Università di Trieste
Trieste, Italy
V. Crescenzi*
R. Rizzo
Istituto di Chimica Fisica
Università di Roma
Rome, Italy

ABSTRACT

Analytical expressions are obtained to evaluate the electrostatic enthalpy of mixing the univalent-salt form of a polyelectrolyte with a neutral salt of the same valency in aqueous solution, following the theory developed by G. S. Manning, avoiding restrictive assumptions regarding the ionic parameters. An endothermic effect is always predicted for the enthalpy of mixing (corrected for the heat of dilution of the polyelectrolyte), which is function of the linear charge density parameter, ξ, of the ratio of neutral salt to the monomoles of polymer, and of the amount of salt initially present in the system. In this way a quantitative method is suggested to test the conformational parameters of polyelectrolytes and their dependence on ion-induced changes, provided that the main non-conformational effect of added salt on the polymer is mostly electrostatic. Specific effects (as site-binding), if present, should be additionally accounted for. However, in the case of the pH-induced conformational transition of polygalacturonate the electrostatic approach is able to give a self-consistent reference for the excess heat effects therein involved. The ionic polysaccharides which have been chosen for confronting the theoretical predictions with the experimental microcalorimetric data are polygalacturonate, polyguluronate, polymannuronate and alginate.

INTRODUCTION

The empirical evidence that practically all biopolymers active in life processes bear electrical charges has led to speculate on the possible intimate relationship between the various forms of biological structures, as they have historically developed on earth, and the ionic character of their fundamental units.[1]   In the context, the class of ionic polysaccharides is neither little populated nor of vanishing importance.[2,3] They are, in fact, directly involved in many delicate biological roles such as, for instance, in the control of the intercellular matrix[4] or in the regulation of the clotting properties of blood.[5]   In addition, important applications have been successfully brought about in pharmacological and in various other industrial fields.[6]   One important feature of most ionic polysaccharides is to give rise in aqueous media to ion-dependent conformational transitions between states characterized by different extent of order, often up to the level of interchain ordered association.[3,7]   As one consequence of such single-chain or multi-chains cooperative conformational changes, the formal charge density along the macroions may result in some cases more than doubled with significant changes in their polyelectrolytic behavior.   Therefore, the need for a theory of charged polymer solutions which could satisfactorily account for the conformation dependent polyelectrolytic properties of ionic carbohydrate polymers has increased.   Among the theories which have been developed until now, the so-called counterion condensation theory has gained   much attention and wide   application.   Firstly proposed by F. Oosawa,[8] it has been particularly developed by G. S. Manning both from the fundamental point of view and from that of the experimental-test.[9-12] The relatively simple formalism which the theory provides has led to extensive application, so that the merits and the limitations of the theory start being assessed.   By and large, the formers seem to overcome the latters as far as applications to different polyelectrolytes are concerned, in particular for the class of polynucleotides including DNA.[13]   One possible reason for the particular success with these polymers is the intrinsic rigidity of their chain backbones, which is one of the requisites of the theory itself.   Several polysaccharides do also present such a feature, so that we have decided to make an extensive application of Manning's theory to different ionic polysaccharides.   As a source of experimental evidences we have focussed our interest onto the calorimetric approach, for various reasons.   Calorimetry has already proved to give prominence to long-range effects in biopolymeric systems,[14] which is the case of polyelectrolytic interactions.   The development of modern instrumentation presently enables one to collect accurate experimental data on systems very close to "working" conditions (sometimes even "in vivo"),[15] to which one often cannot apply more conventional techniques (e.g. spectroscopy).   Finally, the experience which has been gained since more than ten years in the field of calorimetry of polymer solutions has induced us to widen the range of macromolecular systems to be investigated to ionic polysaccharides with the collection of a number of relevant data.[16] Before analyzing some of these results, a detailed survey of the energetic aspects of counterion condensation (theory) will be given; different novel derivations will be proposed and a few old ones will be revisited to assess the validity of the underlying assumptions. For sake of simplicity, the selection of experimental results will be limited to the class of polyuronates, linear polysaccharides bearing the weakly acid carboxylic group. A more comprehensive survey including polysaccharide sulfates will follow.

DESCRIPTION OF THE MODEL AND BASIC EQUATIONS

The derivation of the equations for the electrostatic part of the enthalpy function of a polyelectrolyte solution will be given hereafter within the frame of G. S. Manning's counterion condensation theory. It may be useful to recall the main assumptions (and limitations) of the theory,[17] so that further assessment of the results is made easier:

a. the real chain is replaced by an infinitely long cylinder of radius equal to zero (infinite line charge);

b. the actual distribution of discrete charges is replaced by a linear charge density, β . $\beta = Z_p.e.b^{-1}$, where Z_p is the valence of the charged groups on the polyion, $\underline{e}$ is the value of the elementary charge, and b is the average spacing of charged groups along the cylinder axis (i.e. the possible helical distribution of charges on the real chain is replaced by a linear one);

c.   interactions between two or more polyions are neglected regardless of the ratio of polyelectrolyte concentration to that of added simple salt;

d.   the dielectric constant D is taken as that of the pure bulk solvent;

e.   for dilute solutions, a definite amount of counterions will "condense" on the polyion to lower the dimensionless charge density parameter ξ to $\xi_{crit} = | Z_iZ_p |^{-1}$, where Z_i is the valence of the counterion. ξ if defined as:

$$\xi = \underline{e}^2/DkTb ,\qquad\qquad (1)$$

where $\underline{e}$, D and b have been previously defined, T is the absolute temperature and k is the Botzmann constant. (All quantities are to be expressed in the proper CGS units). The following treatment will be restricted   to the   case of monovalent counterions   and   groups ($| Z_i | = | Z_p | = $ 1), so that the critical condition reads: $\xi_{crit} = $ 1; it will be formally referred to the case of a polyanion;

f. the uncondensed mobile ions may be treated in the Debye–Hückel approximation;

g. the reference state is chosen as that of the solution containing an amount of small ions equivalent to the sum of the ions stemming from the added simple salt at concentration c_s and those of the polysalt at concentration c_e. The self contribution of mobile ions to the excess free energy of the solution is therefore neglected.

If the electrostatic (Helmholtz) free energy, F^{el}, of the system is approximately evaluated in terms of ionically screened (Debye–Hückel) potentials of the type $r^{-1}\exp(-\underline{K}r)$, summed over all pairs of charges on the same polyion (thus neglecting interactions betwen polyions), one finds:

$$F^{el} = -n_e \underline{R} \, T \quad \ln(1 - e^{-\underline{K}b}) \tag{2}$$

where n_e is the total number of moles of univalent charged groups on the polyions, and $\underline{K}$ is the Debye screening parameter given by

$$\underline{K}^2 = \lambda \, (c_+ + c_-) \tag{3a}$$

with

$$\lambda = 4\pi e^2 \, N/10^3 DkT \tag{3b}$$

where c_+ and c_- refer to the counterions and the coions, respectively, N is the Avogadro number and all other quantities have been defined previously. Moreover,

$$c_j = n_j \, V^{-1} \quad (j = +, -, e) \tag{3c}$$

where V is the volume of the solution, and

$$c_+ = c_e + c_s \tag{3d}$$

$$c_- = c_s \tag{3d}$$

Neglecting the small difference between the Helmholtz and the Gibbs free energies for a condensed phase, one readily obtains the electrostatic (excess) enthalpy of the polyelectrolyte solution:

$$H^{el} \simeq \left(\frac{\partial (F^{el}/T)}{\partial \, 1/T} \right)_{n_j,b} = -n_e \underline{R} \left[\left(\frac{\partial \, \xi}{\partial \, 1/T} \right)_{n_j,b} \ln(1 - e^{-\underline{K}b}) + \right.$$

$$\xi \left(\frac{\partial \ln(1-e^{-Kb})}{\partial 1/T} \right)_{n_j,b} \Bigg] = -\tfrac{1}{2}n_e RT \ \xi \left(1 + \frac{d \ln D}{d \ln T} \right) \Bigg[2 \ln(1 - e^{-Kb}) +$$

$$\frac{Kb}{e^{Kb} - 1} \Bigg] \tag{4}$$

The following partial derivatives have been used:

$$\left(\frac{\partial \ \xi}{\partial \ 1/T} \right)_{n_j,b} = T \ \xi \left(1 + \frac{d \ln D}{d \ln T} \right) \tag{5}$$

$$\left(\frac{\partial \ \ln(1-e^{-Kb})}{\partial \ 1/T} \right)_{n_j,b} = \frac{1}{2} \ \frac{Kb}{e^{Kb}-1} \ T \left(1 + \frac{d \ln D}{d \ln T} \right) \tag{6}$$

ENTHALPY OF MIXING (DILUTION) OF POLYELECTROLYTE SOLUTIONS

1. The Case of $\xi < 1$. The task is now to obtain an expression for the (electrostatic) enthalpy change upon mixing n_e moles of polymeric salt (in the presence of $n_{s,i}$ moles of simple salt) with an additional amount of simple salt, $n_{s,add}$, so that the ratio of (total) simple salt to polymer equivalents passes from

$$R_i (R_i = \frac{n_{s,i}}{n_e}) \text{ to } R_f (R_f = \frac{n_{s,i} + n_{s,add}}{n_e} = \frac{n_{s,f}}{n_e}).$$

By application of equation (4) to the final and initial states of the mixing process one obtains:

$$\Delta\overline{H}_{mix} = \frac{H_f^{el} - H_i^{el}}{n_e} = \frac{1}{2} \underline{R}T \left(1 + \frac{d \ln D}{d \ln T} \right) \xi \, x$$

$$x \left(2 \ln \frac{1 - e^{-\underline{K}_f b}}{1 - e^{-\underline{K}_i b}} + \frac{\underline{K}_f b}{e^{\underline{K}_f b} - 1} - \frac{\underline{K}_i b}{e^{\underline{K}_i b} - 1} \right) \tag{7}$$

where $\underline{K}_f^2 = \lambda \, c_e^f (2R_f + 1)$, $\underline{K}_i^2 = \lambda \, c_e^i (2R_i + 1)$, λ has been previously defined, and c_e^i and c_e^f refer to the polymer equivalent concentration in the initial and in the final solution, respectively.

Before proceeding further, an explicit word of caution is needed as far as the reference state is concerned. Equation (7) as well as all the equations derived from, or related to it, refer to the heat effect on mixing the polyelectrolyte solution with the solution containing the simple salt corrected for the heat effects due to the mixing two solutions containing equivalent amounts of simple salt (i.e. $n_{s,i} + n_e$, and $n_{s,add}$, respectively).

For $\underline{K}_f b$, $\underline{K}_i b \ll 1$, it can be easily verified that:

$$\lim_{\underline{K}_f b, \underline{K}_i b \to 0} \left(\frac{\underline{K}_f b}{e^{\underline{K}_f b} - 1} - \frac{\underline{K}_i b}{e^{\underline{K}_i b} - 1} \right) = 0$$

$$\text{and} \quad \lim_{\underline{K}_f b, \underline{K}_i b \to 0} \left(2 \ln \frac{1 - e^{-\underline{K}_f b}}{1 - e^{-\underline{K}_i b}} \right) = \ln \frac{\underline{K}_f^2 b^2}{\underline{K}_i^2 b^2} =$$

$$= \ln \frac{c_e^f}{c_e^i} + \ln \frac{2R_f + 1}{2R_i + 1}$$

Under such circumstances, equation (7) will therefore reduce to:

$$\Delta \overline{H}_{mix} = -\tfrac{1}{2} \underline{R}T \; \xi \left(1 + \frac{d \ln D}{d \ln T} \right) x$$

$$x \left(\ln \frac{c_e^f}{c_e^i} + \ln \frac{2 R_f + 1}{2 R_i + 1} \right) \tag{8}$$

Equation (8) coincides with that (equation 5a) already reported by Boyd, Wilson and Manning[18] for the case of $R_i = 0$.

It is to be remarked that if $R_i = R_f$ (i.e. dilution of polymer salt initially in the presence of R_i moles of simple salt per mole of fixed charge against <u>pure solvent</u>) then:

$$\Delta \overline{H}_{mix} = -\tfrac{1}{2} \underline{R}T \; \xi(1 + \frac{d \ln D}{d \ln T}) \; \ln \frac{c_e^f}{c_e^i} = \Delta \overline{H}_{dil} \tag{9}$$

Equation (9) has been already reported by a number of authors[19-22] only for the salt-free case (i.e. $R_i = R_f = 0$): it is shown here that it holds for <u>any mixing</u> (dilution) experiment at constant R(with $\underline{K}_{i,f}b << 1$). Equation (8) can be split into two separate contributions:

$$\Delta \overline{H}_{mix} = \Delta \overline{H}_{dil} + \Delta \overline{H}_{salt} \qquad \text{where}$$

$$\Delta \overline{H}_{dil} = -\tfrac{1}{2} \underline{R}T(1 + \frac{d \ln D}{d \ln T}) \; \xi \; \ln \frac{c_e^f}{c_e^i} \tag{10a}$$

$$\Delta \overline{H}_{salt} = -\tfrac{1}{2} \underline{R}T(1 + \frac{d \ln D}{d \ln T}) \; \xi \; \ln \frac{2R_f+1}{2R_i+1} \tag{10b}$$

$\Delta \overline{H}_{salt}$ is the (molar) enthalpy change associated with the process of addition of salt to polymer solution, at constant polymer concentration. Since it is always true that $c_e^f < c_e^i$ and $R_f > R_i$, and since $(1 + \frac{d \ln D}{d \ln T})$ for water at 298 K is -0.372,[23] it follows that $\Delta \overline{H}_{dil}$ will always assume negative values, while $\Delta \overline{H}_{salt}$ will always be > 0.

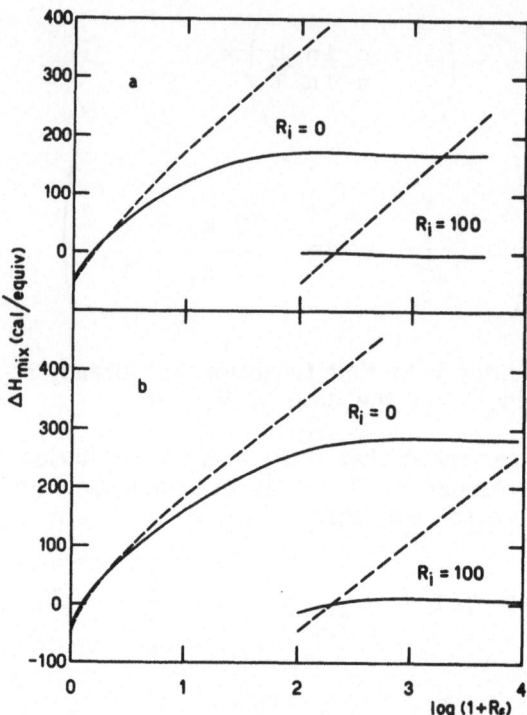

Figure 1.   a. Dependence of $\Delta \bar{H}_{mix}$ on $\log(1 + R_f)$, calculated with $\xi = 0.693$ for the cases $R_i = 0$ and $R_i = 100$ by using equation (7) (full curves) and equation (5a) of Reference 18 (dashed curves). b. Same as in a., but with $\xi = 1.640$. Full curve drawn according to equation (13); dashed curve drawn according to equation (5b) of Reference 18. The value of T is 298 K for this Figure as well as for the following ones.

The dependence of $\Delta \bar{H}_{mix}$ on salt concentration, expressed as $\log(1 + R_f)$, calculated by equation (7) for the case of $\xi = 0.693$ (κ-carrageenan in the random coil form) has been reported in Figure 1a. Examples are given for two different values of initial ratio of salt to polymer (R_i) as full lines.   For $R_i = 0$, the salt-free case, $\Delta \bar{H}_{mix}$ increases from exothermic to endothermic values, reaching an asymptotic value at $R_f >$ 100.   For $R_i = 100$, $\Delta \bar{H}_{mix}$ is practically constant and almost equal to zero. Clearly, the effect of the addition of salt is to suppress all excess effects due to the polyelectrolytic nature of the chain, as expected. The curves calculated according to the approximated equation 5a of Reference 18 have been reported as dashed lines in Figure 1a, too. For the salt-free case, the departure of the behavior of the latter from the predictions of equation (7) becomes significant on increasing R_f. The approximated equation predicts a linear increase of $\Delta \bar{H}_{mix}$ with $\log(1 + R_f)$ above $R \simeq$ 1. For $R_i = 100$ the discrepancy between the predictions of equations (7) and 5a of Reference 18 gets much larger. Boyd et al.[18] reported a deviation of the experimental $\Delta \bar{H}_{mix}$ values from the calculated curve for the case of sodium polystrenesulfonate at the highest R_f values, which approached a constant positive value rather than to increase steadily. Those authors did not discuss such a deviation, which we attribute to the breakdown of the assumption that $Kb \ll 1$, underlying the derivation of the approximated equation. In our opinion, this may be a central point in the discussion of several works concerning the experimental test of G. S. Manning's theory. Much care is always to be given to check if the physical conditions meet the requirements of the assumptions on which the equations to be tested are based.

2. The Case of $\xi > 1$. The detailed derivation of equations so far reported would be of little practical help if it couldn't be extended to the case of values of ξ larger than the critical value of 1. In order to achieve that, some additional considerations are to be made. The total (excess) free energy of polyelectrolytic origin for $\xi > 1$ certainly has to include a purely electrostatic term (similar to that of eq. (2)) and a "mixing" term contributed by the associated counterions, the free counterions and by (negligible) contributions of coions and solvent.[24,25] The latter term is supposed to be of entropic nature only, and independent of temperature: it will therefore be thought not to contribute to the total (excess) enthalpy. To evaluate the former, electrostatic, contribution we can apply eq. (4) to the general case in which the effective value of ξ is permanently reduced to the critical value of unity, while the number of uncompensated fixed charges in solution has changed from n_e to n_e / ξ due to counterions condensation.[22] The equation for H^{el} will therefore read:

$$H^{el} = -\tfrac{1}{2} n_e \, \underline{R}T \, \xi^{-1} \left(1 + \frac{d \ln D}{d \ln T}\right) \left[2 \ln (1 - e^{-\underline{K}b}) + \frac{\underline{K}b}{e^{\underline{K}b} - 1} \right] \tag{11}$$

where $\underline{K}^2$ has accordingly changed to:

$$\underline{K}^2 = \lambda \, (c_+ + c_- - c_{cond}) \tag{12a}$$

and c_+, c_- and c_{cond} stand for the total stoichiometric concentration of counterions and coions, and for the bulk concentration of condensed counterions, respectively. It is to be recalled that:

$$c_+ = c_s + c_e \quad ,$$

$$c_- = c_s \quad ,$$

and
$$c_{cond} = (1 - \xi^{-1}) \, c_e \qquad (12b)$$

so that
$$\underline{K}^2 = \lambda \, c_e \, (2R + \xi^{-1}) \qquad (12c)$$

In deriving equation (11) the weak dependence of the product D.T - in ξ^{-1}, (eq. 12b) - on T has been neglected. An alternative approach to the problem could start from explicitly taking into account the temperature dependence of the fraction of condensed counterions in eq. (12a) as well as the possible dependence of the free energy of mixing on temperature through the dependence of ξ on T. No such a procedure has been yet reported, nor it will be given here, in view of the large formal complications to be introduced in front of the little gain in numerical accuracy for aqueous systems.[1]

Having assessed the assumptions in the derivation of the total excess electrostatic enthalpy of a polyelectrolyte solution with $\xi > 1$, we will proceed further to evaluate $\Delta \bar{H}_{mix}$, by use of equations (11) for the final and initial states, obtaining:

$$\Delta \bar{H}_{mix} = \frac{H_f^{el} - H_i^{el}}{n_e} = -\tfrac{1}{2} \, \underline{R}T \left(1 + \frac{d \ln D}{d \ln T} \right) \xi^{-1} \; x$$

$$x \left(2 \ln \frac{1 - e^{-\underline{K}_f b}}{1 - e^{-\underline{K}_i b}} + \frac{\underline{K}_f b}{e^{\underline{K}_f b} - 1} - \frac{\underline{K}_i b}{e^{\underline{K}_i b} - 1} \right) \qquad (13)$$

where: $\underline{K}_f^2 = \lambda c_e^f (2R_f + \xi^{-1})$,

$\underline{K}_i^2 = \lambda c_e^i (2R_i + \xi^{-1})$.

It can be easily checked that for $\xi = 1$, eq. (7) and eq. (13) correctly give the same value of $\Delta \bar{H}_{mix}$. For $\underline{K}_f b$ and $\underline{K}_i b \ll 1$, eq. (13) reduces to

$$\Delta \bar{H}_{mix} = -\tfrac{1}{2} \underline{R}T \; \xi^{-1} \left(1 + \frac{d \ln D}{d \ln T}\right) x$$

$$(14)$$

$$x \left(\ln \frac{c_e^f}{c_e^i} + \frac{2 \, R_f + \xi^{-1}}{2 \, R_i + \xi^{-1}}\right)$$

which corresponds to equation 5b of Reference 18 for $R_i = 0$.

Equation (14) can be rearranged:

$$\Delta \bar{H}_{mix} = \Delta \bar{H}_{dil} + \Delta \bar{H}_{salt} \qquad\qquad where$$

$$\Delta \bar{H}_{dil} = -\tfrac{1}{2} \underline{R}T \; \xi^{-1} \left(1 + \frac{d \ln D}{d \ln T}\right) \ln \frac{c_e^f}{c_e^f} \qquad (15a)$$

$$\Delta \bar{H}_{salt} = -\tfrac{1}{2} \underline{R}T \; \xi^{-1} \left(1 + \frac{d \ln D}{d \ln T}\right) \ln \frac{2 \, R_f + \xi^{-1}}{2 \, R_i + \xi^{-1}} \qquad (15b)$$

As already shown for the case $\xi < 1$, $\Delta \bar{H}_{dil}$ coincides with that already reported by other authors,[18-22] and furthermore results independent of the amount of salt initially present in the polymer-cointaining solution, provided that the dilution is performed against pure solvent ($R_f = R_i$).

On the other hand, for $R_i = 0$ one obtains:

$$\Delta \bar{H}_{salt} = -\tfrac{1}{2} \underline{R}T \; \xi^{-1} \left(1 + \frac{d \ln D}{d \ln T}\right) \ln(2 \, \xi \, R_f + 1) \qquad (16)$$

For low values of $(2 \, \xi \, R_f)$ the following approximation will hold:

$$\ln(2 \, \xi \, R_f + 1) \simeq 2 \, \xi \, R_f \qquad (17)$$

and therefore:

$$\lim_{2 \, \xi \, R_f \to 0} \frac{\Delta \bar{H}_{salt}}{R_f} = -\underline{R}T\left(1 + \frac{d \ln D}{d \ln T}\right) \qquad (18)$$

independent of ξ. Equation (18) states that the curves of $\Delta\bar{H}_{salt}$ versus R_f of initially salt-free solutions of different polyelectrolytes - with different values of ξ, provided they are larger than 1, - should exhibit a common slope in the initial part.

Figure 1b shows that also in the case of $\xi > 1$ an increasing discrepancy manifests on increasing R_f between the dependence of $\Delta\bar{H}_{mix}$ on R_f as predicted by equation (13) and that predicted by equation 5b of Reference 18.

ENTHALPY OF DISSOCIATION OF WEAK POLYACIDS

An interesting case of application of equations (4) and (11) is to evaluate the (excess) electrostatic enthalpy change upon proton dissociation of a weak polyacid (e.g. a carboxylated polysaccharide).

The following approach represents a novel extension of Manning's theory, which has not received much attention in the literature. The only previous attempt to give a theoretical expression for $\Delta\bar{H}_{diss}$ is that of Crescenzi et al.[26], based on the so-called "cell-model" of polyelectrolyte solutions of Morawetz[27] and Katchalsky[28,29] further developed with regard to the calorimetric aspect by the Yugoslav school.[30-36]

Upon introducing as a new variable the degree of ionization, α, the electrostatic enthalpy per mole of ionizable group, n_p, will read:

$$\bar{H}^{el} = \frac{H^{el}}{n_p} = -\tfrac{1}{2}\underline{R}T\left(1 + \frac{d \ln D}{d \ln T}\right)\alpha\left(\alpha\bar{\xi}\right)\left[2 \ln\left(1 - e^{-\underline{K}\bar{b}\,\alpha^{-1}}\right) + \frac{\underline{K}\bar{b}\,\alpha^{-1}}{e^{\underline{K}\bar{b}\alpha^{-1}} - 1}\right] \tag{19}$$

for $\alpha < \bar{\xi}^{-1}$ (i.e. $\xi < 1$)

with $\underline{K}^2 = \lambda \cdot c_p (2R + \alpha)$, and $R = \dfrac{c_s}{c_p}$;

and :

$$\bar{H}^{el} = \frac{H^{el}}{n_p} = -\tfrac{1}{2}\underline{R}T \left(1 + \frac{d \ln D}{d \ln T}\right) \alpha \left(\alpha \bar{\xi}\right)^{-1} \times$$

$$\times \left[2 \ln \left(1 - e^{-\underline{K}\bar{b}\alpha^{-1}}\right) + \frac{\underline{K}\bar{b}\alpha^{-1}}{e^{\underline{K}\bar{b}\alpha^{-1}} - 1}\right] \qquad (20)$$

for $\alpha > \bar{\xi}^{-1}$ (i.e. $\xi > 1$)

with $\underline{K}^2 = \lambda \; c_p \left(2R + \bar{\xi}^{-1}\right)$, and $R = \frac{c_s}{c_p}$.

Use of the following positions has been made:

$$n_e = \alpha \, n_p \;;\; c_p = \frac{n_p}{V} \;,\; b = \frac{\bar{b}}{\alpha} \;;\; \xi = \alpha \, \bar{\xi} \;; \qquad (21)$$

where $\bar{b}$ and $\bar{\xi}$ refer to the fully ionized chain. For an infinitely long chain the electrostatic part of the enthalpy change on dissociation (per mol H^+) can be identified with the derivative of $\bar{H}^{el}$, above defined, with respect to α:

$$\Delta \bar{H}^{el}_{diss} = \frac{d\bar{H}^{el}}{d\alpha} = -\underline{R}T \left(1 + \frac{d \ln D}{d \ln T}\right) \times$$

$$\times \alpha \bar{\xi} \left[2 \ln \left(1 - e^{-\underline{K}\bar{b}\alpha^{-1}}\right) + \frac{\underline{K}\bar{b}\alpha^{-1}}{e^{\underline{K}\bar{b}\alpha^{-1}} - 1} - \right. \qquad (22)$$

$$\left. - \tfrac{1}{4}\left(\frac{\alpha + 4R}{\alpha + 2R}\right) \frac{\underline{K}\bar{b}\alpha^{-1}}{e^{\underline{K}\bar{b}\alpha^{-1}} - 1}\left(3 - \frac{\underline{K}\bar{b}\alpha^{-1}}{e^{\underline{K}\bar{b}\alpha^{-1}} - 1} e^{\underline{K}\bar{b}\alpha^{-1}}\right)\right]$$

for $\alpha < \bar{\xi}^{-1}$;

for $\underline{K}\bar{b}\,\alpha^{-1} << 1$ it reduces to:

$$\Delta \bar{H}^{el}_{diss} = -\underline{R}T \left(1 + \frac{d \ln D}{d \ln T}\right) \alpha \, \bar{\xi} \quad x \tag{23}$$

$$x \left[\ln \left(\underline{K}\bar{b} \, \alpha^{-1}\right) + \tfrac{1}{2} \left(\frac{\alpha}{\alpha + 2R} \right) \right]$$

Equation (23) for the salt-free case reads:

$$\Delta \bar{H}^{el}_{diss} = -\underline{R}T \left(1 + \frac{d \ln D}{d \ln T}\right) \alpha \, \bar{\xi} \left[\ln(\underline{K} \, \bar{b} \, \alpha^{-1}) + \tfrac{1}{2} \right], \tag{24a}$$

and for the condition of excess salt:

$$\Delta \bar{H}^{el}_{diss} = -\underline{R}T \left(1 + \frac{d \ln D}{d \ln T}\right) \alpha \, \bar{\xi} \quad \ln (\underline{K} \, \bar{b} \, \alpha^{-1}) \tag{24b}$$

Following the same procedure one derives from equation (20) the expression for $\Delta \bar{H}_{diss}$ (for $\alpha > \xi^{-1}$):

$$\bar{H}^{el}_{diss} = \underline{R}T \left(1 + \frac{d \ln D}{d \ln T}\right)(\alpha\bar{\xi})^{-1} \cdot \tfrac{1}{2} \frac{\underline{K}\bar{b}\alpha^{-1}}{e^{\underline{K}\bar{b}\alpha^{-1}} - 1} \left(3 - \frac{\underline{K}\bar{b}\alpha^{-1}}{e^{\underline{K}\bar{b}\alpha^{-1}} - 1} e^{\underline{K}\bar{b}^{-1}}\right) \tag{25}$$

being $\underline{K}^2 = \lambda \quad c_p \left(2R + \bar{\xi}^{-1}\right)$

For $\underline{K} \, \bar{b} \, \alpha^{-1} \ll 1$ it reduces to

$$\Delta\bar{H}^{el}_{diss} = \underline{R}T \left(1 + \frac{d \ln D}{d \ln T}\right) \left(\alpha \, \bar{\xi}\right)^{-1} \tag{26}$$

The most striking result which appears upon comparing equations (22) and (25) at the common value of α corresponding to $\alpha\bar{\xi} = 1$ is that the two calculated values of $\Delta\bar{H}^{el}_{diss}$ do not coincide. The $\Delta\bar{H}^{el}_{diss}$ functions calculated for the case $\xi = 1.640$ and $c_p = 10^{-3}$M by use of equations (22) and (25) have been reported in Figure 2 for $c_s = 0$ and $c_s = 0.05$M, respectively. ΔH^{el}_{diss} values are always exothermic and show a discontinuity at $\alpha = 0.\underline{61}$ corresponding to $\xi = 1$. To our knowledge no similar behavior of $\Delta\bar{H}_{diss}$ has been ever reported in the literature for weak polyacids.

Following the suggestion of a referee, we resorted to the equations that G. S. Manning has recently given[37] to describe the potentiometric behavior of a weak polyacid in term of the variation of α, and hence of ξ. By taking the proper temperature derivatives of the excess free energy term in equations (63) and (60) of Reference 37 we obtained different expressions for $\Delta\bar{H}^{el}_{diss}$ for $\xi < 1$ and for $\xi > 1$, respectively. The former coincides with equation (22), while the latter slightly differs from equation (25) for the case of excess salt and coincides with it for the salt-free case. Therefore it seems likely that the surprising prediction of a marked discontinuity of $\Delta\bar{H}^{el}_{diss}$ as a function of α is to be attributed to some internal inconsistency of the model. A similar discontinuous behavior of pK_{diss} vs. α for $\xi = 1$ has led G. S. Manning to attribute it to an overidealisation of the real chain as it is presently built in the condensation model.[37]

To assess the factors which could modify the above behavior, we have first taken into account the finite length of the chain as opposed to the previous assumption of infinite line charge distribution, still neglecting end effects. A reduction of the chain length reduces the absolute value of $\Delta\bar{H}^{el}_{diss}$, but it only slightly decreases the extent of the discontinuity at $\xi = 1$ (Figure 2). A second source of modification of the $\Delta\bar{H}^{el}_{diss}$ curve can stem in real cases from a non-homogeneous structural distribution of ionizable groups. Numerical evaluation of $\Delta\bar{H}^{el}_{diss}$ for idealized block-copolymers with two different values of b has led to notably reduced extent of the singularity at $\xi = 1$, although it never disappeared. It is not hard to extrapolate that a continuous distribution of b values could lead to a more smooth and continuous $\Delta\bar{H}^{el}_{diss}$ curve. Even in the case of homopolymers, a distribution of $\bar{b}$ values can arise from the fluctuations of the real chain in the conformational space. The additional possibility of a distribution of $\bar{b}$ due to fluctuations of protons over negative sites[38] should be considered. We are aware that, besides the difficulty of quantification, the above arguments can show some inconsistency with a few assumptions of the original condensation theory. Still, we believe that it is worthwhile to refine this model as far as the enthalpic aspects of dissociation are concerned due to their sensitity to physically meaningful polyelectrolyte parameters.

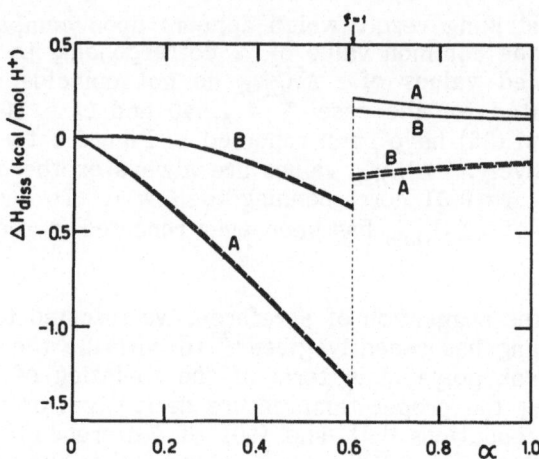

Figure 2.   Dependence of $\Delta \overline{H}_{diss}^{el}$ on the degree of ionization, α, for a weak polyacid with $\xi = 1.64$, calculated according to equations (22) and (25) (full lines). Dashed lines have been calculated for the same system by use of the equation $\Delta \overline{H}_{diss}^{el} = (H^{el})_{\alpha} - (H^{el})_{\alpha - \frac{1}{N}}$, where N is the degree of polymerization of the chain (N = 100); H^{el} is given by equations (19) and (20) for $\xi < 1$ and $\xi > 1$, respectively. Curve A: $c_e = 10^{-3}$ M, $c_s = 0$; Curve B: $c_e = 10^{-3}$ M, $c_s = 0.05$ M.

ELECTROSTATIC ENTHALPY OF CONFORMATIONAL TRANSITION OF POLYELECTROLYTES

The derivation of the equations until here reported has pertained to the rather simple case in which the polyelectrolyte geometry (conformation) was constant and unaffected by the addition of either solvent (dilution), simple electrolyte (mixing, salt) or protons. However, in the real cases, polymer conformations are often affected by such agents: limiting conditions can be envisaged in which the polyelectrolyte chain can assume well defined but different values of b and, hence, of ξ. Conformational transitions can result from variation of bond angles (intramolecular conformational transition), or from side-by-side (or intertwined) alignment of polymer chains (dimerisation, double helix formation), or from both. From the standpoint of the counterion condensation theory, however, the real process of redistribution of molecular masses is immaterial, since the real chains have to be replaced by infinite line charge distributions and the only determining parameter is the "charge-per-unit-length" (ξ). For instance, a dimerisation process will be only supposed to change ξ to 2ξ. An expression is aimed at for the (excess) electrostatic enthalpy of a transition ($\Delta \overline{H}_{tr}^{el}$), ideally supposed to be of the first-order at the given values of T, c_e and c_s. In the initial conformation the value of b is b_i (and that of ξ is ξ_i), while in the final conformation

$b = b_f$ and $\xi = \xi_f$. The following discussion will be referred to the quite common case of an ion-induced conformational transition with an increase of ξ ($\xi_i < \xi_f$). For the "melting" process the subscripts $\underline{i}$ and $\underline{f}$ should be interchanged. In general it will hold:

$$\Delta \ \bar{H}^{el}_{tr} \ = \ (\bar{H}^{el})_f \ \ - \ (\bar{H}^{el})_i \tag{27}$$

Three cases can be identified:

a. $- \xi_i, \xi_f < 1$. Upon insertion of equation (4) into equation (27) one obtains:

$$\Delta \ \bar{H}^{el}_{tr} = -\tfrac{1}{2} \ \underline{R}T \ \left(1 + \frac{d \ln D}{d \ln T}\right) \left\{ \xi_f \left[2 \ln (1 - e^{-\underline{K}_t b_f}) + \right. \right.$$

$$\left. + \ \frac{\underline{K}_t b_f}{e^{\underline{K}_t b_f} - 1} \right] \ - \ \xi_i \left[2 \ln(1 - e^{-\underline{K}_t b_i}) + \frac{\underline{K}_t b_i}{e^{\underline{K}_t b_i} - 1} \right] \left. \right\} \tag{28}$$

where $\underline{K}_t^2 = \lambda c_e(2R_t + 1)$ and R_t is the value of $R(R = \dfrac{c_s}{c_e})$ at which the transition occurs.

b. $- \xi_i, \xi_f > 1$. Combining equations (11) and (27) one can write:

$$\Delta \bar{H}^{el}_{tr} = -\tfrac{1}{2} \ \underline{R}T \ \left(1 + \frac{d \ln D}{d \ln T}\right) \left\{ \xi_f^{-1} \left[2 \ln\left(1 - e^{-\underline{K}_{t,f} b_f}\right) \ + \right. \right.$$

$$\frac{\underline{K}_{t,f} b_f}{e^{\underline{K}_{t,f} b_f} - 1} \right] - \xi_i^{-1} \left[2 \ln\left(1 - e^{-\underline{K}_{t,i} b_i}\right) \ + \tag{29}$$

$$\frac{\underline{K}_{t,i} b_i}{e^{\underline{K}_{t,i} b_i} - 1} \left. \right] \left. \right\}$$

where $\underline{K}_{t,i,f}^2 = \lambda c_e(2R_t + \xi_{i,f}^{-1})$, with R_t previously defined.

c. - $\overline{\xi_i < 1, \ \xi_f > 1}$. Such process can be ideally separated into the following sub-processes:

$$(\xi = \xi_i) \to (\xi = 1) \tag{30a}$$

and
$$(\xi = 1) \to (\xi = \xi_f) \tag{30b}$$

By stepwise application of proceding equations (28) and (29) one gets:

$$\Delta\overline{H}_{tr}^{el} = -\tfrac{1}{2}\,\underline{RT}\left(1 + \frac{d \ln D}{d \ln T}\right)\left\{\xi_f^{-1}\left[2 \ln\left(1 - e^{-\underline{K}_{t,f}\,b_f}\right) + \right.\right. \tag{31}$$

$$\left. \frac{\underline{K}_{t,f}\,b_f}{\exp(\underline{K}_{t,f}\,b_f) - 1}\right] - \xi_i\left[2 \ln\left(1 - e^{-\underline{K}_{t,i}\,b_i}\right) + \right.$$

$$\left.\left. \frac{\underline{K}_t\,b_i}{\exp(\underline{K}_{t,i}\,b_i) - 1}\right]\right\}$$

the terms relative to $\xi = 1$ being cancelled out. Once again one can make the simplifying assumption that $\underline{K}_t b_f, \ \underline{K}_t b_i \ll 1$ thereby obtaining:

$$\Delta \bar{H}_{tr}^{el} = -\tfrac{1}{2}\underline{R}T(1 + \frac{d \ln D}{d \ln T})\left[\xi_f \ln b_f^2 - \xi_1 \ln b_1^2 + (\xi_f - \xi_1)x\right.$$

$$\left. \ln(e \, \lambda \, c_e \, (2R_t + 1))\right] =$$

$$= -\left\{\tfrac{1}{2}\underline{R}T(1 + \frac{d \ln D}{d \ln T}) \left[2(\xi_f \ln b_f - \xi_1 \ln b_1) + (\xi_f - \xi_1) \, x\right.\right.$$

$$\left.\left.\ln(e. \, \lambda \, .2)\right]\right\}_{ISI} -$$

$$-\left\{\tfrac{1}{2}\underline{R}T(1 + \frac{d \ln D}{d \ln T}) \left[(\xi_f - \xi_1) \ln(c_s + \tfrac{1}{2} c_e)\right]\right\} ISD \qquad (32)$$

for $\xi_1, \xi_f < 1$

$$\Delta \bar{H}_{tr}^{el} = -\tfrac{1}{2}\underline{R}T(1 + \frac{d \ln D}{d \ln T}) \left\{\xi_f^{-1}\ln b_f^2 - \xi_1^{-1} \ln b_1^2 + \right.$$

$$\left. + \xi_f^{-1}\ln\left[e \, \lambda \, c_e \, (2R_t + \xi_f^{-1}))\right] - \xi_1^{-1} \ln\left[e \, \lambda \, c_e \, (2R_t + \xi_1^{-1})\right]\right\} =$$

$$= -\left\{\tfrac{1}{2}\underline{R}T(1 + \frac{d \ln D}{d \ln T}) \left[2(\xi_f^{-1} \ln b_f - \xi_1^{-1} \ln b_1) + \right.\right.$$

$$\left.\left. + (\xi_f^{-1} - \xi_1^{-1}) \ln(e \cdot \lambda \cdot 2)\right]\right\} ISI \quad -$$

$$- \left\{\tfrac{1}{2}\underline{R}T(1 + \frac{d \ln D}{d \ln T}) \left[\xi_f^{-1} \ln (c_s + \tfrac{1}{2} \xi_f^{-1} c_e) - \right.\right.$$

$$\left.\left. - \xi_1^{-1} \ln (c_s + \tfrac{1}{2} \xi_1^{-1} c_e)\right]\right\} ISD \qquad \text{(for } \xi_f, \xi_1 > 1) \qquad (33)$$

$$\Delta \bar{H}_{tr}^{el} = -\tfrac{1}{2} RT \left(1 + \frac{d \ln D}{d \ln T}\right) \left\{ \xi_f^{-1} \ln b_f^2 - \xi_1 \ln b_1^2 \right.$$

$$+ \xi_f^{-1} \ln \left[e \, \lambda \, c_e \, (2R_t + \xi_f^{-1}) \right] - \xi_1 \ln \left[e \, \lambda \, c_e \, (2R_t + 1) \right] \right\} =$$

$$= - \left\{ \tfrac{1}{2} \underline{RT} \left(1 + \frac{d \ln D}{d \ln T}\right) \left[2(\xi_f^{-1} \ln b_f - \xi_1 \ln b_1) + \right. \right.$$

$$\left. + (\xi_f^{-1} - \xi_1) \ln (e \cdot \lambda \cdot 2) \right] \right\}_{ISI} -$$

$$- \left\{ \tfrac{1}{2} \underline{RT} \left(1 + \frac{d \ln D}{d \ln T}\right) \left[\xi_f^{-1} \ln(c_s + \tfrac{1}{2} \xi_f^{-1} c_e) - \right. \right.$$

$$\left. - \xi_1 \ln (c_s + \tfrac{1}{2} c_e) \right] \right\}_{ISD} \quad \text{(for } \xi_i < 1 \, ; \, \xi_f > 1) \tag{34}$$

In equations (32), (33) and (34) the subscripts ISI and ISD define the Ionic Strength Independent part and the Ionic Strength Dependent part, respectively; e is the base of natural logarithms and all other symbols have been previously defined.

Figure 3 shows the dependence of ΔH_{tr}^{el} on $-\log c_s$ for an ideal system in which $c_e = 10^{-3}$ equiv. L^{-1} and ξ_i and ξ_f have been chosen both in relation to a case of practical interest, i.e. carrageenans, and to span over different sets of ξ values ($\xi \gtrless 1$). Curves A and B refer to the hypothetical intrachain and to the double-chain conformational transitions of κ-carrageenan, respectively, while for curves C and D the values of ξ of ι-carrageenan have been used for analogous conformational changes. Equations (28), (29) and (31) have been used where appropriate. There is no clear general behavior in the curves shown; $\Delta \bar{H}_{tr}^{el}$ can be either exo- or endothermic depending upon the particular set of ξ values and of c_s, or it may change sign upon changing pc_s. This evidence should induce one to be rather cautious in "predicting" the contribution of the polyelectrolytic term to $\Delta \bar{H}_{tr}$ in biopolymers relying upon "reasonable electrostatic arguments". Conditions can be easily achieved in which a conformation with higher charge density is granted a lower electrostatic internal energy (enthalpy) than that of a less charged conformation, which is contrary to expectation. The only trend which can be noticed is that of a levelling of $\Delta \bar{H}_{tr}^{el}$ values at high values of c_s due to an ionic atmosphere screening effect.

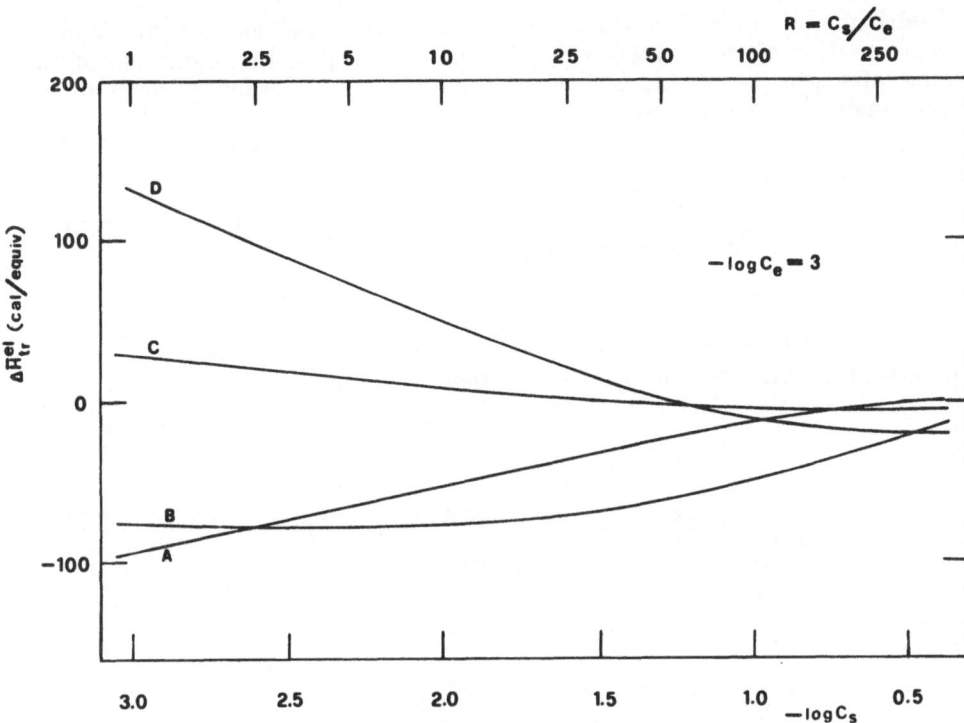

Figure 3.   Dependence of the electrostatic contribution to the enthalpy of transition, $\Delta \bar{H}_{tr}^{el}$, on the concentration of the added salt, c_s.   Curve A: $\xi_i = 0.693$, $\xi_f = 0.870$, (equation (28)); curve B: $\xi_i = 0.693$, $\xi_f = 1.740$ (equation (31)); curve C: $\xi_i = 1.386$, $\xi_f = 1.647$ (equation 29)); curve D: $\xi_i = 1.386$, $\xi_f = 3.293$ (equation 29)). Polymer concentration, c_e, 10^{-3} equiv. L^{-1} in all cases.

The absolute value of $\Delta \bar{H}_{tr}^{el}$ is small ($\sim \pm 10^2$ cal/equiv) and fully comparable with $\Delta \bar{H}_{mix}$ values at corresponding values of c_s. Moreover, the variation of $\Delta \bar{H}_{tr}^{el}$ with c_s is rather small too, for a comparatively large change in c_s as that usually accomplished in the actual measurements. Experimentally determined values of $\Delta \bar{H}_{tr}$ (otherwise called $\Delta \bar{H}_M$) in biopolymer systems – including polysaccharides – normally exceeds such $\Delta \bar{H}_{tr}^{el}$ values by one full order of magnitude. This fact implies that an observed anomalous change in ΔH of the order of 10^3 cal/equiv., corresponding to a conformation transition of a polyelectrolyte, can be largely ascribed to non-ionic contributions (H-bonds, non-polar interactions, etc.) thereby affording a means for their evaluation. The negligible contribution of the ionic-strength dependent part of $\Delta \bar{H}_{tr}$ is the basic assumptions of the treatment due to G. S. Manning of the ionic effects on thermally induced conformational transitions of poly-electrolytes.[39] This states that the dependence of the reciprocal of the "melting" (i.e. mid-transition) temperature of a polyelectrolyte from a conformation characterized by ξ_i to one with ξ_f on the logarithm of the ionic strength (c_s) should be linear as long as $\Delta \bar{H}_{tr}$ is independent of c_s. The analytical expression is:

$$\frac{\Delta \bar{H}_{tr}}{(d \log c_s/d \, 1/T_M)} = -2.288(\xi_f^{-1} - \xi_i^{-1}) \qquad (35)$$

for ξ_i, $\xi_f > 1$, and $\Delta \bar{H}_{tr}$ expressed in cal/equiv. The treatment has been extended to the case of ξ smaller than 1 giving:[40]

$$\frac{\Delta \bar{H}_{tr}}{(d \log c_s/d \, 1/T_M)} = -2.288 \, (\xi_i - \xi_f) \qquad (36)$$

$$\text{for } \xi_i, \; \xi_f < 1$$

and

$$\frac{\Delta \bar{H}_{tr}}{(d \log c_s/d \, 1/T_M)} = -2.288 \, (2 - \xi_f - \xi_i^{-1}) \quad (37)$$

$$\text{for } \xi_i > 1, \; \xi_f < 1$$

(Dealing with a "melting" process, the subscripts i and f refer to conformations opposite to those of equations (28) to (34)).

It is not the purpose of the present paper to discuss the above equations, which have been already applied with success to the conformational changes of nucleic acids[11,13] and of certain ionic polysaccharides.[40] What will be pointed out here is the possibility of using isothermal calorimetry alone for the acquisition of all the experimental parameters to be used in equations (35) to (37). Upon recording the $\Delta \bar{H}_{mix}$ vs. R (i.e. c_s) plots at different temperatures for a system undergoing a conformational change, one should expect a situation like that schematically depicted in Figure 4. For a given temperature, T, a sudden decrease of $\Delta \bar{H}_{mix}$ should usually be recorded at a given value of c_s, between two smoothly increasing tracts. The latters correspond to the $\Delta \bar{H}_{mix}$ function at ξ_i and ξ_f, respectively, while the first drop

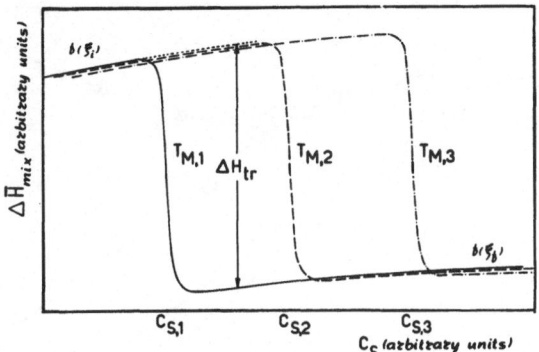

Figure 4. Idealized dependence of the isothermal $\Delta \bar{H}_{mix}$ upon the concentration of simple salt, c_s, for a system undergoing a cooperative conformational transition between states defined by ξ_i and ξ_f, respectively. $c_{s,1}$, $c_{s,2}$, $c_{s,3}$ denote the concentration values of simple salt at the transition midpoint for three different values of temperature, $T_{M,1}$, $T_{M,2}$ and $T_{M,3}$, respectively. $\Delta \bar{H}_{tr}$ is the hypothetical value of the total enthalpy of transition, practically independent of c_s. $f(\xi_i)$ and $f(\xi_f)$ represent the $\Delta \bar{H}_{mix}$ function for the initial and final conformation, respectively; their small dependence on the temperature has been neglected.

corresponds to the change from ξ_i to ξ_f and it gives $\Delta \bar{H}_{tr}$. The set of values of T and c_s of transition therefore gives a data point for the construction of the log c_s vs. $1/T_M$ plot. Different methods have been used until now for collecting the data points for the latter plot, but to our knowledge to the possibility of exploiting isothermal calorimetry has not been given sufficient attention. In conclusion, we want to stress that molecular polyelectrolyte theory is able to give a self-consistent frame of predictions as far as the thermal behavior of polyelectrolyte solutions is concerned. The possibility of separating the purely electrostatic contribution from the non-ionic one in conformational transitions (contrast effect) appears to be a significant achievement, enabling one to evaluate the latter term and to relate it to structural (conformational) parameters.

COMPARISON OF THEORETICAL PREDICTIONS WITH EXPERIMENTAL CALORIMETRIC DATA

1. Enthalpy of Dilution. We shall start comparing the theoretical predictions with the experimental results for the simplest case of mixing, i.e. the dilution, of a polyelectrolyte solution with pure solvent ($R_i = R_f = 0$).

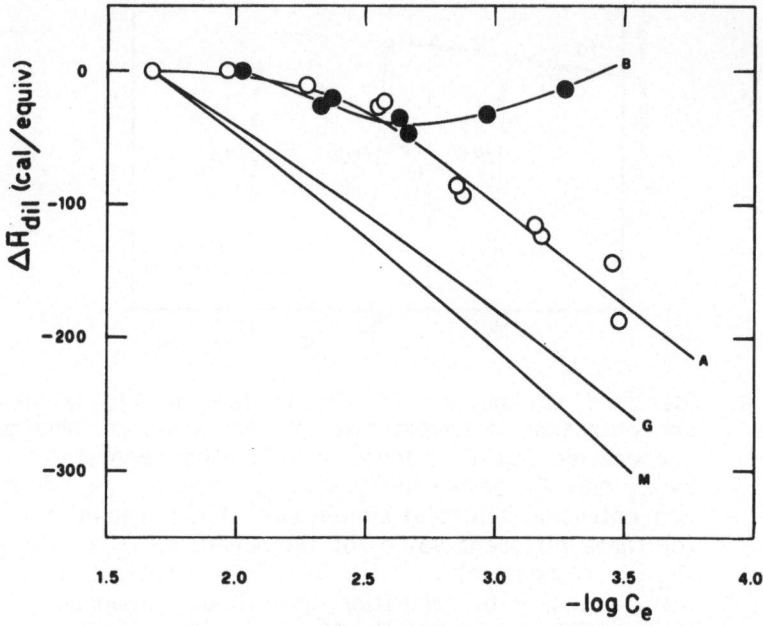

Figure 5.   Dependence of the heats of dilution, $\Delta \bar{H}_{dil}$, on the – log of the final polymer concentration, c_e, for sodium alginates in water.   Curve A: deacetylated bacterial alginate; curve B: native bacterial alginate (~8% acetyl groups).   Curves G and M have been calculated by use of equation (13) for the model compounds poly (sodium guluronate), and for poly (sodium mannuronate), respectively.   The experimental data of the present and of the following Figures have been obtained using either batch – or flow – type 10700 – 2 LKB isothermal microcalorimeters.   For full details, see Reference 45.

Curve A of Figure 5 shows the heats of dilution at 298 K of sodium alginate as a function of -log c_e.[41] The sample used was from a bacterial source (<u>Azotobacter vinelandii</u>), treated with ammonia to eliminate acetyl ester groups.   Alginic acid is a linear copolymer of β-D-mannuronic (ManUA) and α-L-guluronic (GulUA) acids, both linked 1 → 4; the actual distribution of the comonomers is quite complicated, but it can be essentially depicted as given by homopolymeric blocks (either [ManUA]$_n$ or [GulUA]$_n$) and heteropolymeric sequences (ManUA – GulUA).   The monomer composition and the doublet frequences (F) as determined by p.m.r. spectroscopy[42] on our sample are given hereafter:

F_M= 0.37; F_G= 0.63; F_{MM}= 0.24; F_{MG}+ F_{GM} = 0.26; F_{GG}= 0.50. M and G stand for ManUA and GulUA, respectively. $\Delta \bar{H}_{dil}$ values are always exothermic and increasingly so on increasing dilution.   Theoretical curves

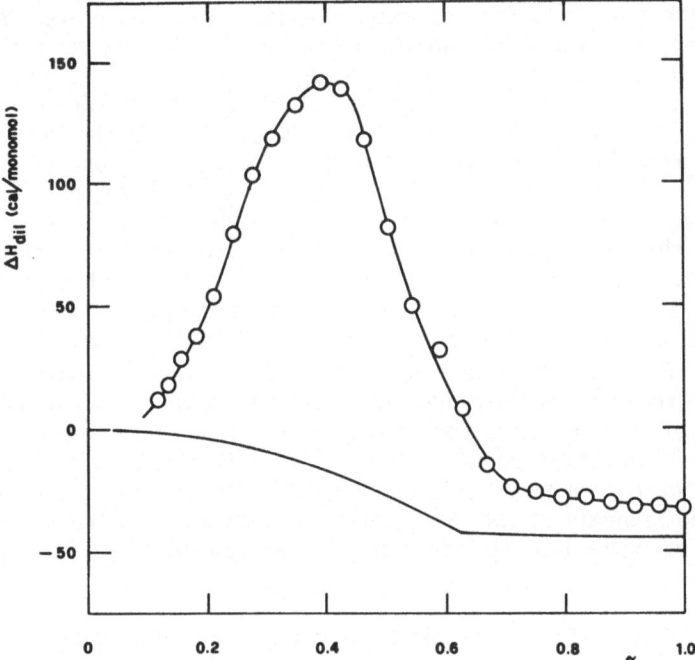

Figure 6. Dependence of the differential heat of dilution, $\Delta\bar{H}_{dil}$, of pectic acid on the degree of ionization, α, in water. Polymer concentrations: $c_e^i = 1.94 \times 10^{-2}$ M, $c_e^f = 9.7 \times 10^{-3}$ M. The full line represent the theoretical curve calculated according to equation (7), $0 < \alpha \lesssim 0.6$, and equation (13), $0.6 \lesssim \alpha < 1.0$, respectively.

calculated on the basis of equation (13), setting $R_i = R_f = 0$, are also reported in Figure 5, for the idealized structures of polymannuronic (M) and polyguluronic acids (G), respectively. The geometrical parameters used have been taken from published X-ray fiber diffraction data,[43] i.e. b = 5.17 Å for M and b = 4.35 Å for G. The corresponding values of ξ are 1.379 and 1.640, respectively. (Use of the approximate form of equation (15a) would give two straight lines of slope -184 cal equiv^{-1} and -155 cal equiv^{-1}, respectively). The experimental curve very closely parallels the theoretical predictions. The behavior of the parent alginate polymer, in which the presence of 8% acetyl groups on C(2) or C(3) of ManUA has been assessed, is significantly different. The $\Delta\bar{H}_{dil}$ vs. -log c_e curve departs from the expected behavior in a way which is suggestive of a concomitant conformational change of the polymer. Both optical activity[41] and viscosity[44] data provide confirmatory evidence for such an interpretation.

Endothermic deviations of experimental $\Delta\bar{H}_{dil}$ data are also shown by poly(α -D-galacturonic acid), pectic acid. Figure 6 reports the experimental calorimetric values as a function of the degreee of ionization, α , of a solution of pectic acid undergoing dilution from c_e = 1.94×10^{-2} M to c_e = 9.7×10^{-3} M at 298 K.[45] The theoretical curve is also reported in the Figure; it has been obtained by use of equations (7) and (13) and the positions: $\xi = \alpha$ $\xi = \alpha \cdot 7.135/\bar{b}(\text{Å})$, $\bar{b} = 4.47$ Å, as obtained from X-ray fiber diffraction data.[46] A bell shaped curve completely shifted to <u>positive</u> value is the notable feature of the plot: a maximum deviation of +150 cal. monomol^{-1} is reached at approximately α = 0.4. A wide variety of experimental techniques have shown that pectic acid undergoes a pH-induced conformational transition in aqueous solution.[45] Furthermore, the heat changes due to non-electrostatic interactions on diluting polysaccharide solutions have been found to be negligible in the absence of changes in coil conformation and of association-dissociation phenomena.[47] The latter one, have been ruled out for the case presently being discussed.[45] We therefore conclude that the reported anomaly in the $\Delta\bar{H}_{dil}$ curve is very likely attributable to the disruption of intrachain bonds, which may parallel a conformational transition.

The heat effect measured on mixing an increasing amount of $NaClO_4$ (aq) with a solution of sodium pectate has been reported as $\Delta\bar{H}_{mix}$ in Figure 7. The dashed curve of Figure 7 has been calculated on the basis of equation (13) using for ξ the value of 1.596 (b = 4.47 Å). The observed behavior parallels the theoretical predictions both in shape and in sign, starting with a negative $\Delta\bar{H}_{mix}$ values of R_f = 0 (dilution) and then turning to positive values. Similar satisfactory agreement between the predictions of the condensation theory and the experimental data have been already reported for some glycosaminoglycans[48] and for carboxymethyl-cellulose.[49]

2. <u>Enthalpy of Dissociation</u>. The experimental data of the enthalpy of dissociation of alginic acid in aqueous solution at 298 K are reported in Figure 8 as a function of α.[41] The sample used, from algal origin, had the following composition parameters[42] (previously defined): F_M = 0.55; F_G = 0.45; F_{MM} = 0.40; $F_{MG} + F_{GM}$ = 0.30; F_{GG} = 0.30; free of acetyl groups. The $\Delta\bar{H}_{diss}$ values for homopolymeric sequences of guluronic acid $(GulUA)_n$ and for mannuronic acids $(ManUA)_n$, isolated from alginic acid samples,[50] also have been reported in the same Figure. The values of $\Delta\bar{H}_{diss}$ are always exothermic, and range from about -1. to -2. kcal(mol H^+)$^{-1}$. As reported in the discussion of the theoretical part, no discontinuity is observed in the trend of the data points. However, the range of variation of the experimental data is encompassed by the limits of variation of the theoretical curves for the pure homopolymeric sequences.

The next example has been chosen to demonstrate the use of the polyelectrolytic interpretation of the calorimetric behavior of biopolymers

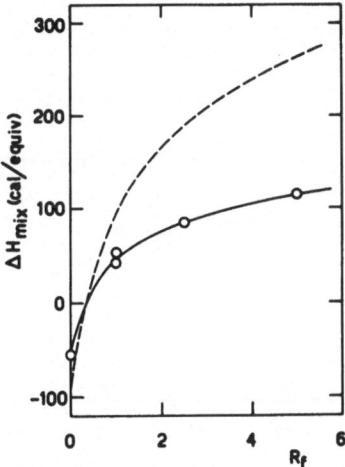

Figure 7.   Dependence of the enthalpy of mixing, $\Delta \overline{H}_{mix}$, of sodium pectate with $NaClO_4$ in water on the final salt-to-polymer concentrations ratio, R_f. Polymer concentrations: $c_e^i = 10^{-2}$ equiv.L^{-1}, $c_e^f = 5 \times 10^{-3}$ equiv.L^{-1}; $R_i = 0$. The dotted curve has been calculated using equation (13), $\xi = 1.596$.

as a "contrast effect" in assessing conformational changes. When the experimental $\Delta \overline{H}_{diss}^{el}$ values of a salt-containing solution of pectic acid are reported as a function of α, the remarkable result of Figure 9 comes out.[45] Starting from exothermic values at low α, the experimental data rapidly move to notably endothermic values on increasing α, to drop again to exothermic values for $\alpha \rightarrow 1$. The bell-shaped curve connecting the experimental points has a maximum at $\alpha \simeq 0.6$. Having already ascertained the presence of a pH-induced intramolecular conformational transition of pectic acid, one could reasonably ask: can the anomalous $\Delta \overline{H}_{diss}$ curve largely stem from electrostatic interactions manifesting themselves on increasing the linear charge density? If not, how can the ionic and the non-ionic contributions be differentiated, so as to allow for a quantitative evaluation of both? A possible answer to both questions is implicitly given in the theoretical $\Delta \overline{H}_{diss}^{el}$ curve for pectic acid, reported as a dashed line in Figure 9. It can be seen that for the given experimental conditions an overall change in ΔH_{diss}^{el} of about 250 calories per mole of H^+ is predicted for the electrostatic contributions. The experimental points span over 2.5 kilocalories per mole of H^+. Therefore, not only the ionic contribution to $\Delta \overline{H}_{diss}$ is small if compared with the non-ionic term, but the predicted behavior of the former provides a "reference" in order to evaluate the latter one. The area between the experimental and the theoretical (electrostatic) curve of Figure 9 can be identified with the (non-ionic) ΔH of transition from the conformation prevailing at low α to that at high α. Such value is +700 cal. monomole[1] ca.[45] The novelty of the procedure is that the "reference" curve has been

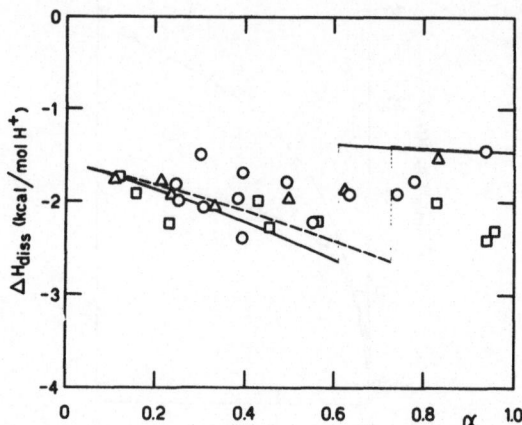

Figure 8.   Dependence of the enthalpy of dissociation, $\Delta\overline{H}_{diss}$, of alginic acid (circles; $c_e = 10^{-3}$ M), $(ManUA)_n$ (squares; $c_e = 3.4 \times 10^{-3}$ M), $(GulUA)_n$ (triangles); $c_e = 4.7 \times 10^{-3}$ M) in water on the degree of ionization, α. The dashed curve and the full one have been drawn each according to equations (22) and (25) for the model compounds $(ManUA)_n$ and $(GulUA)_n$, respectively.

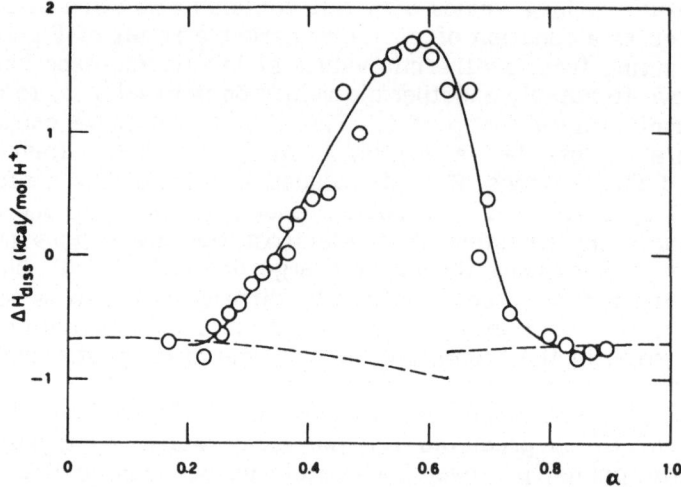

Figure 9.   Dependence of the enthalpy of dissociation, $\Delta\overline{H}_{diss}$, on the degree of ionization, α, of pectic acid in 0.05 M NaClO$_4$ aqueous solution. Polymer concentration: $c_e = 9.7 \times 10^{-3}$ M. Th dashed curve has been calculated according to equations (19) and (20) for poly(galacturonic acid) ($\xi = 1.596$, N = 100) (see Figure 2).

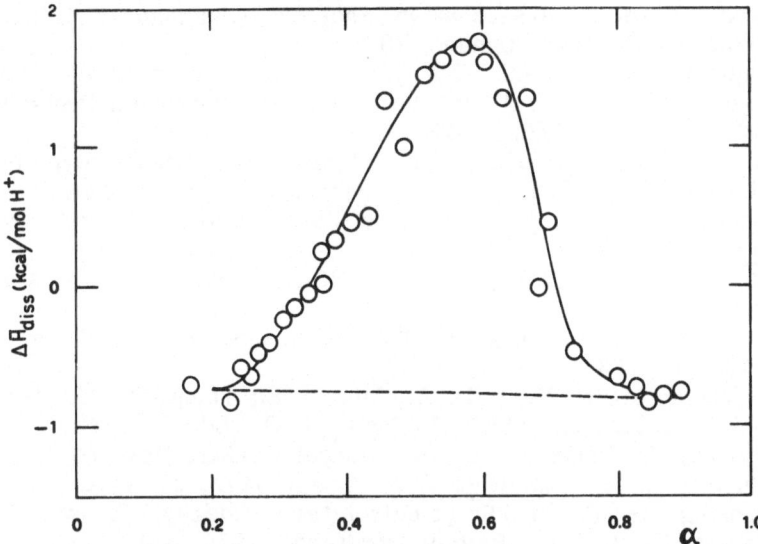

Figure 10. Dependence of the enthalpy of dissociation, $\Delta\bar{H}_{diss}$, on the degree of ionization, α, of pectic acid in 0.05 M NaClO$_4$ aqueous solution. Polymer concentration: C_e = 9.7 x 10^{-3} M; Na$^+$ counterion. The dashed curve has been calculated according to equation (31) and (33) for poly (galacturonic acid) (ξ = 1.596).

chosen on the basis of purely physical parameters and a well-defined theory (though crude it may be), and not according to an alleged "reasonably smooth" behavior of questionable soundness. We have previously called "contrast effect" such a combined use of polyelectrolyte theory and calorimetric techniques; its use should provide a self-consistent scenery on which biopolymers conformational changes can be more easily and neatly described.

ACKNOWLEDGMENTS

Research herein described was supported by grants from the Italian C.N.R. and the University of Trieste. The authors are grateful to G. S. Manning for his suggestions in the derivation of some equations and helpful criticism to the manuscript.

REFERENCES

1. Manning, G. S. In "Ions in Macromolecular and Biological Systems" (Proceedings of the 29th Symposium of the Colston Research Society); Everett, D. H.; Vincent, B., Eds., Scientechnica:Bristol, 1978; p 157.

2. Rees, D. A. "Polysaccharide Shapes"; Chapman & Hall-Outline Studies in Biology: London, 1977.
3. Smidsrød, O. "IUPAC 27th International Congress of Pure and Applied Chemistry"; Varmavuori, A., Ed.; Pergamon Press: Oxford and New York, 1980; p 315.
4. Chakrabarti, B.; Park, J. W. "Glycosaminoglycans: Structure and Interaction"; CRC Review in Biochemistry 1980, 8, 225.
5. Jaques, L. B. "Heparins-Anionic Polyelectrolyte Drugs"; Pharmacological Reviews 1979, 31, 99.
6. Whistler, R. L.; BeMiller, J. N. Eds. "Industrial Gums" Academic Press: New York, 2nd ed., 1973.
7. a. Smidsrød, O.; Haug, A.; Whittington, S. G. Acta Chem. Scand. 1972, 26, 2563
    b. for a review: Rees, D. A.; Morris, E.; Thom, D.; Madden, J. K. In "The Polysaccharides"; Aspinall, G. O., Ed.; in press.
8. Oosawa, F. "Polyelectrolytes", Marcel Dekker: New York, 1971.
9. Manning, G. S. Ann. Rev. Phys. Chem. 1972, 23, 117.
10. Manning, G. S. In "Polyelectrolytes"; Sélégny, E.; Mandel, M.; Strauss, U. P., Eds.; Reidel: Dordrecht, 1974; p 9.
11. Manning, G. S. Q. Rev. Biophys. 1978, 11, 179.
12. Manning, G. S. Acc. Chem. Res. 1979, 12, 443.
13. Record, M. T., Jr.; Anderson, C. F.; Lohman, T. M. Q. Rev. Biophys. 1978, 11, 103.
14. Barisas, B. G.; Gill, S. J. Ann. Rev. Phys. Chem. 1978, 29, 141.
15. Beezer, A. E. "Biological Microcalorimetry"; Academic Press: London, 1980.
16. For a recent survey, see Cesàro, A.; Paoletti, S.; Delben, F.; Crescenzi, V.; Rizzo, R.; Dentini, M. Gazz. Chim. Ital. 1982, 112, 115.
17. Manning, G. S. J. Chem. Phys. 1969, 51, 924.
18. Boyd, G. E.; Wilson, D. P.; Manning, G. S. J. Phys. Chem. 1976, 80, 808.
19. Mita, K.; Okubo, T. J. Chem. Soc., Faraday Trans. 1, 1973, 69, 106.
20. Mita, K.; Okubo, T.; Ise, N. J. Chem. Soc., Faraday Trans. 1, 1976, 72, 504.
21. Boyd, G. E.; Wilson, D. P. J. Phys. Chem. 1976, 80, 805.
22. Paoletti, S.; Delben, F.; Crescenzi, V. J. Phys. Chem. 1981, 85, 1413.
23. Wyman, J. Phys. Rev. 1930, 35, 623.
24. Manning, G. S. Biophys. Chem. 1977, 7, 95.
25. Manning, G. S. Biophys. Chem. 1978, 9, 65.
26. Crescenzi, V.; Delben, F.; Quadrifoglio, F.; Dolar, D. J. Phys. Chem. 1973, 77, 539.
27. Alfrey, T.; Berg, P. W.; Morawetz, H. J. Polym. Sci. 1951, 7, 543.
28. Fuoss, R. M.; Katchalsky, A.; Lifson, S. Proc. Nat. Acad. Sci. U.S. 1951, 37, 579.
29. Lifson, S.; Katchalsky, A. J. Polym. Sci. 1954, 13, 43.
30. Škerjanc, J.; Dolar, D.; Leškovsek, D. Z. Phys. Chem. (Frankfurt am Main) 1967, 56, 208; 1970, 70, 31.

31. Skerjanc, J.; Hočevar, S.; Dolar, D. Z. Phys. Chem. (Frankfurt am Main) 1971, 76, 85.
32. Dolar, D.; Škerjanc, J. J.Chem. Phys. 1974, 61, 4106.
33. Škerjanc, J,; Dolar, D. J. Chem. Phys. 1975, 63, 515.
34. Škerjanc, J. J. Phys. Chem. 1975, 79, 2185.
35. Dolar, D.; Škerjanc, J. J. Polym. Sci., Polym. Phys. Ed. 1976, 14, 1005.
36. Škerjanc, J.; Pavlin, M. J. Phys. Chem. 1977, 81, 1166.
37. Manning, G. S. J. Phys. Chem. 1981, 85, 870.
38. Paoletti, S.; van der Touw, F.; Mandel, M. J. Polym. Sci., Polym. Phys. Ed. 1978, 16, 641.
39. Manning, G. S. Biopolymers 1972, 11, 937.
40. Paoletti, S.; Smidsrød, O.; Grasdalen, H. Biopolymers, 1984, 23, 000.
41. Delben, F.; Cesàro, A.; Paoletti, S.; Crescenzi, V. Carbohydr. Res. 1982, 100, C46.
42. Grasdalen, H.; Larsen, B.; Smidsrød, O. Carbohydr. Res. 1979, 68, 23.
43. Atkins, E. D. T.; Nieduszynski, I. A.; Mackie, W.; Parker, K. D., Smolko, E. E. Biopolymers 1973, 12, 1865; 1879.
44. Paoletti, S.; unpublished results.
45. Cesàro, A.; Ciana, A.; Delben, F.; Manzini, G.; Paoletti, S. Biopolymers 1982, 21, 431.
46. Walkinshaw, M. D.; Arnott, S. J. Mol. Biol. 1981, 153, 1055.
47. Basedow, A. M., Ebert, K. H.; Feigenbutz, W. Makromol. Chem. 1980, 181, 1071.
48. Cleland, R. L. Biopolymers 1979, 18, 2673.
49. Hales, P. W.; Pass, G. J. Chem. Soc., Faraday Trans. 1 1980, 76, 2080; ibid. 1981, 77, 2009; ibid. 1982, 78, 283.
50. Haug, A.; Larsen, B.; Smidsrød, O. Acta Chem. Scand. 1967, 21, 691.

STUDIES ON DILUTE SOLUTIONS OF RODLIKE MACROIONS III. INTEGRATED INTENSITY AND PHOTON CORRELATION LIGHT SCATTERING INVESTIGATION OF ASSOCIATION

Y. Einaga[†] and G. C. Berry
Department of Chemistry
Carnegie-Mellon University
Pittsburgh, PA 15213

ABSTRACT

Integrated intensity and photon correlation experiments are combined to study the effects of intermolecular association in dilute solutions of the rodlike poly(1,4-phenylene-2,6-benzobisthiazole) in dilute solution. The autocorrelation function $g^{(2)}(\tau)$, the dependence of the second factorial moment $n^{(2)}(\Delta\tau)$ on the sampling interval duration $\Delta\tau$, and the integrated polarized and depolarized intensities are analyzed in terms of a quasi-two component mixture comprising unassociated chains and intermolecular aggregates. The latter persist in dilute solutions, even though the polymer exists as a macroion in the strong proton sulfonic acid used as a solvent. The unassociated chains which are 20-25 nm long are found to be predominant in weight fraction, but the aggregates, about 3 nm in diameter and 210 nm in length are large enough to dominate the scattering at small scattering angle.

INTRODUCTION

In previous studies[1-2] it was shown that the rodlike poly(1,4-phenylene-2,6-benzobisthiazole) tends to form metastable aggregates in solution. These persist at the concentration normally used for integrated intensity and photon correlation light scattering even though the polymer is protonated in the strong sulfonic acids used as solvents. In this study polarized and depolarized integrated intensity and polarized photon

---

[†]Permanent Address: Department of Polymer Science, Osaka University, Osaka, Japan

correlation scattering are used to assess the nature of the aggregates and their concentration relative to the dissociated polymer. The analysis employs the intensity correlation function $g^{(2)}(\tau,\Delta\tau)$ and the second factorial moment $n^{(2)}(\Delta\tau)$, both determined in the self-beating photon correlation mode.   Here $\Delta\tau$ is the sampling interval and τ is the correlation time.   For monodispersed solutes,[3-6]

$$g^{(2)}(\tau,\Delta\tau) - 1 = f(A)q(2\gamma)\left|g^{(1)}(\tau,\Delta\tau)\right|^2 \qquad (1.1)$$

$$n^{(2)}(\Delta\tau) - 1 = f(a)h(2\gamma) \qquad (1.2)$$

where $f(A)$ is a coherence factor fixed by the optical system (typically, $0.1 < f(A) < 0.7$), $\gamma = \Delta\tau/\tau_h$ where τ_h is the coherence time at scattering angle θ with wave vector modulus $h = (4\pi/\lambda)\sin\theta/2$, and[3]

$$q(2\gamma) = \left(\frac{\sinh\gamma}{\gamma}\right)^2 \qquad (1.3)$$

$$h(\gamma) = \frac{2}{\gamma} - \frac{2}{\gamma^2}(1 - \exp - \gamma) \qquad (1.4)$$

In obtaining Equations 1.3 and 1.4, the magnitude $|g^{(1)}(\tau,\Delta\tau)|$ of the correlation function has been put equal to $\exp-\gamma\tau/\Delta\tau$; this approximation will be used throughout in the following.   For a collection comprising components characterized by γ_l, each contributing a fraction r_i of the scattered light,

$$g^{(2)}(\tau,\Delta\tau) - 1 = f(A)\ \Sigma\Sigma\ r_i r_j q(\gamma_i+\gamma_j)\exp-(\gamma_i+\gamma_j)\tau/\Delta\tau \qquad (1.5)$$

$$n^{(2)}(\Delta\tau) - 1 = f(A)\ \Sigma\Sigma\ r_i r_j h(\gamma_i+\gamma_j) \qquad (1.6)$$

With neglect of $q(\gamma_i + \gamma_j)$, which is essentially unity unless the γ_i are very large, Equation 1.5 can be written in the more convenient form

$$\left[g^{(2)}(\tau,\Delta\tau) - 1\right]^{\frac{1}{2}} = f^{\frac{1}{2}}(A)\ \Sigma\ r_i\ \exp-\gamma_i\ \tau/\Delta\tau \qquad (1.7)$$

for analysis of data to obtain r_i and γ_i. If the solution comprises unassociated and aggregated components, it may appear to be a quasi-two component mixture since the γ_i for these two components can be expected to differ significantly. In this case, simple methods of data analysis may be used to estimate r_1, γ_1, r_2 and γ_2 for the two "components."   Sophisticated methods (e.g., see ref. 4) have been proposed to deal with more complex situations. In general, if the γ_i are closely spaced and numerous (as with a distribution of chain lengths) estimation of the r_i and γ_i set will be imprecise and may not be unique since the problem is ill-posed.

The r_i are related to the integrated scattered intensity. For vertically polarized scattering with vertically polarized incident light r_i is determined by the Rayleigh ratio $R_{Vv,i}$ for component i at concentration c_i, divided by the total Rayleigh ratio:

$$r_i = \mathbb{R}_{Vv,i}\,(h,c)/\mathbb{R}_{Vv}\,(h,c) \qquad (1.8)$$

$$(Kc_i/\mathbb{R}_{Vv,i})_0^{1/2} = (Kc_i/\mathbb{R}_{Vv,i}^0)_0^{1/2}\,[1 + \frac{A_{2,i}M_i}{(1+4/5\,\delta_i^2)}\,c_i + \ldots] \qquad (1.9a)$$

$$(Kc_i/\mathbb{R}_{Vv,i}^0) = M_i^{-1}\,(1 + \frac{4}{5}\,\delta_i^2)^{-1}\,(1 + \frac{1}{3}\,J(\delta_i)R_{G,i}^2\,h^2 + \ldots) \qquad (1.9b)$$

with δ the molecular optical anisotropy, A_2 the second virial coefficient, M_w the weight average molecular weight, superscript and subscript 0 denoting quantities at infinite dilution and zero scattering angle, respectively, and

$$J(\delta) - \left[1 - (4/5)\,f_1\delta + (4/7)(f_2\delta)^2\right]/\left[1 + (4/5)\delta^2\right] \qquad (1.9c)$$

The depolarized scattering $R_{Hv}(h,c)$ is given by

$$(Kc_i/\mathbb{R}_{Hv,i})_0 = (Kc_i/\mathbb{R}_{Hv,i}^0)_0 + O(c^2) \qquad (1.10a)$$

$$(Kc_i/\mathbb{R}_{Hv,i})_0 = (5/3\ M_i\delta_i^2)[1 + (3/7)f_3^2\ R_{Gi}^2 h^2 + \ldots] \qquad (1.10b)$$

The coefficients f_1, f_2, and f_3 are essentially unity for rodlike chains.[7a]

Data on $n^{(2)}(\Delta\tau)$ and $g^{(2)}(\tau,\Delta\tau)$ obtained in self-beating experiments can also be used to determine a set of average coherence times of the form

$$<\tau_h^{(s)}> \equiv 2\ \Delta\tau\{\Sigma\Sigma\ r_i r_j (\gamma_i+\gamma_j)^{-s}\}^{1/s} \qquad (1.11)$$

Thus, $<\tau_h^{(2)}>$ is equal to $2\Delta\tau_G$, with $\Delta\tau_G$ the value of $\Delta\tau$ for the crossover of $g^{(2)}(\Delta\tau,\Delta\tau) - 1$ from its limiting value $f(A)$ at small $\Delta\tau$ to the function $f(A)(\Delta\tau_G/\Delta\tau)^2$ obtaining at large $\Delta\tau$. Similarly, $<\tau_h^{(1)}>$ is equal to the value $\Delta\tau_N$ of $\Delta\tau$ for the crossover of $n^{(2)}(\Delta\tau) - 1$ from its limiting value $f(A)$ at small $\Delta\tau$ to the function $f(A)\Delta\tau_N/\Delta\tau$ obtaining at large $\Delta\tau$. Values of $<\tau_h^{(s)}>$ with $-s = 1,2$,etc. are calculated from the cumulants[8] K_γ of ln $|g^{(1)}(\tau)|$ given by $(-1)^\gamma (\gamma!)^{-1} \partial \ln |g^{(1)}(\tau)|/\partial\tau$, so that $<\tau_h^{(-1)}> = K_1^{-1}$, and $<\tau_h^{-2)}> = [K_2 + K_1^2]^{-1/2}$; in these expressions, $q(2\gamma)$ is taken to be unity. The latter average coherence times may be put in the form

$$<\tau_h^{(-1)}> = \Delta\tau \{\Sigma \ r_i\gamma_i\}^{-1} \tag{1.12}$$

$$<\tau_h^{(-2)}> = \Delta\tau \{1/2 \ \Sigma \ r_i\gamma_i^2 + 1/2 \ (\Sigma \ r_i\gamma_i)^2\}^{-1/2} \tag{1.13}$$

If, for example, $<\tau_h^{(-1)}>$ is calculated from data extrapolated to infinite dilution, so that $r_i = R_{yy,i}^0(h)/R_{yy,v}^0(h)$, and if $kTh^2\tau_{h,i} = 4\pi\eta_s k_T M_i^\alpha$ (with η_s the solvent viscosity, and k_T a constant), then for a Schulz-Zimm distribution (e.g., see reference 9a) of M_i,

$$kTh^2 <\tau_h^{(-1)}> = 4\pi\eta_s k_T M_w^\alpha Q_{(1-\alpha)} \ z+1) \tag{1.14}$$

$$<\tau_h^{(-2)}> = [1 + \Delta]^{-\frac{1}{2}} <\tau_h^{(-1)}> \tag{1.15}$$

where $Q_{(\mu)}(x) = x^\mu \ (x)/\Gamma(x+\mu)$ with Γ the gamma function, and $2\Delta+1 = Q_{(1-\alpha)}^2(z+1)/Q_{(1-2\alpha)}(z+1)$ with $z^{-1} = (M_w/M_n) - 1$; α equal to $1/2$ and 1 correspond to random-flight coil and rodlike chains, respectively. Since $Q_{(1-\alpha)} \geq 1$, being exactly unity for $\alpha = 1$, and nearly unity for arbitrary α between $1/2$ and 1 (unless z is very small, e.g. $Q_{(1/2)}(1.1) = 1.1$), it is seen that $<\tau_h^{(-1)}>$ varies with M_w. For $\alpha = 1$, $2\Delta+1 = M_w/M_n$ so that

$$\frac{<\tau_h^{(-1)}>}{<\tau_h^{(-2)}>} = \{1+[(M_w/M_n)-1]/2\}^{\frac{1}{2}} \tag{1.16}$$

or in other terminology, $M_w/M_n = 1+2K_2/K_1^2$ for $\alpha = 1$. Since $Q_{(0)} = 1$, $\Delta = (Q_{(1/2)}^2-1)/2$ and is nearly zero for $\alpha = 1/2$, even for a very broad molecular weight distribution (e.g. $\Delta(z = 0.1) = 0.12$ and $\Delta(z = 10) = 0.005$) so that $<\tau_h^{(-2)}> \sim <\tau_h^{(-1)}>$ for random-flight chains. Values of $<\tau_h^{(1)}>$ and $<\tau_h^{(2)}>$ are less easily evaluated, but if the distribution is not too broad, for random-flight chains $kTh^2 <\tau_h^{(1)}>$ is approximately equal to $4kT\eta_s k_T M_z^{1/2}$, and $<\tau_h^{(2)}> \sim <\tau_h^{(1)}>$.

EXPERIMENTAL

Details are given elsewhere for the light scattering apparatus[1] and methods.[1,7b] Briefly, a photometer constructed in our laboratory is coupled with a Birnboim Autocorrelator and Data Acquisition System. The latter provides the means to collect data automatically at selected scattering angle θ and sampling interval $\Delta\tau$; the number of photon counts per interval $<n>$ may be as large as $2.^{15}$ Values of $\Delta\tau$ used are given by $\Delta\tau = 50 \times 2^m$ ns, with m between 2 and 23. The full auto-correlation function $G^{(2)}(\tau, \Delta\tau)$ is computed over the primary data base of $T = 2^{12}$ intervals for $m > 6$ to give a correlation with up to 512 points spaced at intervals $\Delta\tau$; $G^{(2)}(\tau, \Delta\tau)$ is averaged over M such experiments until TM $<n> \sim 10^6$. The photon count distribution among the TM total intervals is used to compute the (unnormalized) factorial moments $N^{(s)}(\Delta\tau)$, s = 1-5. The normalized correlation function $g^{(2)}(\tau, \Delta\tau)$ is computed as $G^{(2)}(\tau, \Delta\tau)/G_N$ for $\tau > 0$ with G_N equal to $[N^{(1)}]^2$, the average $<<n>^2>$ of $<n>^2$ over the M separate experiments, or $G^{(2)}(\tau', \Delta\tau)$ where τ' is chosen to be very large compared with the coherence times of the sample. Usually, these estimates of G_N were close enough that differences among $g^{(2)}(\tau, \Delta\tau)$ calculated with the various G_N were unimportant. Typically, $g^{(2)}(\tau, \Delta\tau)$ was obtained for 16 to 32 correlation points, and with $\Delta\tau$ in the range 3 to 48 μs. The normalized second factorial moments $n^{(s)}$ is computed as $N^{(s)}(\Delta\tau)/[N^{(1)}(\Delta\tau)]^2$. Values of $\Delta\tau$ ranged from 3 μs to 0.15 s in these experiments. When the average photon count rate $<n>$ is small, the contribution of dark counts from the photomultiplier to $n^{(2)}(\Delta\tau)$ and $g^{(2)}(\tau, \Delta\tau)$ becomes significant. The experimental estimators are given by

$$\hat{n}^{(2)}(\Delta\tau) - 1 = \left[n^{(2)}(\Delta\tau) - 1\right](1-\varepsilon)^2 + \left[n_d^{(2)}(\Delta\tau) - 1\right]\varepsilon^2 \quad (2.1)$$

$$\hat{g}^{(2)}(\tau) - 1 = \left[g^{(2)}(\tau) - 1\right](1-\varepsilon)^2 + \left[g_d^{(2)}(\tau) - 1\right]\varepsilon^2 \quad (2.2)$$

where $n_d^{(2)}$ and $g_d^{(2)}$ are the second factorial moment and the correlation function for dark counts, and ε is the fraction of dark counts in the total counts registered. Usually, $n_d^{(2)}-1$ and $g_d^{(2)}-1$ are not very large in the range of time scale corresponding to macromolecular diffusion. The correction terms in Equations 2.1 and 2.2, therefore, only increase the experimental values of $\hat{n}^{(2)}$ and $\hat{g}^{(2)}$ to larger values, but do not affect evaluation of $<\tau_h^{(s)}>$. On the other hand, distortion by the dead time of photomultiplier or computer become appreciable if $<n>/\Delta\tau$ is very large. A correction for this effect is given in the literature.[10]

An additional source of error to $n^{(2)}$ is introduced by the truncation of the experimental count distribution $P(n; \Delta\tau)$ by the digital nature of the experiments. Thus, for m such that $TM<n>P(m; \Delta\tau) < 1$, the experimental estimator $\hat{n}^{(s)}$ given by

$$\hat{n}^{(2)} = \sum_{n}^{m} \frac{n!}{(n-s)!} P(n, \Delta\tau) \tag{2.3}$$

may be evaluated by use of the approximate photon distribution (see below)

$$P(n, \Delta\tau) = \frac{\Gamma(n+k)}{n!\Gamma(k)} (1 + \frac{k}{<n>})^{-n} (1 + \frac{<n>}{k})^{-k} \tag{2.4a}$$

$$k = [n^{(2)}(\Delta\tau) - 1]^{-1} \tag{2.4b}$$

for which $n^{(s)} = \Gamma(k+s)/\Gamma(k)k^s$; Equation 2.4 is expected to be precise for large and small $<n>$.[11] The sum is truncated when $n > m$. For small $<n>$, which is the only situation for which the error $(\hat{n}^{(s)} - n^{(s)})/\hat{n}^{(s)}$ is important, m ln TM n /ln n $^{-1}$. Consequently, m may be increased, and the error made negligible by sufficient signal averaging. With experiments reported here, TM n was greater than 10^6. The function f(A) is fixed by the detector optics. With the optical system used in this study, f(A) is principally determined by pinholes. Pinholes from 100 to 1000 μm diameter are used. Typical data for f(A) as a function of the pinhole diameter are given in Figure [1], along with a theoretical estimate for f(A).[4]

The polarized 514.5 nm wavelength line from an argon-ion laser (Lexel 85) was used for photon correlation measurements, and the polarized 632.8 nm wavelength line from a He-Ne laser (spectra Physics 120) was used for integrated intensity measurements.

One of the polymer samples studied is an anionic polystyrene supplied from Pressure Chemicals Co. The weight-average molecular weight M_w and the ratio of M_w to the number-average molecular weight M_w/M_n of this sample are 4.0×10^5 and 1.06, respectively, according to the supplier. The polystyrene sample was dissolved in tert-butyl acetate, which was distilled prior to use. The solutions were filtered into the light scattering cells, sealed under vacuuum, and then centrifuged for 24 hours before light-scattering measurements while suspended by flotation in centrifuge tubes in a swinging-bucket rotor.

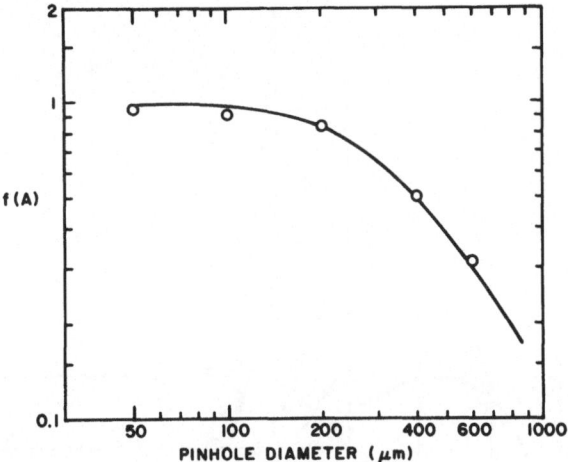

Figure 1.   The coherence factor f(A) determined for several values of
the pinhole diameter D.   The curve is a theoretical
estimate, fitted to the data by use of an arbitrary
proportionality between D^2 and the coherence area A.

Solutions of the rodlike macroion were prepared by dissolution in
methane sulfonic acid (MSA) of deuterated poly(1,4-phenylene-2,6-
benzobisthiazole)(d-PBT), which was supplied by Dr. J. R. Wolfe, SRI
International. Solutions were prepared in a glove box with air circulated
over drierite.   The solvent MSA was distilled under vacuum.   The
solutions were filtered several times with a teflon millipore filter into
the light scattering cells in a glove box filled with dry nitrogen. Cells
were sealed with flame under vacuum and then centrifuged for 24 hours
before measurements.

RESULTS

Polystyrene Solutions

Data on the polystyrene solutions provide a useful comparison as a
system devoid of intermolecular association and with a narrow molecular
weight distribution.   The gradual variation of P(n, $\Delta\tau$) with increasing $\Delta\tau$ is illustrated in Figure 2; the solid curves were constructed using
Equation 2.4 and the experimental values of <n> and $n^{(2)}(\Delta\tau)$. The gradual
change in P(n, $\Delta\tau$) with $\Delta\tau$ provides the desired information on the
coherence time.   Typical data on $n^{(2)}(\Delta\tau)$ obtained over a range of θ from
30 to 135 deg are shown in Figure 3 along with values of $g^{(2)}(\Delta\tau,\Delta\tau)$ for
the same solution.   These data are well fitted by Equations 1.1 and 1.2,
respectively, and permit estimates of $h^2 <\tau_h^{(2)}>$ and $h^2 <\tau_h^{(1)}>$ that

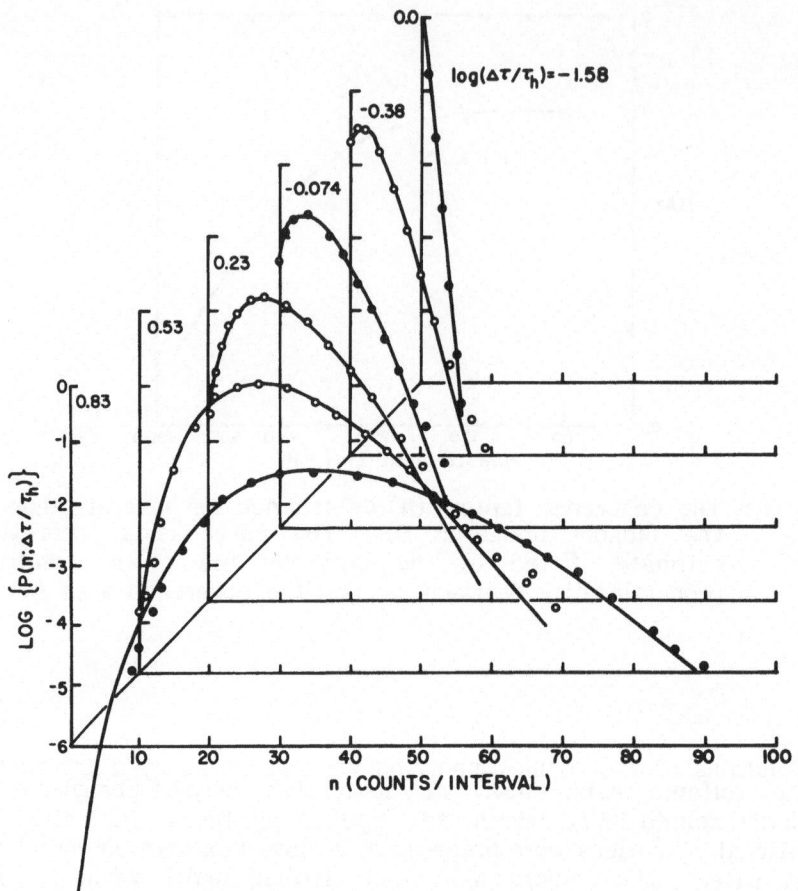

Figure 2.   The distribution $P(n,\gamma)$ determined for a polystyrene solution
             for several values of γ.   Numbers on each curve give log
             $\Delta\tau/\tau_h$.   The solid curves were constructed with Eqn. 2.4 using
             the experimental $n^{(2)}$ to compute k.

do not depend on h, (e.g., for the data shown, $<\tau_h^{(s)}>\sin^2\theta/2$ is 40 μs for
s = 1 or 2).   Data for $g^{(2)}(\tau,\Delta\tau)$ are given in Figure 4 over a range of
(from 6.4 to 51.2 μs) and θ (from 30 to 135 deg).   Use of the first
cumulant gives $<\tau_h^{(-1)}>\sin^2\theta/2$ = 37.9 μs for the solution with $<\tau_h^{(1)}>\sin^2\theta/2$
= 40 μs.   Values of $<\tau_h^{(s)}>$ with s = -1, 1 and 2 are given as a function
of θ in Figure 5, and data determined at three temperatures are given
in Figure 6.   In general, $<\tau_h^{(1)}>/<\tau_h^{(-1)}>$ was found to be about 1.04-1.06 for
these solutions, and $<\tau_h^{(-2)}> \sim <\tau_h^{(-1)}>$, so that on the scale used in Figure
6 it does not matter which estimate is used.   Values of $<\tau_h^{(-1)}>$ plotted
in Figure 6 have been normalized by $kTh^{2}{}^{-1}$ where the sovlent viscosity
is 0.740, 0.495, and 0.323 m Pa's at 19.6, 50.6, and 89.8° C, respectively.

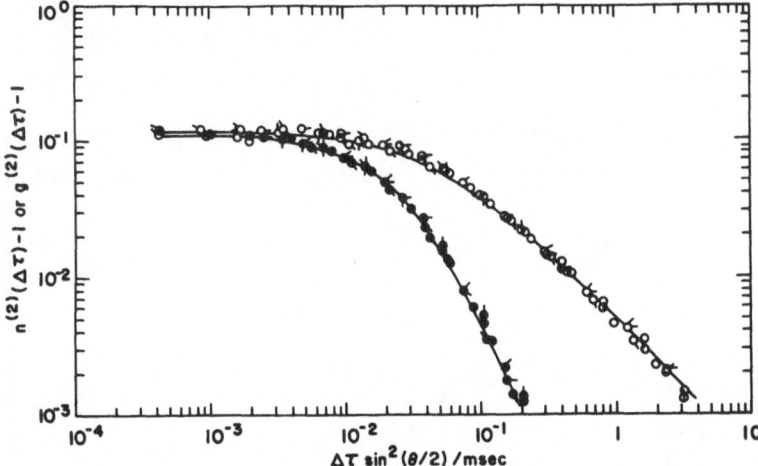

Figure 3. $\log[n^{(2)}(\Delta\tau)-1]$, O, and $\log[g^{(2)}(\Delta\tau)-1]$, ●, versus $\Delta\tau \sin^2\theta/2$ for solutions of polystryene in t-butyl acetate at 50.6°C; C = 10 g/L, 30 < θ > 135° (pip up at 30, rotating clockwise with increasing θ).

Values of $n^{(3)}(\Delta\tau)$ were also determined with these solutions. Data obtained with different optical arrangements for the detector are given in Figure 7. For a monodisperse solute,[12]

$$\frac{n^{(3)}(\Delta\tau)-1}{3\,n^{(2)}(\Delta\tau)-1} -1 = \frac{g(A)}{f(A)}\frac{2\left[\gamma-1+(\gamma+1)\exp-2\gamma\right]}{2\gamma^2-1+\exp-2\gamma} \qquad (3.1)$$

where g(A) is a coherence factor set by the optical arrangement. The data shown are in reasonable accord with Equation 3.1 for each A, and are fitted by the empirical relation $g(A) \sim (4/3)\,f^2(A)$.

D-PBT Solutions

Photon Correlation Scattering Examples of correlation functions $g^{(2)}(\tau,\Delta\tau)$, $n^{(2)}(\Delta\tau)$, and $g^{(2)}(\Delta\tau,\Delta\tau)$ are given in Figures 8 and 9. In contrast to the data for polystyrene solutions, the plot of ln $[g^{(2)}(\tau)-1]$ vs τ given in Figure 9 is not linear. Curvature becomes more severe as the scattering angle θ increases. It is, however, not difficult to evaluate the initial slope or the first cumulant. The data sets of $n^{(2)}(\Delta\tau)-1$ and give $g^{(2)}(\Delta\tau,\Delta\tau)-1$ versus $h^2 \Delta\tau$ are not independent of the scattering angle; the curves shift to shorter Δτ region with increasing scattering angle. These results indicate that the system d-PBT/MSA has a broad distribution of

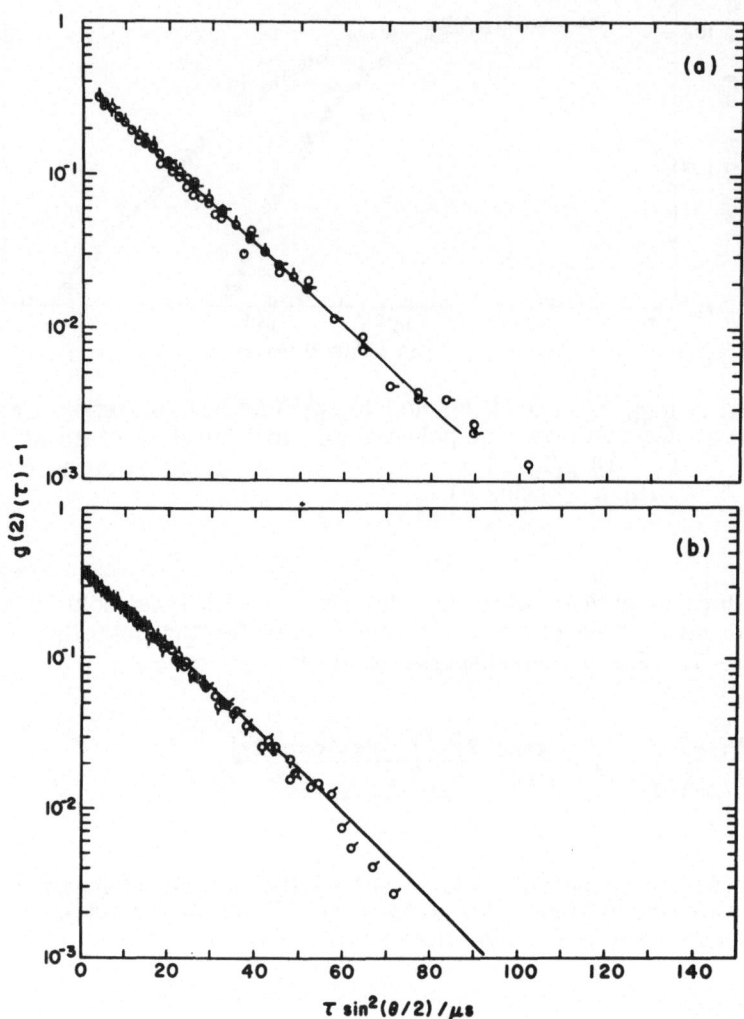

Figure 4a.   log[g$^{(2)}$(τ)-1] versus τ sin^{2}θ/2 for polystyrene solution in t-butyl acetate at 50.2°C, C = 5 g/L, θ = 60°.

Figure 4b.   log[g$^{(2)}$(τ)-1] versus τ sin^{2}θ/2 for a polystyrene solution in t-butyl acetate at 50.2°C, C = 5 g/L, 30 < θ > 135°.

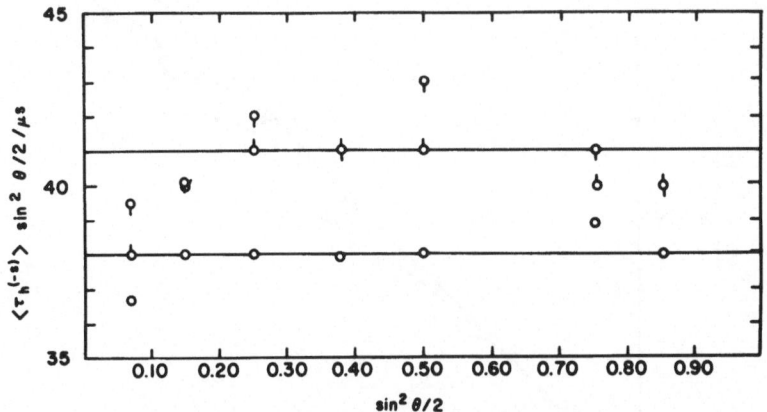

Figure 5. $< \tau_h^s > \sin^2\theta/2$ versus $^2\theta/2$ for polystyrene in t-butyl acetate at 50.6°C; C = 10 g/L; S = -1, O; 1, O; and 2, O.

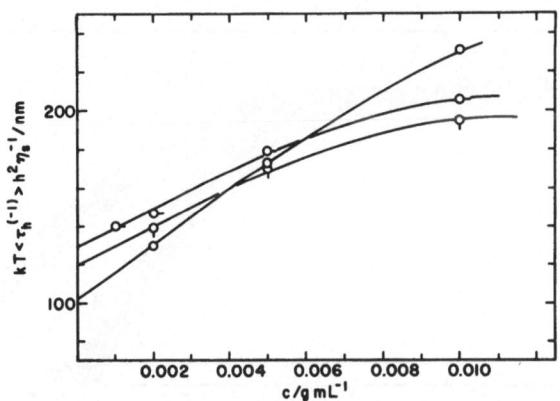

Figure 6. $kT < \tau_h^{(-1)} > h^2 \eta_s^{-1}$ versus C for solutions of polystyrene in t-butyl acetate at three temperatures: O, 19.6°C; O, 50.6°C; O, 89.8°C.

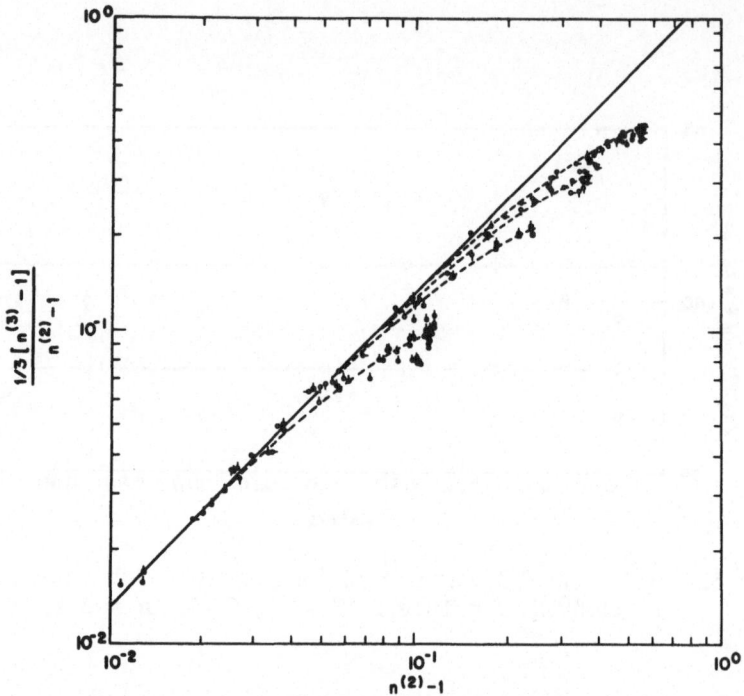

Figure 7.   Relation between $n^{(3)}(\Delta\tau)$ and $n^{(2)}(\Delta\tau)$ for four optical
arrangements. Dashed lines represent Eqn. 3.1 -- solid lines
represent asymptotic behavior.

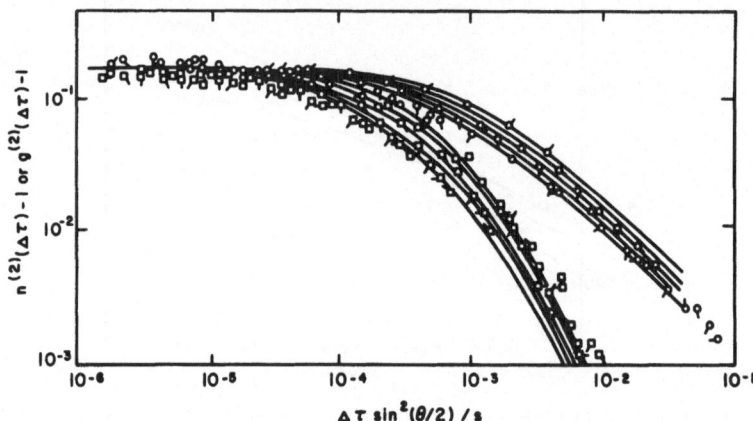

Figure 8. $\log[n^{(2)}(\Delta\tau)-1]$, O, and $\log[g^{(2)}(\Delta\tau)-1]$, □, versus $\Delta\tau \sin^2\theta/2$
for a solution of d–PBT in MSA at 30°C; C = 0.65 g/L,
45 < θ < 135° (pip up at 45, rotating clockwise with
increasing θ). The solid curves are calculated with Eqns. 1.5
and 1.6 using $\tau_{h,1}$, $\tau_{h,2}$, and r_1 given in Fig. 10.

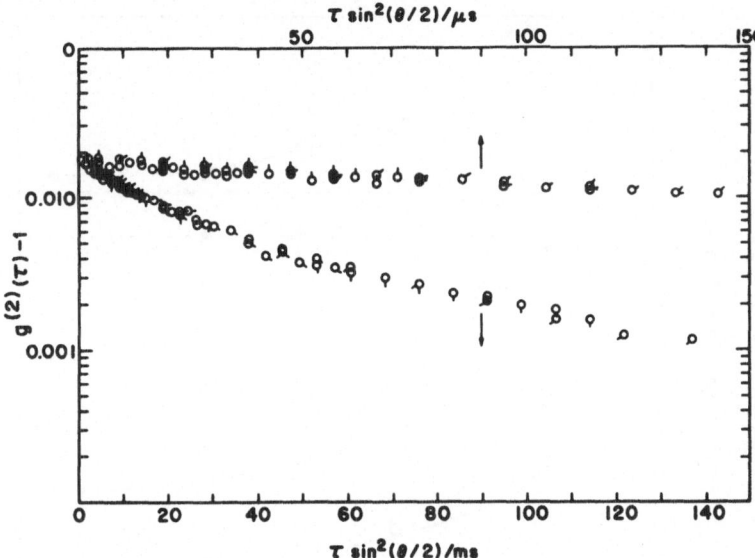

Figure 9.   log[g$^{(2)}$(τ)–1] versus τ sin$^2\theta$/2 for solutions of d–PBT in MSA
at 30°C; C = 0.65 g/L, θ = 75°, for several values of $\Delta\tau$.

molecular (or particle) sizes, or modes of relaxation other than
translational diffusion as a whole molecule.

In Figure 10, the average relaxation times $<\tau_h^{(s)}>$ sin$^2\theta$/2 for a
solution with c = 6.5 g/L are plotted against sin^2(θ/2) for s = –1 and 1.
Data points of each average relaxation time give a curve convex
downward and are not proportional to sin^2(θ/2). Values of $<\tau_h^{(1)}>$ are
slightly smaller than those of $<\tau_h^{(2)}>$, and much larger than $<\tau_h^{(-1)}>$ values.
The ratio $<\tau_h^{(1)}>/<\tau_h^{(-1)}>$ is about 2.0, which indicates that the distribution
of molecular sizes is broad, if the distribution of relaxation time
detected is due to the heterogeneity of molecular size.

The extreme curvature observed with g$^{(2)}$(τ,$\Delta\tau$) invites the use of
Equation 1.7 to obtain r$_i$ and γ_i for a few widely different components.
Equation 1.7 is reminiscent of relations encountered in other applications
in which the measured quantity represents a weighted sum of exponential contributions (e.g., stress relaxation, small-angle X-ray scattering,
absorption spectroscopy, total intensity light scattering, etc.). In this
case, the relation between r$_i$ and the concentration c$_i$ and molecular
weight M$_i$ of the components with coherence time $\tau_{h,i}$ is complex, but
need not be known to attempt assessment of a resolution of g$^{(2)}$(τ,$\Delta\tau$) into
representative components. If the $\tau_{h,i}$ are sufficiently widely spaced
then [g$^{(2)}$(τ)–1]$^{1/2}$ may be decomposed into a sum of exponentials by

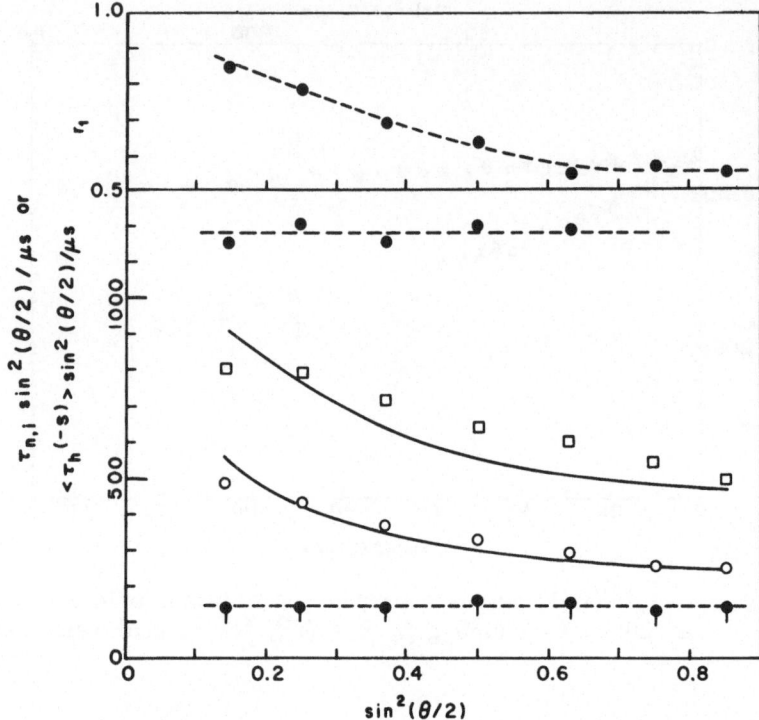

Figure 10.   (a) $\langle\tau_h^{(s)}\rangle \sin^2\theta/2$ or $\tau_{h,i}\sin^2\theta/2$ versus $\sin^2\theta/2$ for a solution
of d-PBT in MSA at 30°C for S = -1, 0 and 1, $\square$; i =
1, o, and 2; C = 0.65 g/L.   Solid lines calculated with
Eqn. 1.11 using values of $\tau_{h,i}$ and r_1 given in the figure.

(b) r_1 versus $\sin^2\theta/2$ for solutions of d-PBT in MSA at 30°C;
C = 0.65 g/L.

simple graphical methods.   Thus, as may be seen in Figure 9,
$\ln[g^{(2)}(\tau)l]$ versus τ is a linear function (within experimental error) for
large τ, permitting assessment of the longest coherence time $\tau_{h,i}$ and the
weight r_1.   Subsequent analysis of $\ln\{[g^{(2)}(\tau)-1]^{1/2}-f^{1/2} r_1 \exp - \tau/\tau_{h,1}\}$
versus τ permits evaluation of $\tau_{h,2}$ and r_2, etc., until the residual
between $[g^{(2)}(\tau)-1]^{1/2}$ and $f^{1/2} \Sigma r_i\exp - \tau/\tau_{h,i}$ at small τ is within
experimental error -- this method is exactly that of "Procedure X" used
to represent stress relaxation data.[13]   This procedure provides a method
to represent the component contributions to $g^{(2)}(\tau,\Delta\tau)$ with minimal
computational effort, and will usually provide a reasonable representation within the limits of the data.   The graphical method is not suitable
to represent components with closely spaced $\tau_{h,i}$ in a meaningful way
(e.g., to accurately represent the molecular weight distribution of a
polymer). The uniqueness of the representation is strongly dependent on
the behavior of large τ and the accuracy to which $\tau_{h,1}$ and r_1 may be
obtained.

Application of successive approximations to $g^{(2)}(\tau,\Delta\tau)$ is illustrated in Figure 11 with data that are fitted by two terms with coherence times differing by about ten-fold. All of the data for d-PBT could be fitted by Equation 1.7 with two components. Typical results are shown in Figure 10, for one concentration. As seen in Figure 10 $h^2\tau_{h,1}$ and $h^2\tau_{h,2}$ are independent of h, as expected, even though $h^2 <\tau_h^{(-1)}>$, etc., depend markedly on h. The curves given in Figure 8 for the angular dependence of the average coherence times calculated with Equations 1.5 and 1.6 and the values of γ_1, γ_2, and r_1 are in good agreement with the observed data. A plot of $h^2\tau_{h,1}$ and $h^2\tau_{h,2}$ versus the overall concentration c is given in Figure 12; r_1 was independent of c, and followed the behavior given in Figure 10.

The significance of the two components found by the preceding analysis will be considered in the following, after the results for the integrated intensity light scattering are presented.

Integrated Intensity Scattering Reciprocal scattering intensities for V_V and H_V modes are plotted against $\sin^2(\theta/2)$ in a form of square root plot in Figure 13. Values of $(Kc/\mathbb{R}_{VV})^{1/2}$ for each solution indicate significant deviation at small scattering angles (ca. 30% at 30°) from the straight line fitted with data at higher angles, while those of $(Kc/\mathbb{R}_H)^{1/2}$ deviate from constant value at high angles to less extent (ca. 12% at 0°). Thus, the scattering from large particles (presumed to be aggregates of d-PBT molecules) contribute to V_V scattering intensity more severely than to H_V intensity. This means that most of the aggregated particles are optically isotropic. H_V scattering is, therefore, mainly attributed to non-associated molecules and V_V scattering includes components from both single molecules and aggregates. The fast and slow modes found above correspond to these two species. Based on this conjecture, the V_V scattering data in Figure 13 may be separated into two components, corresponding to the fast and the slow modes, using r_1 in Figure 10. Values of $[Kc/\mathbb{R}_{VV}^{(F)}]^{1/2}$ and $[Kc/\mathbb{R}_{VV}^{(S)}]^{1/2}$ thus obtained are plotted against $\sin^2(\theta)$ in Figure 14. Here $\mathbb{R}_{VV}^{(F)} = r_2\,\mathbb{R}_{VV}$ and $\mathbb{R}_{VV}^{(s)} = (1-r_2)\mathbb{R}_{VV}$ indicate the fast and slow components, respectively. Then, the data were analyzed with Equations 1.9 and 1.10 on the assumption that both $\mathbb{R}_{HV}$ and $\mathbb{R}_{VV}^{(F)}$ result from unassociated d-PBT molecules. The results give δ equal to 0.53, which is close to that estimated previously for PBT.[1] The weight average molecular weight M_W is calculated to be 4600 on the basis that the weight fraction $x = c_2/c$ of the unassociated molecules is essentially unity so that $c_2 \approx c$, see below.

In contrast with $\mathbb{R}_{VV}^{(F)}$, $\mathbb{R}_{VV}^{(s)}$ is very dependent on the scattering angle, indicating a large size for the aggregates. The slopes of the straight lines in Figure 17 give the apparent radius of gyration of the aggregates in each solution, which increases with increasing original concentration. The apparent radius of gyration obtained for each solution is extrapolated to zero concentration to give about 60 nm, if we

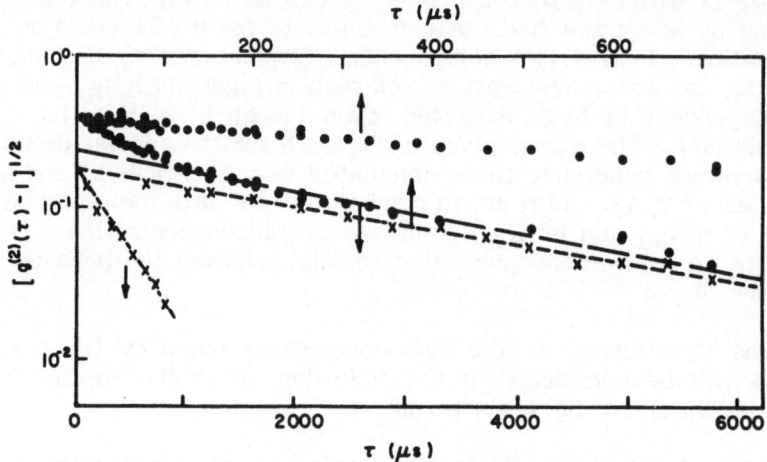

Figure 11. $\log[g^{(2)}(\tau)-1]^{1/2}$, ●, and $\log\{[g^{(2)}(\tau)-1^{1/2} - f^{1/2} r_1 \exp - \tau/\tau_{h,1}\}$, X, versus τ for a solution of d-PBT in MSA at 30°C; $C = 0.65$ mg/ml; $\theta = 90°$.

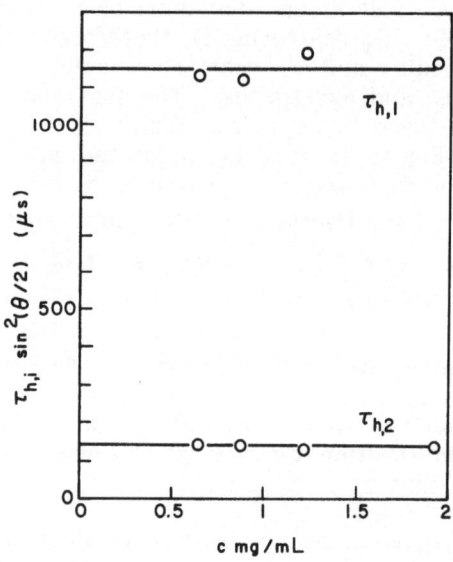

Figure 12. $\tau_{h,1} \sin^2\theta/2$ and $\tau_{h,2}\sin^2\theta/2$ versus C for solutions of d-PBT in MSA at 30°C

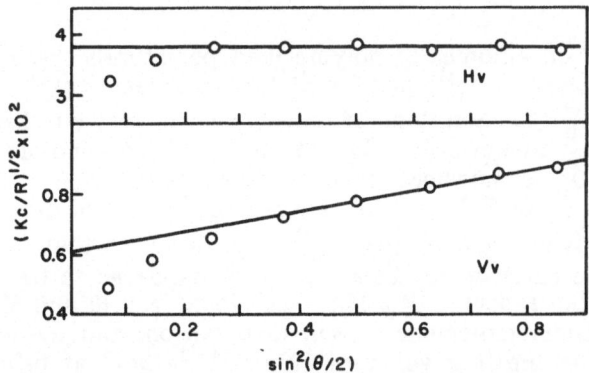

Figure 13. The polarized (Vv) and depolarized (Hv) scattering as a function of $\sin^2\theta/2$ for a d-PBT solution in MSA at 30°C; C = 0.65 g/L.

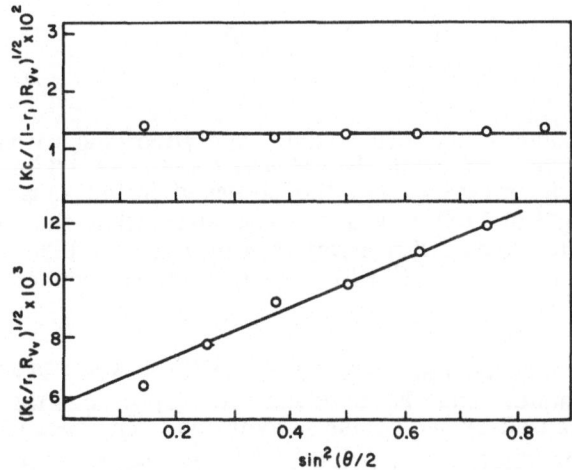

Figure 14. The components of the Vv scattering in Fig. 13 attributed to the moieties with coherence times $\tau_{h,1}$ (lower) and $\tau_{h,2}$ (upper).

can assume that the concentration of the aggregates is proportional to the original polymer concentration.

DISCUSSION

The data on anionically polymerized polystyrene result in a small difference between $<\tau_h^{(1)}>$ and $<\tau_h^{(-1)}>$, whereas $<\tau_h^{(2)}> \sim <\tau_h^{(1)}>$ and $<\tau_h^{(-2)}> \sim <\tau_h^{(-1)}>$. Solutions of polystyrene in tert-butyl acetate have an upper theta temperature Θ_U at about 13° $\overline{C}$ and a lower theta temperature Θ_L somewhere near 110° C (e.g., $A_2 = 0$ when the temperature is Θ_U or Θ_L), with A_2 equal to a maximum near 70°C. With the relations given above, $<\tau_h^{(1)}>/<\tau_h^{(-1)}> \sim (M_z/M_w)^\alpha$ so that in accord with experiment, the ratio of coherence times is expected to be about 1.04 to 1.06 based on the reported $M_w/M_n = 1.10$ (e.g., z = 10 and $M_z/M_w = 1.09$ for a Schulz-Zimm molecular weight distribution) and $\alpha \sim 0.6$. As seen in Figure 6, the limiting value of $(\partial \ln <\tau_h^{(-1)}>/\partial c)^0$ at infinite dilution decreases from 140 mLg^{-1} to 75 mLg^{-1} as T increases from 19.6 to 50.6° C. This trend is in accord with the usual behavior which can be expressed in the form[9b] (suppressing the superscript s).

$$<\tau_h>h^2 = (\Xi^0/kT)\left[1+(d_1[\,\eta\,]-d_2A_2M+\overline{v})c+...\right] \qquad (4.1a)$$

so that the molecular friction coefficient at infinite dilution is

$$\Xi^0 = kT <\tau_h>^0 h^2 \qquad (4.1b)$$

Here v is the polymer specific volume, and according to one treatment d_1 and d_2 are constants, equal to 0.2 and 1.2, respectively. Since $[\eta]$ is expected to be 50 mLg^{-1} at Θ_L, the observed initial tangent $(\partial \ln <\tau_h>/\partial c)^0$ of 140 mLg^{-1} at 19.8° C is seen to be larger than the value of about 10 mLg^{-1} predicted at Θ_U with Equation 4.1a with $d_1 = 0.2$. This behavior and the absolute value of Ξ^0 will receive further comment in a future communication.

The difference between $<\tau_h^{(1)}>$ and $<\tau_h^{(-1)}>$ is more pronounced with the d-PBT sample, and is believed to represent the effects of heterogeneity caused by molecular association. The result $\tau_{h,2}^0 \sin^2\theta/2 = 135$ µs at infinite dilution is attributed to the unassociated species such that the molecular friction coefficient is given by $\Xi^0/\eta_s = 70.7$ nm (the solvent viscosity η_s is 9 mPa·s at 25^0). For rodlike chains, the weight average contour length L_w is given by[1]

$$L_w = (H\ \Xi^0\ /3\pi\eta_s L_w^{0.2})^{5/4} \qquad (4.2)$$

where H is about equal to $1.887 \ (L_W/d)^{0.2}$, with d the hydrodynamic diameter of the chain, taken to be 1.5 nm. The result, $L_W = 25$ nm, is in reasonable accord with the estimate 21 nm calculated as M_W/M_L, using the value $M_W = 4600$ estimated from the integrated intensity light scattering data; $M_L = 220$ dalton/nm is the mass per unit chain contour length.

The values of $R_{G,1}$ and $\tau^0_{h,1}$ are attributed to a large aggregate, which will be taken to be ellipsoidal such that

$$R_G = \left(\frac{1+2P^2}{3}\right)^{\frac{1}{2}} \frac{L}{2} \tag{4.3}$$

$$\Xi^0/3\pi\eta_s = L \ (1-P^2)^{\frac{1}{2}}/\ln P^{-1}\left[1+(1-P^2)^{\frac{1}{2}}\right] \tag{4.4}$$

where L and LP are the major and minor diameters, respectively. The results $R_{G,2} = 60$ nm and $\Xi^0_2/\eta_s = 600$ nm correspond to $P^{-1} = 13$, LP = 16 nm and L = 210 nm for the aggregated moities, or an ellipsoidal aggregate with volume $\pi L^3 P^2/6 = 29 \times 10^3$ nm^3 per particle. If it is assumed that the aggregate particle have the partial specific volume $v^0_{sp} = 0.48$ cm^3/g, then this particle volume corresponds to a particle molecular weight of 3.6×10^7, in comparison with $(Kc/r_1R^0_{V_V})^{-1} = (1-x)M_W$ 6.2×10^4, so that 1-x is seen to be 0.002, in accord with the premise that the weight fraction c_1/c of the aggregate is small. This analysis indicates that the aggregates in the d-PBT solution are elongated, being about tenfold the average molecular length and diameter, as might be expected if the molecules are arrayed with their axes in (nearly) parallel array.

ACKNOWLEDGEMENT

This study was supported in part by the Air Force Materials Laboratory, Polymers Branch, contract number F33615-79-C-5034, and the National Science Foundation, Polymers Program, grant number DMR-8009226. Ms. R. Furukawa assisted with aspects of the light scattering measurements.

REFERENCES

1.   C. C. Lee, S.-G. Chu, and G. C. Berry, J. Poly. Sci., Poly. Phys. Ed. (in press).

2.   P. Metzger Cotts and G. C. Berry, J. Poly. Sci., Poly. Phys. Ed., submitted.

3. E. R. Pike and E. Jakeman, Adv. Quantum Electron, 2, 1 (1974).
4. J. M. Schurr, CRC Crit. Rev. in Biochem. 4, 371 (1977).
5. B. J. Berne and R. Pecora, Dynamic Light Scattering, Wiley-Interscience, New York, 1976, Chapter 8.
6. B. Chu, Laser Light Scattering, Academic Press, NY, 1974.
7. a) G. C. Berry, J. Poly. Sci., Poly. Symp. 65, 143 (1978); b) G. C. Berry, Discuss. Farad. Soc. 49, 121 (1970).
8. D. E. Koppel, J. Appl. Phys. 42, 3216 (1971).
9. H. Yamakawa, Modern Theory of Polymer Solutions, Harper and Row, New York, 1971, a) p 218, b) p 314.
10. E. Jakeman, in Photon Correlation and Light Beating Spectroscopy, H. Z. Cummins and E. R. Pike, eds., Plenum Press, New York, NY 1974.
11. G. Bedard, J. C. Chang, and L. Mandel, Phys. Rev. 160, 1496 (1967).
12. A. K. Jaiswal and C. L. Mehta, Phys. Rev. A2, 168 (1970).
13. A. V. Tobolsky, Properties and Structures of Polymers, Wiley, New York, NY (1960).

STUDIES OF PSM IN AQUEOUS SOLUTION NEAR THE OVERLAP

CONCENTRATION

R. L. Shogren, A. M. Jamieson, and J. Blackwell
Department of Macromolecular Science
Case Western Reserve University
Cleveland, Ohio 44106

ABSTRACT

Porcine Submaxillary Mucin (PSM) is a glycoprotein whose primary structure consists of a protein core with frequent, short oligosaccharide side chains. Previous static and dynamic light scattering data indicate, in dilute aqueous 0.1M NaCl, PSM exists as quasi-spherical, internally branched, highly hydrated, polydisperse aggregates which slowly dissociate till a stable species of weight average molecular weight (M_W) 7.4 x 10^6 is reached. In 6M GdnHCl the non-covalent bonds between PSM molecules are apparently broken giving a highly elongated molecule of M_W 2.0 x 10^6. The forces which bond PSM into its native aggregate form in 0.1 M NaCl seem to be specific in nature since the dissociation is irreversible. Upon reduction of PSM with mercaptoethanol, polydispersity decreases substantially, and M_W decreases to 9 x 10^5. A discrete change in the solution properties of PSM in 0.1 M NaCl at a concentration of 2 mg/ml, manifested by a sudden decrease in the translational diffusion coefficient, increase in viscosity number and decrease in slope of the osmotic compressibility, was observed. We tentatively propose that a weak secondary association process occurs at this concentration although it cannot be ruled out that hydrodynamic interaction of the congested macromolecules are responsible for the observed effect.

INTRODUCTION

We are investigating the solution properties of mucin glycoproteins in order to understand the relationship between their chemical structure and viscoelastic properties[1]. Gel-forming mucins such as tracheo-

211

bronchial mucin (TBM) serve, for example, as mechanical couplers between particles stuck in mucus and beating cilia[2]. To understand how the unique mucus gel is formed, we need to characterize the interactions which occur between mucin molecules. In the present paper we describe studies of porcine submaxillary mucin (PSM), which forms viscous solutions and serves as a lubricant, etc., in saliva. This macromolecule has a similar chemical structure to TBM but is in many ways simpler and better understood.

Partial chemical structures of many mucins have only recently become available. All mucins possess a protein backbone rich in serine, threonine, glycine, proline, alanine, glutamic acid (glutamine) and aspartic acid (asparagine)[3]. Oligosaccharide side chains of 1-20 sugar residues are linked via the Cl of galactosamine to the hydroxyl groups of serine and threonine[4]. Galactose, glucosamine, fucose and sialic acids are the only other saccharide components. Usually one out of every 3-5 amino acid residues is a serine or threonine with an oligosaccharide substitution. Interestingly, gel-forming mucins isolated from tracheobronchial, gastro-intestinal and cervical mucus have longer oligosaccharides (1-20 residues) than many submaxillary mucins[5] (1-5 residues). Structures of oligosaccharides have been determined for many mucins[3]. Since the sugars are sequentially linked to the protein core by different enzymes, there is a wide distribution in oligosaccharide lengths. Some TBM species have chains with short repeating sequences of some sugars[4]. The protein core is probably also a source of variation between different mucins.

A partial sequencing of ovine submaxillary mucin by Hill et al[6] revealed no recognizable repeat pattern. Bhushana Rao et al[7] and Pigman et al[8], however, obtained only a small number (2) of low molecular weight glycoproteins after extensive proteolysis of bovine cervical mucin (BCM) and bovine submaxillary mucin (BSM) respectively, suggesting a repeating structure. For porcine gastric mucin (PGM) Allen[3] found a bare peptide region at the end of each molecule which contained many cysteine residues. Similarly, evidence has been presented using light scattering[9], water vapor sorption[10] and enzymatic digestion data[11] to suggest that areas of the protein core of PSM are also exposed.

Silberberg et al[12] have classified two types of interactions between mucins: type I crosslinks that involve both secondary bonds and disulphide bonds linking the subunits into linear or slightly branched chains and type II crosslinks that are secondary bonds leading to the formation of the transient mucus network. The presence of intermolecular disulphide bonds has been clearly demonstrated for a number of mucins. The molecular weight of cervical, tracheobronchial and gastric mucins decreases from 2×10^6 to 5×10^5 upon reduction of the disulphide bonds[12]. Recently, Pearson et al[13] found that a 70,000 molecular weight protein is released upon reduction of PGM and may

Table 1. Molecular weight and size of PSM in different solvents.

	0.1M NaCl		6M GuHCl		0.1M Mercapto-ethanol + 6M GuHCl
	Exper.	Calc.	Exper.	Calc.	Exp.
M_w (g/mol)	$7.4 \pm 1.0 \times 10^6$	7.4	$2.0 \pm 0.2 \times 10^6$	2.0×10^6	$9.4 \pm 1.0 \times 10^5$
R_G(Å)	1420 ± 100	1130	1080 ± 80	830	721 ± 70
D_t^o (cm^2/s)	$1.75 \pm 0.3 \times 10^{-8}$	1.97	$3.9 \pm 0.2 \times 10^{-8}$	4.3×10^{-8}	$5.3 \pm 0.3 \times 10^{-8}$
R_S(Å)	1230 ± 200	1090	550 ± 50	500	400 ± 50
$[\eta]$ (cm^3/g)	840 ± 50	1070	400 ± 40	546	-
D_r (sec^{-1})	1.16 ± 25	85	-	-	-
a(Å)	-	1950	-	1500	
b(Å)	-	650	-	150	
$(\bar{\nu}_2 + \delta_1 \bar{\nu}_1^0)$ (cm^3/g)	-	280	-	42.4	
a/b	-	3	-	10	

represent a "link" protein. Most submaxillary mucins contain either very small or negligible amounts of cysteine[5]. Holden et al[14] found, however, that reduced PSM migrates much faster than the oxidized form during gel electrophoresis. Type II bonds are labile and are slowly broken by shearing or by the addition of guanidine hydrochloride or other denaturing agents.

Our recent work on PSM has not yet been published[1] so we briefly describe parts of it here. Table 1 summarizes experimental data, hydrodynamic radius R_h, radius of gyration, R_G, intrinsic viscosity [ ] and weight average molecular weight, M_w, for PSM particles observed in two solvent systems. Also shown are calculated values for an equivalent prolate ellipsoid of axial ratio a/b. PSM dissolved in 6M guanidine hydrochloride has an M_w of 2×10^6. All non-covalent bonds are apparently broken in this species since the same M_w is obtained at 20°C, 40°C and 60°C. Disulphide bond reduction further reduces M_w to 9×10^5 and reduces the polydispersity as measured by the second moment of the correlation function μ_2/Γ^2 (see experimental section). PSM in 6M GnHCl is a highly extended molecule (axial ratio ~ 10/1) presumably reflecting the stiffening effect of the oligosaccharide side chains on the

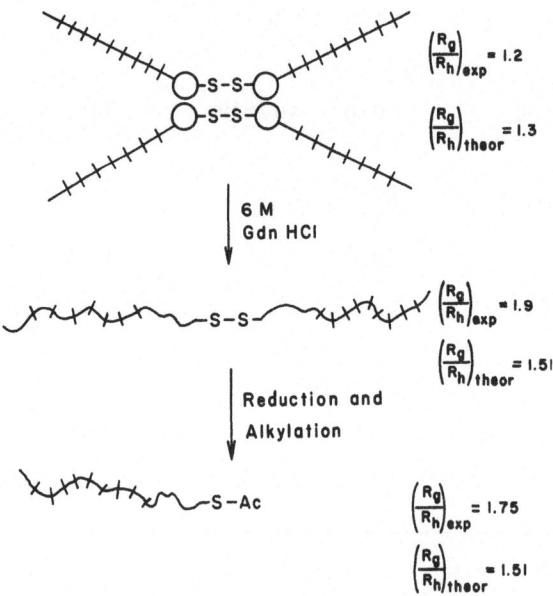

Figure 1.   Hierarchical structures of primary PSM aggregates in 0.1M NaCl at infinite dilution (top) and after dissolution by 6M GnHCl and finally reduction and alkylation.

protein backbone. In 0.1M NaCl, PSM forms a quasi-spherical aggregate (axial ratio ~ 3/1) with an M_w of 7×10^6 which is highly hydrated (300 cm^3/g). Since the ratio R_g/R_h is much larger for the species observed in GdnHCl in comparison to the aggregate species in 0.1M NaCl (Table 1), it is proposed[1], based on theory[15], that the former are linear while the latter exhibit a branched internal structure. A structural model for the primary aggregates is shown in Fig. 1. Since the polydispersity decreases upon dispersion of the primary aggregates in dissociative solvents, their size distribution is probably due to the presence of aggregates formed by different numbers of subunits. The primary aggregates dissociate at different rates depending on the solvent; the solvents, listed in order from good to poor are: 6M GnHCl, 6M urea, 1% SDS, 0.1M NaCl, 33% DMSO in H_2O. Increasing temperature greatly increases the rate of dissolution. Once dissociated by GnHCl, for example, reassociation does not occur.

In the following we present a characterization of the physical properties of solutions of PSM aggregates in 0.1M NaCl in the dilute and semi-dilute regimes. Diffusional properties were measured using quasielastic laser light scattering, osmotic compressibilities via integrated light scattering and shear dependent behavior via capillary viscometry.

EXPERIMENTAL

Sample Preparation

Porcine submaxillary mucin (PSM) was generously supplied by Dr. Pi-Wan Cheng of the Department of Pediatrics, CWRU (present address: Department of Pediatrics, University of North Carolina, Chapel Hill, N.C.), and was prepared by a method similar to that of De Salegui and Plonska[16] but without the hydroxylapatite treatment. The structures of the oligosaccharides of PSM have been determined by Carlson[17].

A stock solution of 0.1% PSM in 0.01M NaCl and 0.002% sodium azide (to retard bacterial growth) was prepared by dissolving a weighed portion of lyophilized mucin in the above salt solution by heating at $37^\circ C$ for 4 days. This heating was found to be necessary to disperse undissolved mucin which otherwise would sediment out during centrifugation (see below). The pH of the solution was next adjusted to 6.4 by addition of a small amount of concentrated NaOH. To remove dust particles, the solution was centrifuged for 4 hours at 20,000 x g. The top 4/5 of the centrifuged solution was removed and evaporated under reduced pressure for two weeks at $23^\circ C$ to about 1/10 the original volume. (The concentrations of PSM in the upper 4/5 and lower 1/5 of the centrifuge tube were measured as described below and found to be equal.) Successive dilutions of the concentrate were made with 0.1M NaCl and 0.02% NaN_3.

The mucin solutions were dialyzed against the solvent for one week at 5°C after which the refractive index increment at constant chemical potential of solvent, $(dn/dc)\mu_3$, was determined using a differential refractometer model RF-600 from C. N. Wood M.F.G. Co. Newtown, Pa. Concentrations were also checked by this method assuming $\Delta n = (dn/dc)\mu_3 \Delta C$.

Light Scattering

The instrumentation used for quasielastic light scattering has been described previously[18]. Solution specimens in small glass tubes were transferred to a larger tube containing decahydronapthalene which has a refractive index close to that of glass thereby minimizing the elastic scattering. The instrument provided a direct determination of the normalized intensity autocorrelation function, $g^{(2)}(\tau)$, which, for a dilute solution of identical, spherical particles is a single exponential:

$$g^{(2)}(\tau) = 1 + f(A)e^{-2\Gamma\tau} \tag{1}$$

where τ is time, $f(A)$ is an optical constant and $1/\Gamma$ is a characteristic relaxation time. Γ is proportional to the translational diffusion coefficient, D_t:

$$\Gamma = D_t K^2 \tag{2}$$

$$K = \frac{4\pi n \sin\theta/2}{\lambda} \tag{3}$$

where n is the refractive index of the solution, θ the scattering angle (40° was used) and λ is the wavelength of the laser light in vacuo (6328Å).

For polydisperse systems, the correlation function is a sum of exponential terms having different relaxation times. The simplest analysis uses the method of cumulants[19] for which the first two terms are:

$$\frac{1}{2}\ln\left[g^{(2)}(\tau) - 1\right] = -\overline{\Gamma}\tau + \frac{1}{2!}\frac{\mu_2}{\overline{\Gamma}^2}(\overline{\Gamma}\tau)^2 \tag{4}$$

Here the first moment, $\overline{\Gamma}$, is related to the z-average diffusion coefficient, $D_{t,z}$:

$$\overline{\Gamma} = D_{t,z}K^2 \tag{5}$$

when $KR_{G,z} \ll 1$, and $\mu_2/\bar{\Gamma}^2$ is a measure of the breadth of the distribution of relaxation times and hence of the polydispersity. In practice, $\bar{\Gamma}$ and $\mu_2/\bar{\Gamma}^2$ are calculated for a number of different correlation times and the results extrapolated to $\tau = 0$ where the higher order terms of eq. (4) can be neglected.

Viscosity

A Cannon–Ubbelohde #100 four bulb viscometer, immersed in a water bath controlled to $0.1^{\circ}C$ by a Haake circulator, was used to determine viscosities which were extrapolated to zero shear rate by plotting the logarithm of the reduced viscosity versus the logarithm of the shear rate. (The above plots were linear in the accessible range of shear rates: $50-1000 \ sec^{-1}$).

RESULTS AND DISCUSSION

Figures 2–4 show the concentration dependence of $D_{t,z}$, viscosity number and concentration divided by excess scattering intensity, $c/\Delta I$, for PSM in 0.1M NaCl.   There is a sigmoidal decrease in $D_{t,z}$ and a corresponding increase in viscosity number, each with a midpoint at a concentration of ~ 2 mg/ml; at the same concentration, the rate of change of c/I, which is, in the limit of zero angle $\theta = O$, proportional to the osmotic compressibility $(\partial \pi/\partial c)_{T,P}$, undergoes a sudden decrease. These events occur at a concentration which is comparable to the overlap concentration defined by deGennes[20] $c^* = M/N_A R_G^3 = 4.1$ mg/ml. Near c^*, well-defined changes in the concentration dependence of D_t[20], n_r[21] and $(\partial \pi/\partial c)_{T,P}$ are expected and observed due to the influence of chain entanglements on thermodynamic and transport properties of polymer solutions.   It should be noted that the above expression for c^* corresponds approximately to the concentration c_h at which a hexagonally close-packed   arrangement of the equivalent   hydrodynamic spheres of linear polymer chains occurs[22],   (c.f. using the Kirkwood-Riseman equation $R_S = 0.665 \ R_G$, $c_h \sim (\ 2/8)M/N_A R_S^3 = 0.60 \ M/N_A R_G^3$). From Table 1, R_G/R_S for PSM in 0.1M NaCl is 1.2 and thus $c_h \sim (2/8)M/N_A R_S^3 = 0.31 \ M/N_A R_G^3$.   This suggests that changes in solution properties of PSM in 0.1M NaCl as a result of molecular congestion may be observed at a concentration around one half of the value where such changes are observed for linear flexible coil molecules, i.e. at c ~ 2 mg/ml.   We therefore propose to interpret the transitions in solution properties evident in Figures 2–4 based on possibilities inherent in a congested system of interacting Brownian particles.

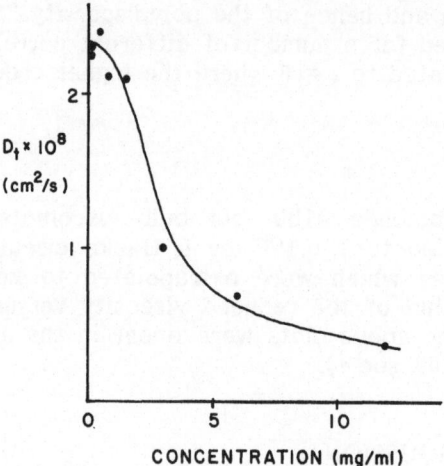

Figure 2.   Concentration dependence of translational diffusion coefficient for PSM in 0.1M NaCl as determined by the method
of cumulants.

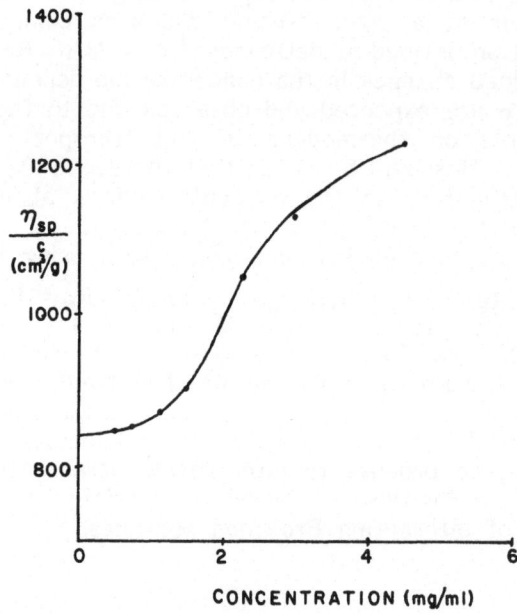

Figure 3.   Concentration dependence of the viscosity number, η_{sp}/c,
for PSM in 0.1M NaCl.

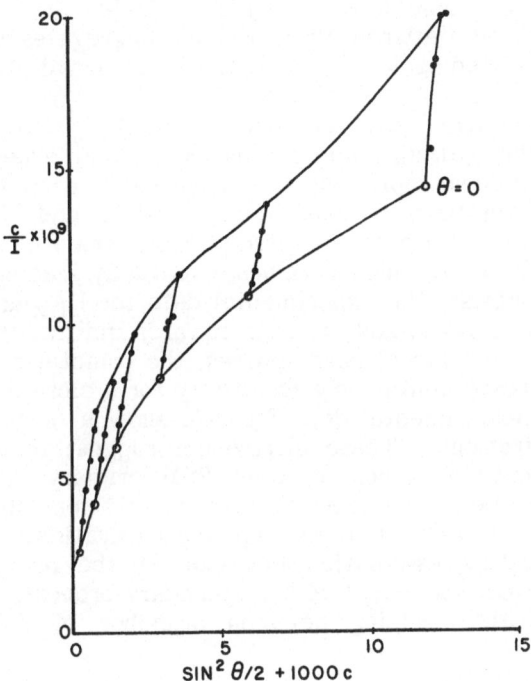

Figure 4.   Zimm plot of PSM in 0.1M NaCl.

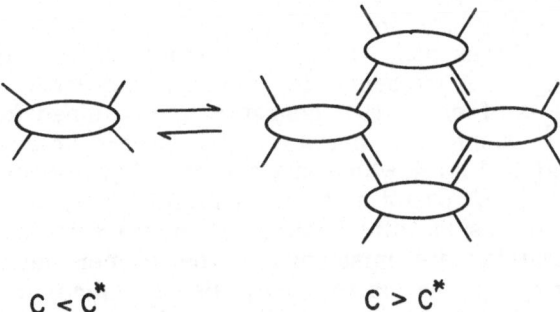

Figure 5.   Schematic representation of proposed secondary association of primary PSM aggregates into a bridged network at concentration above c*.

Two specific interaction schemes are considered: a) the particles interact by predominantly hard-sphere repulsive forces; b) a short range attractive interaction between particles exists, such that a weak tendency for self association results. Likely candidates for the attractive potential between PSM primary aggregates are hydrophobic and/or hydrogen-bonding interactions of the carbohydrate side chains[23].

If PSM particles are regarded as hard spheres with a radius equivalent to the hydrodynamic radius, we should expect translational diffusion to slow down rapidly at a concentration near $c \sim 0.3 \ c^*$ and $D_{t,z}$ to decrease continuously to even smaller values and, finally, to zero. Likewise, we would expect a corresponding smooth increase in the viscosity number to even higher values, eventually leading to a solid-like response. In contrast, the experimental data for $D_{t,z}$ and η_{sp}/c appear, respectively, to decrease and increase towards limiting values at highest concentrations. For model hard spheres, the osmotic compressibility is expected to increase continuously to infinity for a close packed system.[24] In contrast the experimental data for $c/\Delta I$ show a decrease in slope at highest concentrations. These differences suggest that in addition to hard sphere repulsive forces between PSM primary aggregates, weak attractive forces become important at concentrations above $c^*/2$. Thus we are led to suggest that at high concentrations a weak, easily reversible secondary association behavior of the previous aggregates occurs. A hypothetical model of the secondary process, shown in Figure 5, schematizes the bridging between branches of different primary aggregates.

Some reservations about interpreting our data as an association process should be pointed out. First, the modelling of highly swollen PSM aggregates as hard spheres is a reasonable approximation since the few experimental or theoretical results reported regarding the transport properties of semi-rigid branched polymers or swollen colloids show behavior in semi-dilute solutions that appears rather similar to that observed for spheres near the locking concentration c_h. However, a substantial degree of molecular overlap will be possible and it should be noted that plots of c/I versus concentration obtained for flexible linear polymers[25-27] which can pervade each others volume show effects similar to that in Fig. 4 which are expected because of the concentration-dependence of osmotic compressibility in such systems. It is possible that the asymptotic leveling off of viscosity number and D_t at high concentrations are misleading: the former may not be in the Newtonian region, the latter may be complicated by unexpected relaxational contributions in the PCS decay function. Thus, based on the present results alone we are not able to unambiguously differentiate between the potential effects of hydrodynamic and weak associative interactions in congested solutions of PSM. Further work, utilizing techniques such as freeze-fracture electron microscopy and including a comprehensive analysis of the shear dependence of viscosity will be necessary to confirm the presence of secondary aggregation of PSM in solution.

ACKNOWLEDGEMENT

Support for this research by the Northeast Ohio Chapter of the Cystic Fibrosis Foundation is gratefully acknowledged.

REFERENCES

1. Shogren, R., Jamieson, A. M. and Blackwell, J. (1983) Biopolymers, in Press.
2. Meyer, R. A. and Silberberg, A. (1980) Biorheology 17, 163-168.
3. Allen, A. (1978) Br. Med. Bull. 34, 28-33.
4. Boat, T. F. and Cheng, P.-W. (1980) Fed. Proc. 39, 3067-3074.
5. Gottschalk, A. and Bhargava, A. S. (1972) in Glycoproteins: Their Composition, Structure and Function (Gottschalk, A. Ed.) pp. 810-829, Elsevier, New York.
6. Hill, H. D., Reynolds, J. A. and Hill, R. L. (1977) J. Biol. Chem. 252, 3791-3798.
7. Bhushana Rao, K.S.P., Van Roost, E., Masson, P. L., Heremans, J. F. and Andre, F. (1973) Biochim. Biophys. Acta 317, 286-302.
8. Pigman, W., Moschera, J., Weiss, M. and Tettamanti, G. (1973) Eur. J. Biochem. 32, 148-154.
9. Bettelheim, F. A. and Dey, S. K. (1965) Arch. Biochem. Biophys. 109, 259-265.
10. Bettelheim, F. A. and Block, A. (1968) Biochim. Biophys. Acta 165, 405-409.
11. Payza, N., Rigvi, S. and Pigman, W. (1969) Arch. Biochem. Biophys. 129, 68-74.
12. Silberberg, A., Meyer, F. A., Gilboa, A. and Gelman, R. A. (1977) in Adv. in Exp. Med. & Biol. 89 (Elstein, M. and Parke, D., eds.) pp. 171-181, Plenum, New York.
13. Pearson, J. P., Allen, A. and Parry, S. (1981) Biochem. J. 197, 155-162.
14. Holden, K. G., Yim, N. C. F., Griggs, L. J. and Weisbach, J. A. (1971) Biochem. 10, 3105-3112.
15. Burchard, W., Schmidt, M. and Stockmayer, W. H. (1980) Macromolecules 13, 1265-1272.
16. De Salegui, M. and Plonska, H. (1969) Arch. Biochem. Biophys. 129, 49-56.
17. Carlson, D. M. (1968) J. Biol. Chem. 243, 616-626.
18. Lee, H. S., Jamieson, A. M. and Simha, R. (1979) Macromolecules 12, 329-332.
19. Koppel, D. E. (1972) J. Chem. Phys. 57, 4814-4820.
20. deGennes, P. G. (1976) Macromolecules 9, 587-598.
21. Klein, J. (1978) Macromolecules 11, 852.
22. Weissberg, S. G., Simha, R. and Rothman, S. (1951) J. Res. Nat'l. Bur. Standards 47, 298-314.

23. Suggett, A. (1975) in Aqueous Solutions of Amphiphiles and Macromolecules (Franks, F., ed.) Chp. 6, Plenum, New York.
24. Calje, A. A., Agterof, W. G. M. and Vrij, A. (1976) in Micellization, Solubilization and Microemulsions V.2 (Mittal, K. L., ed.) pp. 779–790, Plenum, New York.
25. Hyde, A. J. (1972) in Light Scattering From Polymer Solutions (Huglin, M. B., ed.) pp. 385–396, Academic Press, New York.
26. Hyde, A. J. and Taylor, R. B. (1963) Makromol. Chem. 62, 204–207.
27. Benoit, H. and Picot, C. (1966) Pure Appl. Chem. 12, 545–561.

PART IV

MICRODOMAINS IN NONAQUEOUS MEDIA

ION DISTRIBUTION AND POLYION CONFORMATION DISPLAYED BY AMPHIPHILIC POLYACIDS IN AQUEOUS AND ORGANIC MEDIA

R. Varoqui and E. Pefferkorn
Centre de Recherches sur les Macromolecules
CNRS 6, rue Boussingault
67083 Strasbourg-Cedex, France

ABSTRACT

The properties of aqueous and n-octanol solutions of amphiphilic polyelectrolytes - the polysoaps of maleic acid and cetyl (or decyl) vinyl ether - are described. Some important differences between the conformation and the counter-ion distribution in the two media are discussed. In water, the occurrence of intra-molecular hydrophobic interactions results in a compact conformation, which is however not that of a spherical mono-molecular micelle with no water incorporated. From conductivity and tracer self-diffusion measurements, it is shown that more than 20% of the alkali metal carboxylate groups are ionized. The polysoaps in n-octanol behave quite differently: although metal carboxylate groups can only be accommodated in organic media with a considerable free energy expense, an "inverted" compact intra-molecular micelle structure is not observed. It is assumed that a redistribution of metal carboxylate groups along the chain enables the partially neutralized form of the polymer to be solubilized in a low dielectric constant medium without undergoing a conformational collapse. The idea that carboxylate (R-COOA) and carboxylic (R-COOH) monomeric groups are not randomly mixed along the chain is inferred from the determination of the excess free energies of these groups and from conductivity measurements. A theoretical model in which the polymer is treated as an uncharged coil with all small metal ions "site bound" and non-homogeneously distributed, yields realistic pair contact energies and overall qualitative agreement with the experimental findings.

INTRODUCTION

Polysoaps are macromolecules in which the chemical structure of a monomeric unit bears great resemblance with that of a low-molecular weight soap or ionic amphiphile. Because of the van der Waals attraction between alkyl groups, the molecular dimensions of polysoaps in water are strikingly small compared to other polyelectrolytes. A globular mono-molecular micelle structure has been established by Strauss et al. in a number of fundamental studies involving poly(vinyl-pyridinium salts)[1] or anionic polysoaps (the alternating copolymer of maleic acid and vinyl ether).[2,3] Some of the polysoaps are also soluble in organic media.[4,5] Solution properties of polymeric amphiphiles in organic media have thus far not been the subject of extensive investigations and in the present paper we shall describe some important differences between conformational and counterion distribution proper-ties of polysoaps in aqueous and n-octanol medium. Polymeric amphiphiles soluble in organic media are currently important from a practical point of view for dispersion properties such as needed in microemulsion formation, or in the deflocculation of oil-suspensions of solids, etc. Unusual electrochemical properties in liquid/liquid ion-exchange or liquid membrance processes were already reported[6,8] and the reversible transition from intramolecular micelle to inverted mono-micelle conformation at a critical composition of a binary polar/non polar solvent mixture was discussed.[9,10]

EXPERIMENTAL SECTION

Polymers

The 1-1 copolymer of maleic anhydride and methylvinyl ether were commercial products and were used without fractionnation. The copolymers of maleic anhydride and decyl (or) hexadecyl vinyl ether were obtained by radical polymerization. For hydrolysis of the polymers with a long alkyl side chain, heat treatment was necessary (60°C for several days in pure water). This resulted however in a reduction of average DP and probably in an enlargement of the initial distribution. Therefore, after hydrolysis, the polymer sample was fractionated with 2:1 acetone:benzene as solvent and methanol as precipitant.

Polysoap $PS16_1$, $M_w=1.02 \times 10^5$ was used in aqueous medium. Its supply was however not sufficient to permit completion of measurements in octanol and the sample of $M_w=2 \times 10^5$ was used in the latter study.

Techniques

The mobility of counter-ions Na^+ and C_s^+ in self-diffusion experiments were determined by the capillary tube method of Anderson and Saddington, using radioisotopes [22]Na and [137]Cs.[11] Electrical con-ductances were determined with a Wayne Kerr bridge and sedimentation data with an ultracentrifuge "Beckman-Spinco model E". Quantitative analysis of ion concentrations was most often performed using trace amounts of alkali metal radioisotopes of [36]Cl. The densities of the solution were measured by means of a digital precision densimeter

("Digital Densimeter DMA 002" of H. Stabinger, H. Leopold and O. Kratky). The temperature was kept at 25 $\pm$ 0.1°C, in all experiments.

RESULTS AND DISCUSSION

1.   Polysoaps in aqueous medium

1.1. Dimensions

Polysoaps PS10 and PS16, when partially in salt form, disperse monomolecularly in aqueous medium. Light scattering measurements, not shown here, proved that for $\alpha > 50\%$, no colloidal aggregates of large molecules are formed. We denote by α, the stoichiometric degree of neutralization, i.e.

$$\alpha = \frac{[\ (AOH) + (H^+) - (OH^-)]}{2\ C_p} \tag{1}$$

where (AOH), (H$^+$) and (OH$^-$) are the molarities of added alkali metal hydroxyde, free hydrogen and free hydroxyl ions respectively and C_p is the polymer concentration expressed in mono-moles per liter.

In Figs. 1 and 2, we have reported the inverse of the sedimentation coefficient and the reduced viscosity as a function of the concentration of the polymers (salt form, $\alpha =1$) in 0.1 M NaCl aqueous solution. Using

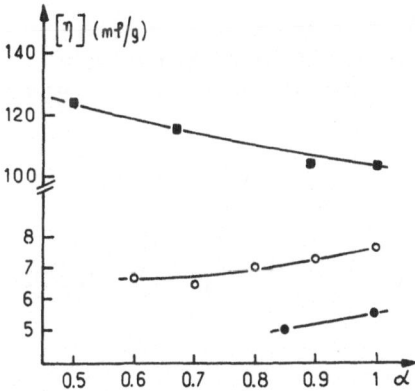

Figure 1.   Reduced viscosity and inverse of sedimentation coefficient as a function of the polymer concentration C (g/ml)   in 0.1 M NaCl aqueous solution, $\alpha = 1$: η_{sp}/C, G119 (Δ), PS10 (+), PS16$_1$ (o).s^{-1}, PS10 (+), PS16$_1$ (o).

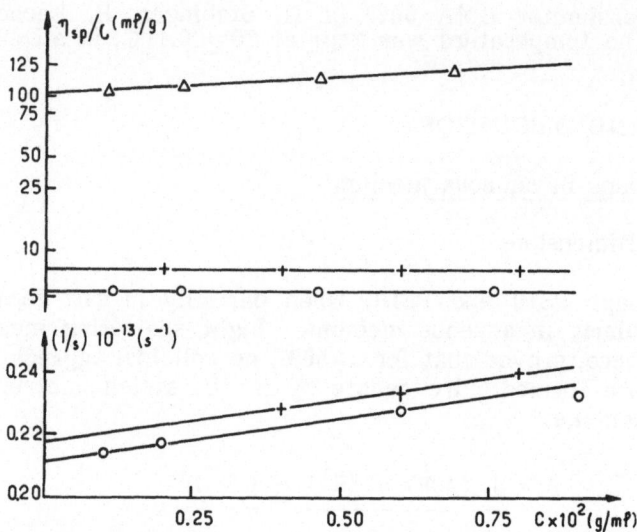

Figure 2. Intrinsic viscosity as a function of the degree of neutralization α in 0.1 M NaCl aqueous solutions: G119 (■), PS10 (o),
PS16$_1$ (●).

the concept of a solvated rigid particle[12] one can determine from
sedimentation data extrapolated to zero concentration an equivalent
Stokes radius $R_E(S)$:

$$R_E(S) = \frac{M_w/N_{AV} - v_H^o}{6\pi\eta S} \qquad (2)$$

where v_H is the volume of the particle in solution, ρ and η are the
density and viscosity of the solvent respectively. The partial specific
volume $\bar{v}_2$ of the solute particle in solution is given by:

$$\bar{v}_2 = \frac{N_{AV}V_H}{M} - \delta_1 v_1^o \qquad (3)$$

assuming that the particle contains δ_1 grams of water of specific
volume $\bar{v}_1^o$. The radius R_E (s) may be compared to the radius R_o given
by Eq. (4) :

$$R_o = [(3/4\pi) \ (M_w/N_{AV})\bar{v}_2]^{1/3} \qquad (4)$$

Data according to Eqs. (2-4) are reported in Table II. The ν values (sixth
column) were obtained from viscosity data and the Einstein-Simha
equation[13]:

$$[\eta] = \frac{\nu v_H \, N_{AV}}{M} \tag{5}$$

By inspection of Table II, one may infer that the polysoaps are probably not of spherical shape. The shape factor ν departs significantly from the 2.5 Einstein figure. Since $R_E(S)$ and R_o are different it is likely that a certain amount of solvent is an inherent part of the macromolecule, i.e. δ_1 is not zero. Looking at the variation of $[\eta]$ with α in Figure 2, we note a decrease in $[\eta]$ with α for the polyelectrolyte, whereas an increase is observed for the polysoap. For the latter, the swelling with α probably originates from electrostatic repulsions between neighboring "surface" located ionized groups (This is in the line of the counter-ion diffusion and conductivity behavior). Invoking topological constraints, one may also conjecture that since part of the polar groups are necessarily incorporated in the micelle, substitution of sodium for hydrogen (at high α) should contribute a positive volume increment.

1.2. Ion association

The variation of the self-diffusion coefficient of sodium and cesium counter-ions as a function of α is reported in Figure 3. The equivalent conductances of the same solutions are reported in Figure 4.

Table I: Polyelectrolyte and polysoaps

$\{CH(COOH)-CH(COOH)-CH_2-CH(OR)\}_n$		
Alkyl chain R $\quad$ CH_3	$C_{10}H_{21}$	$C_{16}H_{33}$
$\overline{M}_w$ $\quad$ 202,000(G119)	98,000(PS10)	102,000(PS16$_1$)
		200,000(PS16$_2$)

Table II:

	$S{\times}10^{13}(sec^{-1})$	$\overline{v}_2(ml/g)$	$R_E(S)$ (Å)	R_o(Å)	ν
PS10	4.2	0.68	60.4	30.2	11
PS16$_1$	4.4	0.78	44.1	31.6	7.1

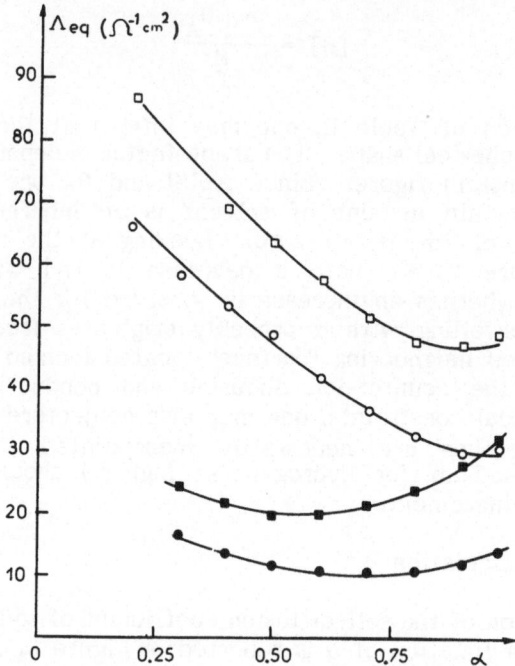

Figure 3. Self-diffusion coefficient of C_s^+ and Na^+ counterions in aqueous solutions as a function of neutralization. G119 (●), PS10 (o), PS16$_1$ (□).

Both experiments were conducted in the absence of added salt at a polymer concentration C_p=0.01 (monomoles/liter). In the polysoap solutions, the counter-ion mobility reduction is enhanced in a manner dependent on alkyl chain length. Furthermore, minima are observed. The reduction in mobility originates from: i) electrostatic interactions between free dissociated charges; and ii) counterion-polyion association (site-binding or ion condensation). Resolution of these effects appears complex and we shall not endeavor a detailed analysis.[14] Lumping both effects together in terms of an apparent degree of ionization α_a=D_S/D_S^0 (D_S^0 being the limiting stoichiometric self-diffusion coefficient at =0), we find α_a to be of the order of 20% for PS16, and 30% for PS10. The effective degree of ionization should be larger than these lower limiting values. To account for the minimum in the D and Λ_E curves, an expansion of the polysoaps at large α must be assumed (as already suggested). This results in a lower charge density and a concomitant decrease of electrostatic interactions. Consequently a higher counterion mobility is expected.

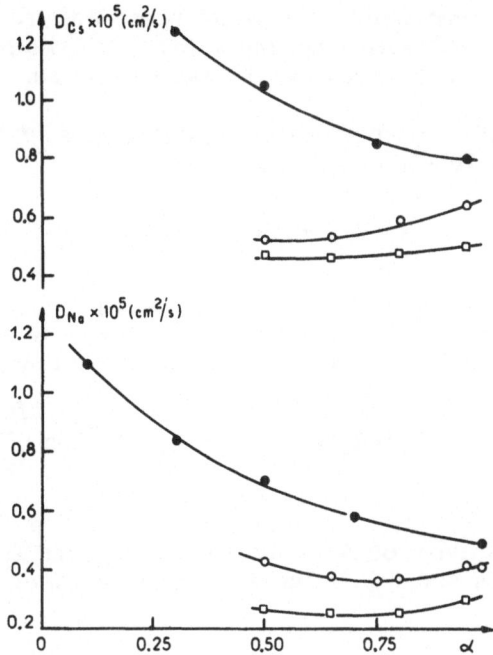

Figure 4. Equivalent conductance Λ_{eq} ($\Omega^{-1}cm^2$) as a function of α: G119 Cs salt ($\square$), Na salt ($\circ$), PS16$_1$ Cs salt ($\blacksquare$) Na salt ($\bullet$).

2.   Polysoaps in non-polar medium

2.1. Thermodynamic and ionization properties

Polysoap PS16$_2$ (M_w = 2x10^5) was dissolved in its acid form in n-octanol saturated with water.   The 1% wt polymer solution was transferred to an aqueous solution containing hydrogen and alkali metal chlorides. The partitioning of ions between the two phases was examined according to equilibria (6) and (7):

$$A^+ \;\; + \;\; \overline{H}^+ \;\; \not\rightleftarrows \;\; \overline{A}^+ \;\; + \;\; H^+ \tag{6}$$

$$RH \;\; + \;\; \overline{ClA} \;\; \not\rightleftarrows \;\; RA \;\; + \;\; \overline{ClH} \tag{7}$$

Barred symbols refer to the octanol phase, and RA and RH represent one equivalent of polymer group plus accompanying counterion, respectively. Before proceeding to the significance of Equation 7, it should be noted that water dissolves only a small amount of octanol ($\approx 0.5\%$ wt at saturation). Octanol is a good solvent for the polymer in its acid form. (The Mark-Houwink relationship was found to be [η] = 8.05x10^{-3} $M_w^{0.70}$

in water-saturated n-octanol). The water phase was free of polysoap in the entire pH range when the potentiometric two-phase acid-base titration, as expressed by Equation (6) was performed.

In Figure 5, pH (aqueous phase) is plotted as a function of α. The aqueous phase was 10^{-2}M in ACl (A$\equiv$Na, K, C$_S$)

$$\alpha = \bar{N}_A - \bar{N}_{Cl} \simeq \bar{N}_A$$

N_A, N_{Cl} are the concentrations (moles per unit polymer acid group) of cation A and anion Cl in n-octanol. We always found $\bar{N}_{Cl} \ll \bar{N}_A$. Exchange reaction (6,7) may be expressed in the following form:

$$pH + \log \left(\frac{1-\alpha}{\alpha}\right) + \log a_A = \log K_A^H + \log \frac{f_{RA}}{f_{RH}} \qquad (8)$$

where a_A is the activity of A in the water. f_{RA} and f_{RH} are rational activity coefficients and K_A is the equilibrium constant.

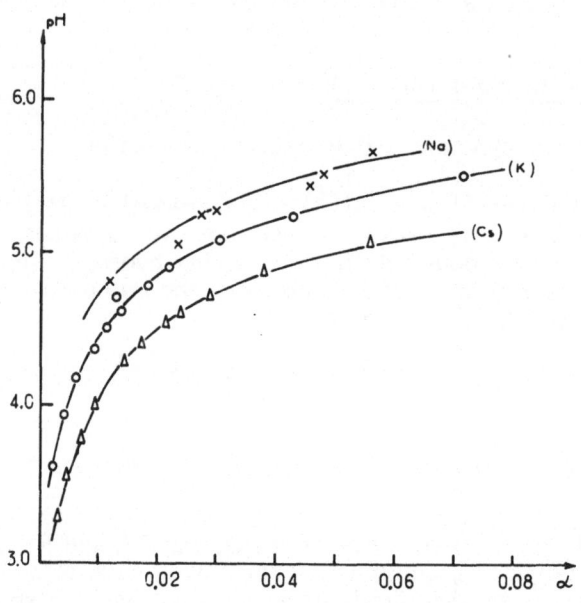

Figure 5.   Two-phase potentiometric acid-base titration data for PS16$_2$ in n-octanol: Na$^+$ (x), K$^+$ (o), Cs$^+$ (Δ) in the presence of 10^{-2} M NaCl, KCl, CsCl (aqueous phase).

Let us first note that the experimental ion-exchange isotherms are in satisfactory accord with Equation (8). This is seen in Figure 6, where we have reported the (H,Na) ion-exchange for different NaCl concentrations of the water phase, by plotting the left hand side of Equation (7) vs. (bottom curves). According to Equation (7), one expects all points to fall on the same curve. The observed divergence is considered to correspond to experimental precision (note also that a_A was not corrected or the presence of octanol in the water phase). It is important to note that the isotherms depart from an ideal behavior and the excess free energy terms cannot be neglected.

Formulation (7,8) supposes implicitly that the polysoap is unionized in n-octanol (cf to Appendix I). In order to substantiate this point further, the conductance of the organic phase was also measured as a function of α as shown in Figure 7. The degree of ionization at different α's for hydrogen ($\alpha_{i,H}$) and for metal ($\alpha_{i,A}$) was determined from the equivalent conductance (see Appendix II). Table III summarizes these results.

Table III: Degree of dissociation of alkali metal ions on the partially neutralized polycarboxylic acid in water satured octanol solutions ($\alpha_{i,H}$ (α = o) = 2.04x10^{-5})

$\alpha\%$	$\alpha_{i,Na}$	$\alpha_{i,K}$	$\alpha_{i,Cs}$
3	4.29x10^{-3}	4.20x10^{-3}	2.27x10^{-3}
6	3.50x10^{-3}	3.25x10^{-3}	1.55x10^{-3}
9	2.80x10^{-3}	2.80x10^{-3}	1.73x10^{-3}

The transference from the aqueous phase to the organic phase of a polar ionic solute is accompanied by a certain amount of water which acts as solvating agent for that species in the organic phase. The differential amount of water adsorbed or desorbed during the ion-exchange process was obtained by measuring the specific volume of the octanol phase as a function of polymer concentration. From the measure of the density of the organic solutions, the number of moles of water $dn_W/d\alpha$ which accompanies the substitution on the polymer of one hydrogen by a metal was derived (see Appendix III for more details) and is reported in Table IV.

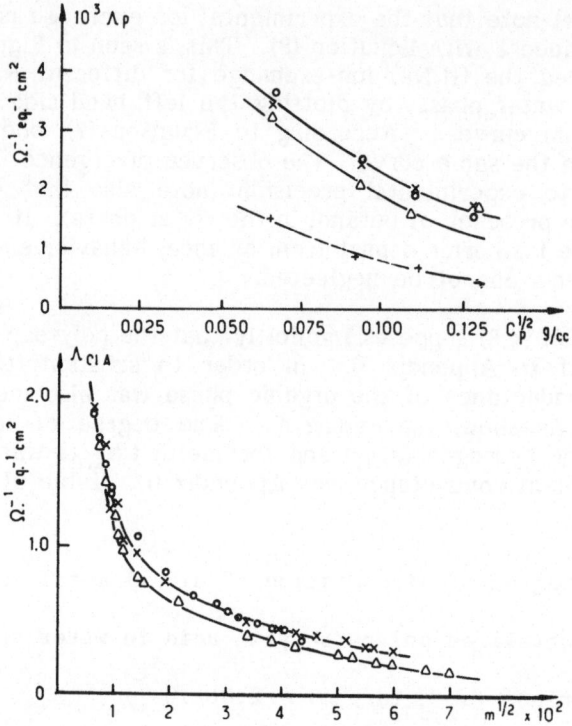

Figure 6. Two–phase potentiometric acid–base and titration data for PS16$_2$, Na form in the presence of 10^{-2} (+), 10^{-1} (o) and 1 M (Δ), NaCl aqueous solutions (upper curves) lower curves according to Eq. 8.

Table IV: Water sorption data

	(H,Na)	(H,K)	(H,C$_s$)
$dn_w/d\alpha$	+5.6±1.2	−0.94±0.2	−2.84±0.20

The exchange of Na$^+$, K$^+$, C$_s^+$ with hydrogen involves +5, −1, and −3 mole of water respectively. While Na gives rise to a positive adsorption of water, the hydration of the COOH group exceeds that of COOK and COOC$_s$ by about one and three water molecules. dn_w/d was found approximately constant with α.

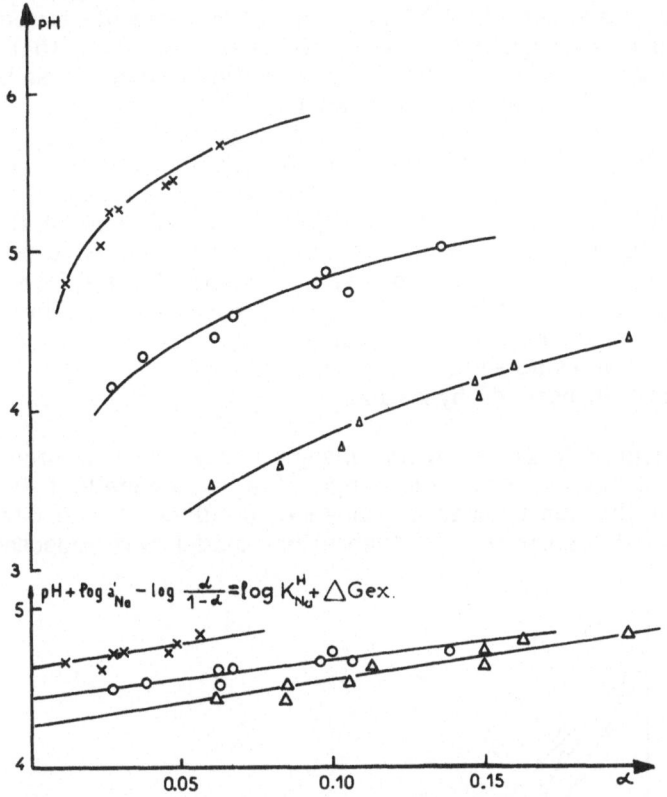

Figure 7.   Upper curves show equivalent conductances of the polymer in
            the organic phase against the root square of polymer
            concentration at α = 9%, Na$^+$ (X), K$^+$ (o), Cs$^+$ (Δ), polymer in
            the acid form (α = 0) (+).   Bottom curves show equivalent
            conductances of NaCl (X), KCl (o), CsCl (Δ) in water-
            saturated octanol solutions against the root square of elec-
            trolyte molarity.

        Summarizing the findings of this study, we can state the following
observations.   (1) Polyamphiphiles such as polysoaps display in organic
media selective interactions with alkali metal ions (see, for example, in
Figure 5, the pH shift from Na$^+$ to Cs$^+$).   (2) Non-ideality is observed
and the amount of free charges is negligible.

        Deviation from ideality means that ion-pairs RH, RA cannot be
considered without interactions among them on the polymer chain.   Such
interactions are obviously correlated to the chain conformation.   For
example a segregated structure — an inverted micelle in this case —
would favor contacts, in the same way as in aqueous media separation

between polar and non-polar parts of the macromolecule is requisite for intramolecular hydrophobic bonds. We will now show that than the intramolecular micelle structure, a more refined model must be devised to account for the results in n-octanol.

2.2. Polymer conformation and ion distribution

The intrinsic viscosity $[\eta]$ (ml/g) of the polysoap in the organic phase is given as a function of α in Figure 8. A marked dependence of $[\eta]$ on the nature of the ion and opposite to the trend of the thermodynamic constant K_A^H (which increases in the direction Na>K>Cs, cf. Figure 5) is displayed. Qualitatively, one may say that the occurrence of intramolecular polymer-polymer bonds is greater for the ions the most in need of hydration.

Although a large viscosity change occurs with α, the order of magnitude of $[\eta]$ is out of the range of a monomicelle behavior, and furthermore the quasi-linear dependence, precludes the existence of a conformational transition. A theoretical model was proposed by the

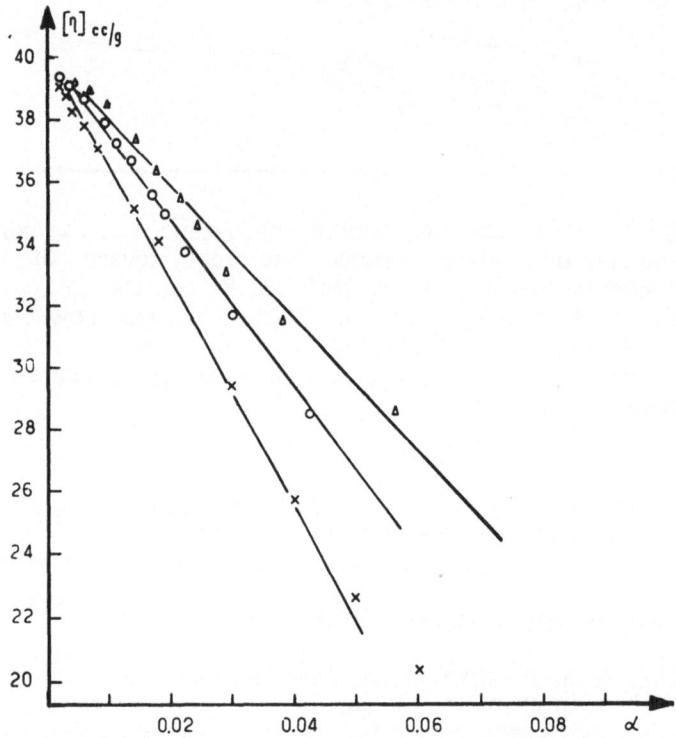

Figure 8. Intrinsic viscosities of PS16$_2$ in n-octanol as a function of fractional metal occupation: experimental points Na$^+$ (x), K$^+$ (o), Cs (Δ); full lines according to Eqs. (8-12) With $\alpha=\alpha(r)$.

authors to interpret the behavior of $[\eta]$ [5]. Let us imagine a polymer coil composed of αx monomers of kind RA and $(1-\alpha)x$ of kind RH, with a gaussian coil segment distribution:

$$\rho = X(\frac{3}{2\pi\gamma^2 \overline{R}_0^2})^{3/2} \exp(-3r^2/2\gamma^2 \overline{R}_0^2) \qquad (8)$$

ρ is the number of segments per unit volume at r, γ the expansion coefficient and $(\overline{R_0^2})^{1/2}$ the radius of gyration in the unperturbed state. The free energy ΔG of mixing polymer segments and solvent is given by[15]:

$$\Delta G = \int_0^\infty \delta(G)4\pi r^2 dr + kT[3(\gamma^2-1)/2 - 3\ln\gamma] \qquad (9)$$

$$\delta(\Delta G) = kT\{[(1-\overline{v}\rho)(\chi-1)-\overline{v}\rho/2] + \alpha\ln\alpha + (1-\alpha)\ln(1-\alpha)\} \qquad (10)$$

With respect to the original Flory theory, $\delta(\Delta G)$ includes here the additional entropy of mixing RA, RH groups along the chain and furthermore χ must be a quadratic function of the local composition α (the number of pair contacts being proportional to the volume fraction), i.e.:

$$\chi = \chi_{S,RH} + \chi_{S,RA} - (\chi_{S,RH} + \chi_{RH,RA})\alpha + \chi_{RH,RA}\alpha^2 \qquad (11)$$

The χ's account for the excess bond energy in a heterobond RA<—>RH over that of the average of homobonds RA<——>RA and RH<——>RH. Minimizing G with respect to γ allows one to express γ (theoretical) in terms of the χ's. Since furthermore

$$[\eta] = \Phi(R_0^2)^{3/2} \gamma^3/M \qquad (12)$$

a fitting of experimental and theoretical $[\eta]$ values can be achieved with a particular choice of $\chi_{S,RA}$ and $\chi_{RA,RH}$.

This process has been applied assuming $\alpha = \alpha$ (stoich.) and also assuming α to be a function $\alpha(r)$ of r. (In the latter case one must minimize ΔG with respect to γ and $\alpha(r)$ with a normalization condition). Skipping all mathematical details (see ref. 5), we simply report the χ values which, as seen in Figure 8, show agreement over the investigated α range (values of relevant parameters for our solvent-polymer system are given in ref. 5).

Table V: Monomer-monomer and monomer-solvent interaction

         energies

metal carboxylate	$\alpha = \alpha_{(stoich.)}$		$\alpha = \alpha(r)$	
	$\chi_{S,RA}$	$\chi_{RA,RH}$	$\chi_{S,RA}$	$\chi_{RA,RH}$
RCOONa	-5,98	-8.62	2.64	-2.13
RCOOK	-2.62	-4.66	1.87	-1.80
RCOOCs	-1.28	-3.00	1.6	-1.38

Since metal carboxylate groups can only be accommodated in organic media with a positive free energy expense, the large negative values for the $\chi_{S,RA}$ reported above for $\alpha = \alpha_{(stoich.)}$ are unrealistic and obviously the hypothesis $\alpha = \alpha_{(stoich.)}$ fails. On the other hand the assumption that α is a function of the metal location leads to plausible thermodynamic interaction parameters (in sign and in magnitude); while interactions S,RA are repulsive (as should be), interactions RA, RH are attractive. We have reported $\alpha(r)$ in Figure 9. It is seen that the location of a metal group on the chain does not proceed at random, but rather owing to the relatively greater occurrence of polymer bonds in the central region of the coil, metal ions should have a tendency to cluster about the center of the coil.

Some simplifying assumptions are necessarily used in the theory. In the model, monomers RA and RH are supposed of dimensions equal to that of the solvent. In view of the diacid structure of our polymer, compliance with this cannot be expected. Furthermore, the water molecules which make up the hydration shell of a carboxylic acid or carboxylate metal group must be considered as part of a monomer, in that way the χ parameters are unambigously defined with respect to a water-saturated n-octanol solvent. In spite of some unavoidable simplificatons it is nevertheless gratifying that the introduction of an $\alpha(r)$ distribution accounts for realistic pair contact energies.

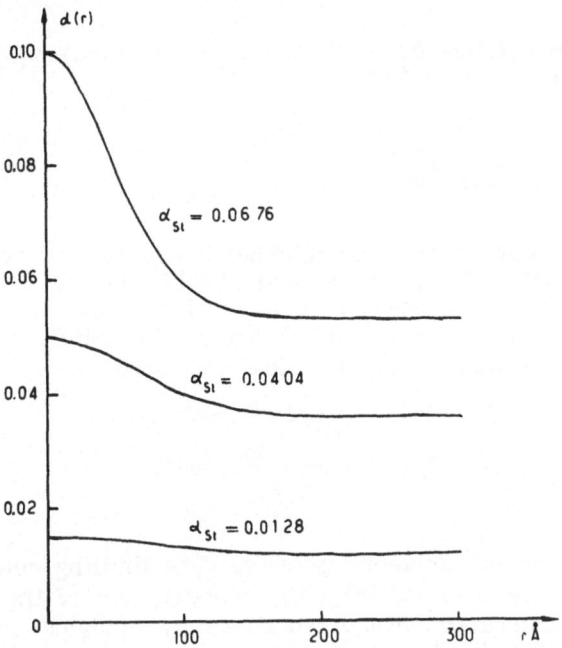

Figure 9. Metal ion distribution as a function of location for α_{stoich} (Na$^+$) equal to 6,8%, 4% and 1.3%.

APPENDIX

1. Assuming the presence of space charges, we would then write as usual:

$$RH \rightleftharpoons R^- + H^+ \qquad (A1)$$

and find the isotherm in the form:

$$pH + \log\left(\frac{1-\alpha}{\alpha}\right) + \log \overline{N}_{Cl}/N_{Cl} = \text{constant} + \log \frac{f_{R^-}}{f_{RH}} \qquad (A2)$$

Since the partition $\overline{N}_{Cl}/N_{Cl}$ of chlorides is almost constant up to 1M, clearly this would not account for the experimental data.

2. The equivalent conductance Λ_p of the polymer is given by the following equation:

$$\Lambda_p = \frac{F^2}{F_p} \left[(1-\alpha)\ \alpha_{i,H} + \alpha,\alpha_{i,A} \right] + (1-\alpha)\ \alpha_{i,H}\ \Lambda_H^o +$$

$$+ \alpha,\alpha_{i,A}\ \Lambda_A^o \tag{A3}$$

The first bracket on the right-hand side of the equation is the polyion contribution, the second bracket is the free metal and hydrogen ion conductance contribution. F_p is the frictional coefficient of the polymer in octanol, which is related to the intrinsic viscosity by the classical expression:

$$F_p = \frac{6^{3/2}\pi\eta_o}{3.7}\ \{[\eta]M/\Phi\}^{1/3} \tag{A4}$$

where η_o is the solvent viscosity. The limiting conductance $_H^o$ and Λ_A^o were obtained from the conductances of the alkali halides in water-saturated octanol solutions by applying the method of Fuoss and Skedlovsky in ref. 16. The two unknown parameters $\alpha_{i,H}$ and $\alpha_{i,A}$ are readily obtained by combining $\Lambda_p(\alpha)$ and $\Lambda_p(\alpha=0)$.

3. The differential amount of water absorbed or desorbed during the ion-exchange process was obtained by measuring the specific volume of the octanol phase as a function of polymer concentration. The significance of partial specific volumes in multi-component systems has been discussed by Casassa and Eisenberg.[17] It has been shown that if equilibrium experiments at constant chemical potential of the solvents are performed (in our case the macromolecular solution at varying polymer concentration was equilibrated with the water phase at constant composition), it is possible to define from the partial specific volume data an excess or deficiency of the diffusible solvent components which can be attributed to "preferential solvation" or hydration of the macro-molecular species. More precisely, if we call Γ the weight of excess water in grams to be removed from the solution per gram of polymer added at constant chemical potential μ_w, μ_o of water and octanol, respectively, then

$$\Gamma = (\partial C_w/\partial C_p)\ \mu_o,\mu_w \tag{A5}$$

the "adsorption" or "membrane distribution coefficient" is obtained from density measurements according to the expression

$$= \frac{(\partial\rho/\partial C_p)_{\mu_o,\mu_w} - (\partial\rho/\partial C_p)_{n_w,n_o}}{1 - \overline{V}_w\rho_s^0} \tag{A6}$$

$(\alpha\rho/\alpha C_p)_{\mu_o,\mu_w}$ is the density increment of the organic phase with respect to the polymer concentration C_p after equilibrium with the water phases is attained, $(\alpha\rho/\alpha C_p)_{n_w,n_o}$ is the density increment of the organic phase at constant solvent composition i.e., when the amount of water with respect to octanol in the organic phase is kept constant at varying polymer concentration, ρ^0 is the density of the organic phase extrapolated to zero polymer concentration, and $\overline{V}_w$ the partial specific volume of water in the oil phase was taken to be 0.997.

Experimental values of $(\alpha\rho/\alpha C_p)_{\mu_o,\mu_w}$ and $(\alpha\rho/\alpha C_p)_{n_w,n_o}$ are reported in Table VI.

Table VI: Differential increments of densities at constant chemical composition and constant "solvent" molar composition.

	α	0.00	0.03	0.06	0.09
$(\frac{\partial\rho}{\partial C_p})_{\mu_o,\mu_w}$	H	0.216	–	–	–
	Na	–	0.222	0.228	0.235
	K	–	0.223	0.231	0.238
	Cs	–	0.229	0.247	0.260
$(\frac{\partial\rho}{\partial C_p})_{n_o,n_w}$	H	0.189	–	–	–
	Na	–	0.189	0.195	0.202
	K	–	0.198	0.203	0.210
	Cs	–	0.204	0.221	0.240

REFERENCES

1. U. P. Strauss and N. L. Gershfeld, J. Phys. Chem., 58, 747 (1954)
2. R. Varoqui, U. P. Strauss, J. Phys. Chem., 72, 2507 (1968)
3. P. L. Dubin, U. P. Strauss, J. Phys., 74, 2842 (1970)
4. K. Ito, Y. Yamashita, J. Colloid Sci., 19, 152 (1964)
5. R. Varoqui, E. Pefferkorn, J. Phys. Chem., 79, 169 (1975)
6. E. Pefferkorn, R. Varoqui, J. Coll. and Polym. Sci., 255, 222 (1977); 255, 543 (1977).
7. E. Pefferkorn, R. Varoqui, J. Coll. and Interface Sci., 52, 89 (1975)
8. E. Pefferhorn, R. Varoqui and H. Benoit, J. Applied Polymer Sci., 19, 2929 (1975).
9. F. M. Fowkes, M. J. Schick, A. Bondi, J. Coll. Science, 15, 531 (1960)
10. Z. Wojtczak, C.Strazielle, R. Varoqui, H. Benoit, European Polym. J., 6, 247 (1970)
11. E. Pefferkorn, R. Varoqui, Eur. Polym. J., 6, 663 (1970)
12. C. Tanford, "Physical Chemistry of Macromolecules", John Wiley, New York 1961 p. 339
13. R. Simha, J. Phys. Chem., 44, 25 (1940)
14. A. Schmitt, R. Varoqui, J. Chem. Soc., Faraday Transactions II, 69, 1087 (1973)
15. P. J. Flory, "Principles of Polymer Chemistry", Cornell University Press, Ithaca, N.Y., 1953, Chapt. XIV
16. R. M. Fuoss and T. Shedlovsky, J. Am. Chem. Soc., 71, 1496 (1949)
17. H. Eisenberg and E. F. Casassa, J. Polymer Sci., 47, 29 (1960)

ASSOCIATION OF THE ION PAIR END-GROUPS OF HALATO

TELECHELIC POLYMERS IN NONPOLAR SOLVENTS

R. Jerome, G. Broze, and Ph. Teyssie
Laboratory of Macromolecular Chemistry and
Organic Catalysis
University of Liege
Sart Tilman - 4000 Liege - BELGIUM

ABSTRACT

Halato telechelic polymers (HTP) result from the complete ionization of both ends of telechelic prepolymers. In nonpolar solvents, they form homogeneous gels which upon dilution generally transform into more or less viscous solutions unless they display a phase separation process. In the former case, critical gel concentrations (C_{gel}) as low as 1.5 g dL^{-1} are observed. This process is dependent on the molecular weight of the prepolymer according to the general relationship $C_{gel} = k.\bar{M}_n^{-1/2}$. Solutions of 10 g dL^{-1} α,ω alkaline earth dicarboxylato polybutadiene ($\bar{M}_n$ = 4,600) in xylene exhibit an attractive thermo-rheological simplicity and obey a deformation mechanism controlled by the stability and the mean size of multiplets formed by the carboxylate end-groups. A secondary relaxation characteristic of these multiplets is clearly evidenced. Furthermore, the shear-thickening character of these solutions is noteworthy. Nature of ion pairs, polarity of the medium, and temperature significantly influence the general behavior of HTP solutions and prove its electrostatic origin. At high concentrations (> 50 g dL^{-1}), the multiplets can possibly aggregate and form periodic structures with a lamellar character, as evidenced by the SAXS spectra of α,ω - Ba and K dicarboxylatopolybutadienes in toluene.

INTRODUCTION

The dispersion of mineral fillers in thermoplastics and rubbers is probably the first example of a new technical development based on heterophase polymeric materials.[1,2]

Block polymers and polymer blends deserve now a great interest because of their multiphase character and their related properties.[3,4] The thermodynamic immiscibility of the polymeric partners gives rise indeed to a phase separation, the extent of which controls the detailed morphology of the solid and ultimately its mechanical behavior. The advent of thermoplastic elastomers and high impact resins (HIPS or ABS type) illustrates the importance of the industrial developments that this type of materials can provide. In selective solvents, and depending on molecular structure, concentration and temperature, block polymers form micelles which influence the rheological behavior and control the morphology of the material.

Today, increasing attention is paid to a third type of heterophase materials, i.e., ion-containing polymers. The general behavior of nonpolar macromolecules is drastically influenced indeed by the random distribution of even modest amounts of ionic substitutents (carboxylate or sulfonate groups). The investigation of the resulting polymers or ionomers leads to the conclusion that microphase-separated ionic domains are formed. It is suggested that ion pairs, formed by the ionic sites on the polymeric backbone and the counterions, aggregate into multiplets of a few ion pairs, or into larger entities called clusters. The ionomers might be regarded as composed of clusters of ionically concentrated material immersed in a polymeric matrix containing mainly multiplets. The multiplets act as moderately strong cross-links, whereas clusters play the role of both cross-links and strongly interacting fillers. However, uncertainties still remain about the shape and size of the clusters, as well as about the geometrical arrangements of ions and polymer chains.[5,6] The solution behavior of ionomers has been rarely considered because of problems of limited solubility, especially in nonpolar solvents. That behavior results again from the ion pair interactions.[7-9] The situation is quite different from conventional polyelectrolytes in water, because the solution properties are controlled by free ions. In that respect, it is worthwhile to mention the investigation by Ise of the "ordered" distribution of electrically charged solutes, such as macroions, proteins and ionic micelles, in dilute solution.[10] The author suggests that ordered and disordered regions might coexist and fluctuate with time in these diluted ionic systems.

In ionomers, the ion aggregation process is hindered to some extent by the entanglements and/or the rigidity of the polymeric backbone onto which the ion pairs are randomly attached. A better insight into the mechanism of ion aggregation should be obtained from macromolecules carrying ion pairs largely independent of each other. Halato-telechelic polymers (HTP) satisfy largely this condition since the salt groups are now selectively attached at both ends of linear, usually flexible polymers.[11,12] For sake of clarity, the Greek word "halato" means salt-like,[13] whereas the word "telechelic" derives from the Greek "tele," far off, and "chele," claw, and commonly designates polymer molecules having two reactive terminal groups. Several years ago, Pineri et al.[14,15]

and Otocka et al.[16] considered similar materials, but the synthetic pathways they used were not fully satisfying. In the present work, a search for an efficient synthetic route to well-defined HTP was undertaken. The chain length, the polydispersity and the nature of the polymeric carrier as well as the nature of the salt groups were modified systematically so that a family of materials with carefully matched molecular characteristics could be prepared. The behavior of these model ion-containing polymers was investigated in mainly nonpolar solvents, over a large range of concentrations.[11,12,17] This paper aims to report a review of that research.

SYNTHESIS OF HALATO-TELECHELIC POLYMERS

This work focuses on HTP resulting from the neutralization of carboxy telechelic polymers (eq. 1), although polymers with sulfonic end-groups have also been investigated.

$$n \; HOOC\text{-}P_x\text{-}COOH + \frac{2n}{v} MeA_v \rightleftharpoons \left\{ OOC\text{-}P_x\text{-}COO\text{-}Me_{\frac{2}{v}} \right\}_n + 2 \; nHA \qquad (1)$$

To prepare well-defined HTP, stoichiometric amounts of the neutralizing agent should be used and the reaction equilibrium should be displaced as completely as possible towards HTP formation. For that purpose, metal alkoxides (methoxide, isopropoxide) and alkyl-metals are the most convenient reagents, thanks to their high reactivity and the formation of HA subproducts (low molecular weight alcohol or hydrocarbon) easily eliminated from the reaction medium.

Pineri et al.[14,15] had previously performed the neutralization of carboxy telechelic polybutadiene (PBD) by dispersing hydroxides of alkaline metals, acetates of divalent metals or aluminum acetylacetonate in bulk PBD within a Brabender Plastograph at 150°C. Only a fair control of the reaction was allowed in that way.[14] The highest neutralization degree was obtained by using quantities of metal reagent exceeding 2-5 times the stoichiometric amounts. Furthermore, the absence of thermooxidative degradation could not be ascertained. Otocka et al.[16] neutralized carboxy telechelic PBD in solution with methoxides of group I and II metals (in methanol solution) or with aluminum isopropoxide. However, the authors did not eliminate completely the alcohol from the solution before characterization. That point has however a critical importance as assessed by the results relative to the neutralization of carboxy telechelic PBD in toluene by magnesium methoxide in methanol (Table I). The relative viscosity of the products is quite different depending on whether the methanol is removed from the solution or not. Practically, methanol can be displaced by continuous distillation of solvent (100 ml/hr). When about 20% of toluene is recovered, the initial volume is completed by addition

TABLE I.   Relative viscosities at different concentrations in toluene at 25°C, for carboxy telechelic polybutadiene (Hycar CTB from Goodrich) mixed with stoichiometric amount of magnesium methoxide.

Conc. g dL^{-1}	Before methanol distillation	After methanol distillation
1.0	1.4	2.9
1.5	2.0	41
1.8	2.5	224

of pure toluene and the distillation is resumed further. Depending on the concentration, a sharp increase of the relative viscosity is observed at the elimination of the alcohol and no further modification is noted after two distillation runs (Table I).   It must be stressed that this characteristic is much more sensitive than most of the usual analytical techniques.   Polar compounds, such as alcohols, have therefore a deep influence on the solution behavior of HTP;   that point will be discussed more thoroughly in a next section.   When the neutralizing agent is used in nonpolar solvent (aluminum isopropoxide or diethylzinc in toluene), the formation of particles of gelatinous material is generally observed. However, the addition of anhydrous alcohol followed by its elimination through solvent distillation is sufficient to yield a completely homogeneous system, at least at high concentration (20-30%).   An overnight refluxing has usually the same effect.

In conclusion, the procedure reported by Otocka on the synthesis of HTP is efficient only when the alcohol formed (eq. 1) is totally removed from the reaction medium.   In that way, the alcohol interacting with the ion pair aggregates is eliminated and the reaction equilibrium is simultaneously displaced in favor of the HTP formation.

The carboxy telechelic PBD commercialized by BF Goodrich under trade mark "Hycar CTB 2000 x 156" was largely used in this work ($\overline{M}_n$ = 4,600; $\overline{M}_w/\overline{M}_n$ = 1.8;   functionality = 2.01 and microstructure cis/trans/vinyl = 20/65/15).   Carboxy telechelic polyisoprene (PIP), poly α-methylstyrene (PMS) and poly tert.butylstyrene (PTBS) were prepared by anionic polymerization in THF at -78°C.

α-Methylstyrene sodium tetramer was used as the difunctional initiator, and excess anhydrous CO_2 as the deactivating agent. The polydispersity of all these polymers did not exceed 1.2 and their functionality was better than 1.95. Poly α-methylstyryl dianions were also deactivated on propanesultone[18] and 1-chloro-3 dimethylaminopropane[19] with formation of α,ω-disulfonic acid and α,ω-di tert.amino PMS, respectively. Carboxy telechelic polymers were neutralized by metal methoxides (Mg, Ca, Ba), Al isopropoxide or di n.butyl Be. The metal methoxides were freshly prepared just before use by reaction of the pure metal with anhydrous methanol. Be $(nC_4H_9)_2$ was obtained by deactivating n.C_4H_9Li onto $BeBr_2$ prepared in refluxing THF according to the procedure reported by Richards et al.[20] The prepolymers were previously dried by three successive azeotropic distillations of benzene and finally dissolved in dry solvent (toluene, xylene, decaline, chloroform). After the addition of the metal alkoxide, 2 or 3 distillation runs of solvent were sufficient to observe the complete disappearance of the absorption at 1700 cm^{-1} (carboxylic acid) in favor of a new absorption at 1560 cm^{-1} (carboxylate) as well as a stabilization of the viscosity. α,ω-di tert.amino PMS was quaternized by a five fold excess of benzyl-chloride in chloroform solution at 25°C, and finally purified by three successive precipitations into hexane.

SOLUBILITY OF HALATO-TELECHELIC POLYMERS

The solubility of ion-containing polymers is quite a general problem at least in nonpolar solvents. Otocka et al.[16] mentioned that α,ω-sodium dicarboxylato PBD was not soluble in benzene but soluble with difficulty in trichlorobenzene. Lenz et al.[9] studied the solubility of poly(isoprene-co-sodium styrenesulfonate) ionomers in usual hydrocarbon solvents of isoprene homopolymer. Although insolubility was the rule, the mixing with methanol did result in almost complete solubilization. The polar cosolvent was believed to preferentially solvate the ion pair aggregates,[8,9] but its concentration in the solvent mixture has a decisive influence on the ultimate balance between the weakening of the ion pair interactions and the decrease of solubility of the backbone polymer. That means that the addition of a solvating agent of the ion pairs cannot systematically be used to dissolve any ion-containing hydrophobic polymer in nonpolar solvent. Furthermore, these observations largely support our recommendations for the synthesis of reliable and representative HTP, according to Otocka's procedure.

In nonpolar solvent, the neutralization of the carboxy telechelic polymers considered in this study generally results in the formation of homogeneous gels (10 wt% of solids), which can however suffer demixing upon subsequent dilution. For instance, a phase separation occurs systematically when PBD (Hycar CTB from B. F. Goodrich) is neutralized with Be and Al derivatives, respectively. This phenomenon is quite reversible since a homogeneous gel is again observed when the excess nonpolar solvent is distilled off. Furthermore, the mixing of small amounts of alcohol homogenizes the two-phase system, which forms again at the elimination of the alcohol while keeping constant the overall polymer concentration. Increasing amounts of nonpolar solvent are expected to increase the degree of swelling of a homogeneous gel of HTP. For weak ion pair interactions and metal cations of low valency ($\leqslant 2$), this process can ultimately lead to the complete disruption of the network with formation of a more or less viscous solution. Sufficiently strong and/or multiple (metals of valency > 2) ion pair interactions can however impede the complete dissolution, and beyond a given degree of swelling the gel cannot accomodate a further amount of solvent without increasing its free energy. A thermodynamic transition (demixing) then takes place, the occurrence of which can be delayed in some circumstances, just as supercooling is observed in melted semicrystalline polymers (metastable situations). In a theoretical approach, Joanny[21] predicted a demixing transition in ionomer solutions in nonpolar solvent. This transition was expected at constant concentration and decreasing temperature (below θ), whereas the HTP solutions in nonpolar solvent demixes at constant temperature (above θ) and decreasing concentrations.

Once completely dried, HTP require a prohibitively long time to dissolve or to swell again. For this reason they are usually prepared in the solvent wherein studied.

GELATION OF SOLUTIONS OF HALATO-TELECHELIC POLYMERS IN NONPOLAR SOLVENT

Figure 1 illustrates the dilute solution behavior of a series of HTP in toluene at 25°C. Compared to the relative viscosity (η rel) curve of the non-neutralized PBD (Hycar CTB), that of the neutralized polymer increases more or less abruptly with concentration. Ultimately a gelation phenomenon occurs, which is the obvious consequence of the intermolecular interactions of the ion pairs that the metal carboxylate end-groups form in toluene at 25°C. In the alkaline-earth cations series (Ba, Ca, Mg), the sharp increase of the relative viscosity appears at decreasing concentrations as the cation size decreases (Figure 1). As already mentioned, the smallest alkaline earth cation (Be) exhibits systematically a phase separation at high dilution which prevents significant measurements in that range with a capillary viscometer. These results can be explained by the equation (eq. 2) relating the attractive force between anion and cation to their charge (e_A and e_C, respectively), the square of their distance (r) and the dielectric constant (ε) of the medium, respectively.

$$f = \frac{1}{\varepsilon} \cdot \frac{e_A \cdot e_C}{r^2} \tag{2}$$

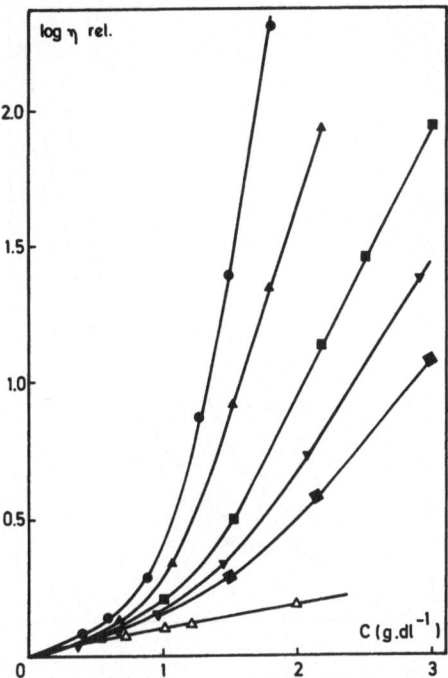

Figure 1.   Relative viscosity of the original α,ω-dicarboxy PBD ($\triangle$) and
the α,ω-divalent metal dicarboxylato PBD versus concen-
tration in toluene at 25°C: (●) Mg, (▼) Cu, (▲) Ca, (■) Ba,
and (◆) Mn.

When f is low, the carboxylate groups are negatively charged and set up
an electrostatic repulsion, well known as the polyelectrolyte effect. At
sufficiently high f, net charges tend to disappear on the chain-ends in
favor of dipoles, the interaction of which can stabilize a chain network.
For alkaline earth carboxylates (e_A = 1, e_C = 2) in toluene (ε = 2.38), f
is high and the ion pair interactions increase in the order Ba (ionic radius
r_i:1.35 Å), Ca (r_i:0.99 Å), Mg (r_i:0.65 Å) and Be (r_i:0.31 Å). Cu (r_i:0.69
Å) and Mn (r_i:0.80 Å), although smaller than Ba, are responsible for a
delayed gelation (Figure 1); this may be the consequence of the less
ionic, more coordinative character of the carboxylate-transition metal
bonds.   This bonding feature results in a decrease of the effective
charges. e_A and e_C can also be modified by capping both ends of an
anionically prepared poly α-methylstyrene ($\overline{M}_n$:6,000) with a carboxylic
acid PMS-$(COOH)_2$, a sulfonic acid PMS-$(SO_3H)_2$ and a tertiary amine
PMS-$(NMe_2)_2$, respectively.   After complete neutralization by magnesium methoxide, the dicarboxylic and disulfonic acid-terminated PMS
have practically the same behavior (Figure 2), the gelation appearing just
earlier for the sulfonate HTP.   When quaternized with benzyl chloride,
the α,ω-di tert.amino-PMS exhibits a greater trend to gelation (Figure 2).
Ionic bonding is prevailing in the ammonium salt end-groups, whereas the

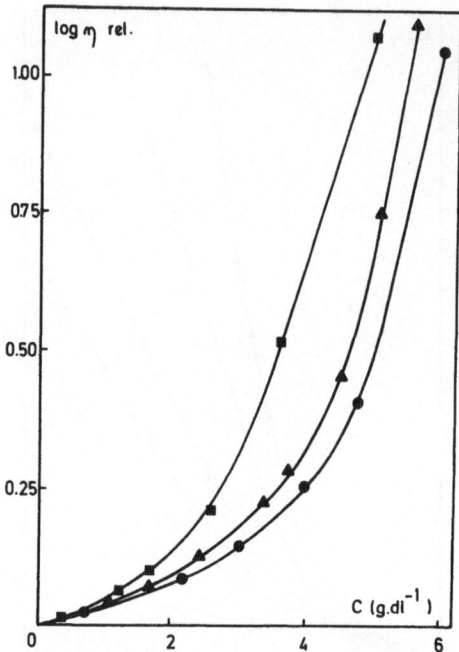

Figure 2.   Relative viscosity-concentration plots in toluene at 25°C for poly α-methylstyrene ($\overline{M}_n$:6,000) α,ω end -capped with: (●) Mg carboxylate, (▲) Mg sulfonate and (■) dimethyl benzyl ammonium chloride.

ionicity of the sulfonate end-groups is still higher than that of the carboxylate salts.   Indeed, the higher the effective charges are, the higher their attractive force is, and the more efficient ion pair interactions are.

The electrostatic origin of gelation is unambiguously assessed by the effect of solvent on the solution behavior of the α,ω-quaternary ammonium chloride PMS (Figure 3). In dimethylformamide (DMF, $\epsilon = 37$), at around 5 g.dL^{-1}, this HTP behaves rather as the non-quaternized prepolymer, whereas in toluene ($\epsilon = 2.38$) the gel is already well-formed. Tetrahydrofuran (THF, $\epsilon = 7.4$) is of course an intermediate case. Clearly, the dissociation of the salt end-groups is enhanced in solvents of high dielectric constant and the intermolecular interactions are accordingly depressed. The deleterious effect of trace polar compounds (1% V/V) on the ion pair interactions is illustrated in Figure 4 and substantiates the caution recommended in the preparation of representative HTP. The addition of THF, triethylamine (NEt$_3$) and propylene carbonate (PC) to toluene is responsible for a shifting of the viscometric curve toward higher concentrations.   Upon addition of methanol to

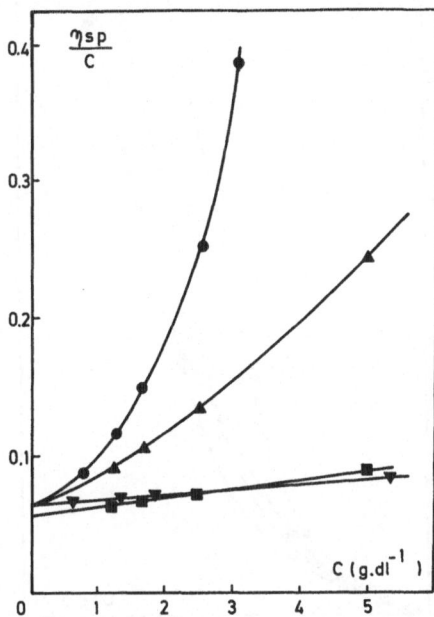

Figure 3.   Effect of solvent on the reduced viscosity-concentration plots for α,ω-quaternary ammonium chloride-PMS at 25°C: (●) toluene, (▲) THF, (■) DMF and (▼) non-quaternized α,ω-di tert-amino PMS in DMF.

toluene, the curve is deformed in such a way that the alcohol seems to interact with the ion pairs more strongly than the other polar compounds (THF, NEt_3, PC), and to affect more deeply the gelation process. This effect might be ascribed to additional hydrogen bonding.[8,9]

An increase in temperature as expected decreases the relative viscosity of α,ω-magnesium dicarboxylato PBD in toluene, especially in the range of concentrations where the ion pairs are interacting together (Figure 5).   The excellent thermoreversibility of these solutions is noteworthy.   After a heating run up to 80°C, the measurement of η rel is repeated at 25°C with a reproducibility higher than 95%.

In conclusion, the carboxylate end-groups behave as dipoles in nonpolar solvent. When their mutual intermolecular interactions engage the prepolymer chains in the tridimensional network, a sharp increase in the solution viscosity and ultimate gelation result. The occurrence of this phenomenon depends on all the parameters that control the formation and the interaction of the ion pairs, i.e., ionicity and geometry ("r" value) of the salt end-groups, nature of solvent and ligands, and temperature.

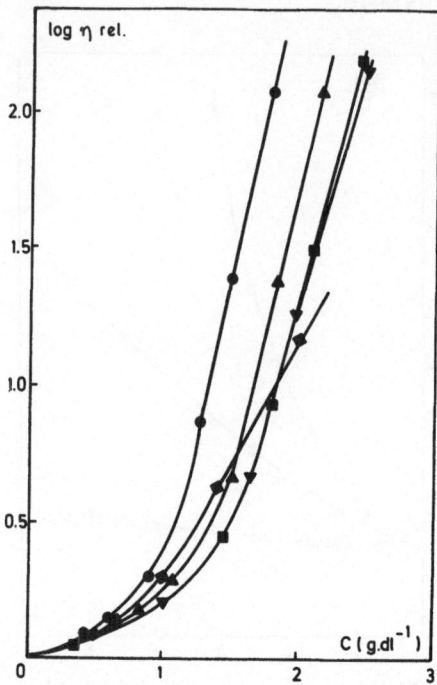

Figure 4.   Relative viscosity-concentration plots for α,ω -Mg di-
            carboxylato PBD (Hycar CTB) in toluene at 25°C. Effect of
            the addition of 1 % polar compounds to toluene: (●) pure
            toluene, (▲) addition of THF, (■) propylene carbonate, (▼)
            triethylamine and (♦) methanol.

EFFECT OF THE MOLECULAR WEIGHT AND THE HETERODISPER-
SITY OF THE PREPOLYMER ON THE GEL FORMATION

    When the molecular weight distribution of the original carboxy
telechelic polymer is very narrow, the increase of the solution viscosity
is very sharp and the critical concentration for gelation (Cgel) is
accurately determined by the vertical asymptote of the log η rel vs.
concentration curve.

    Figure 6 shows this pattern for polyisoprene samples of different
molecular weight ($\overline{M}_w/\overline{M}_n$ < 1.2), and the comparison with the commercial
PBD (Hycar CTB, $\overline{M}_w/\overline{M}_n$ ≃ 1.8, Figure 1) is convincing in that respect.
From Figure 7, the inverse dependence of C_{gel} on the prepolymer
molecular weight is unambiguous and means that the higher the average
end-to-end distance of the prepolymer and the easier the gelation even
if the ion pair end-groups concentration is lower.

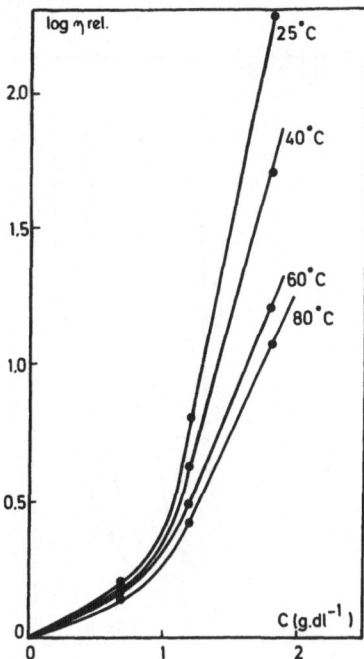

Figure 5.   Temperature effect on the relative viscosity- concentration plot for α,ω -Mg dicarboxylato PBD (Hycar CTB) in toluene.

For Mg carboxylate end-groups,in toluene at 25°C, a general relationship - $C_{gel} = k.\overline{M}_n^{-0.5}$ - is experimentally observed regardless of the nature of the polymeric backbone. The value of k is 800, 490, 407, 370 and 136 for poly (tert. butylstyrene), poly α-methylstyrene, polyisoprene, polystyrene and polybutadiene (85% 1,4-), respectively.

Since C_{gel} is in close relationship with $\overline{M}_n$, it is not surprising that a broad molecular weight distribution of the prepolymer causes the gelation to occur gradually as the concentration of the original solution increases. Indeed, the ability of the chains to form a tridimensional network depends on their length. When chains with a well-defined length have reached their critical gel concentration, they are able to form multiplets with a size just sufficient to provide for a continuous network. However, the other chains, the concentration of which is still lower than the critical value, participate also to the multiplet formation and preferentially form loose ends or possibly cyclic subchains. This means that even though the size of the multiplets (i.e. the mean number of cations per multiplet) is unchanged, all the subchains emerging from these entities are no longer active for building up a single giant molecule. It is only when all the chains will satisfy their own critical condition of gelation that the original solution will be completely gellified.

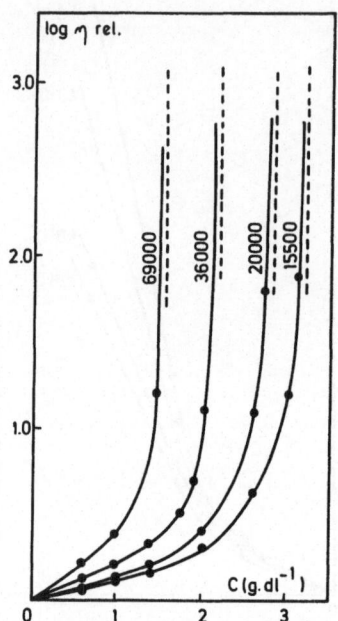

Figure 6.   Relative viscosity-concentration plots in toluene at 25°C, for α, ω -Mg dicarboxylato polyisoprenes with different molecular weights.

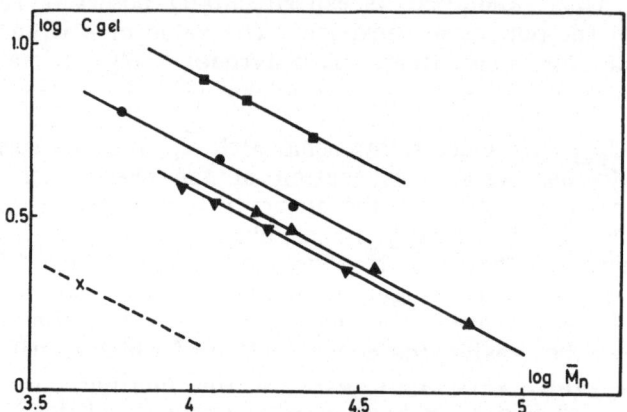

Figure 7.   Dependence of the critical gel concentration (C_{gel}) upon the prepolymer molecular weight in α, ω -Mg dicarboxylato polymers (toluene, 25°C): (●) PMS, (■) PTBS, (x) PBD, (▲)PIP and (▼) PS.

In media of low dielectric constant, ion aggregation into multiplets lowers significantly the potential energy as calculated for both alkaline and alkaline-earth carboxylates.[22]   Tables II and III show that a very large potential energy is released upon doublet formation, whereas the gain in energy decreases as the multiplet size increases. The order of magnitude of the involved energies may be given by the value of $e^2/\varepsilon r$ which amounts to 44 kcal/mol at 300°K, for $\varepsilon = 3$ and $r = 2.5$ Å. When 2 triplets formed by alkaline earth carboxylate aggregate into one sextet, the gain in energy (10.7 kcal/mol, Table III) is still very high compared to the thermal energy at 300°K (RT = 0.6 kcal/mol). By assuming that the multiplets occupy well-defined positions within ideal space lattices, it is possible to calculate the dependence of the multiplet size on the concentration (C) and the molecular characteristics of the carboxy telechelic polymer.[22] For alkaline earth carboxylates, the following relationship is established.

$$\bar{n} = 7.4 \times 10^{-3} \cdot a_{(\bar{n})} \cdot \left[\overline{r_0^2} / M \right]^{3/2} \cdot \alpha^3 \cdot C \cdot \bar{M}_n^{1/2} \tag{3}$$

where $\bar{n}$ is the mean number of cations per multiplet, $a_{(\bar{n})}$ is a parameter depending on the coordination number of the corresponding lattice, $(\overline{r_0^2}/M)$ is the intrinsic flexibility and α is the expansion coefficient of the prepolymer for the selected solvent and temperature.

Under defined experimental conditions (solvent, temperature, ion pair) the critical mean size of the multiplets $(\bar{n}_{gel})$ has a characteristic value which may be calculated by applying the critical conditions to equation 3.

$$\frac{\bar{n}_{gel}}{7.4 \times 10^{-3} \cdot a_{(\bar{n}_{gel})}} = (\overline{r^2}/M)^{3/2} \cdot C_{gel} \cdot \bar{M}_n^{1/2} = K \tag{4}$$

$(\overline{r^2}/M)$ is largely independent of M in a limited range of molecular weight (about 50,000).[23]   In these conditions and for given alkaline earth carboxylates, solvent and temperature, the result is:

$$C_{gel} = k \cdot \bar{M}_n^{-1/2} \tag{5}$$

in agreement with our experimental observations. From equations 4 and 5, it is obvious that:

$$k^{-1/3} = K^{-1/3} (\overline{r^2}/M)^{1/2} \tag{6}$$

where $(\overline{r^2}/M)^{1/2}$ can be estimated from the Flory relationship

$$[\eta] = \emptyset \cdot (\overline{r^2}/M)^{3/2} \tag{7}$$

TABLE II. Potential energy of alkali metal carboxylate multiplets.

Multiplets	$(\Delta E)_{min}$ $X\left(\dfrac{e^2}{\varepsilon r_d}\right)^{-1}$	Mean number of cation per multiplet $(\bar{n})$	$(\Delta E)_{min}$ reduced to one cation $X\left(\dfrac{e^2}{\varepsilon r_d}\right)^{-1}$
Doublet	-0.855	1	-0.855
Triplet	-1.221	1.5	-0.814
Quartet	-2.053	2	-1.026
Sextet	-3.227	3	-1.076

TABLE III. Potential energy of alkaline earth metal carboxylate multiplets.

Multiplets	$(\Delta E)_{min}$ $X\left(\dfrac{e^2}{\varepsilon r_d}\right)^{-1}$	Mean number of cation per multiplet $(\bar{n})$	$(\Delta E)_{min}$ reduced to one cation $X\left(\dfrac{e^2}{\varepsilon r_d}\right)^{-1}$
Doublet + free anion	-1.710	1	-1.710
Doublet + quartet	-5.156	1 + 1	-2.578
Triplet	-2.925	1	-2.925
Sextet	-6.336	2	-3.168

TABLE IV. k values for α,ω –Mg dicarboxylato polymers and $(\bar{r}^2/M)^{1/2}$ of the corresponding non–neutralized prepolymers in toluene at 25°C.

Prepolymer	$\bar{M}_n$	$\dfrac{k}{g^{3/2}dl^{-1}mol^{-1/2}}$	$(\dfrac{\bar{r}^2}{M})^{1/2}$
PTBS	15.000	800	0.714
PMS	6.000	490	0.740
PIP	70.000	407	0.816
PS	11.500	370	0.846
PBD	4.600	136	1.145

The intrinsic viscosity [η] of the non–neutralized carboxy telechelic prepolymers has been measured in toluene at 25°C and the calculated values of $(\bar{r}^2/M)^{1/2}$ are reported in Table IV, together with the experimental values of k.

Figure 8 clearly shows that the experimental results are in agreement with the theoretical equation 6.   From the slope of this graph, K is estimated to 227($\pm$29) and $\bar{n}_{gel}$ to ca. 2.3.

The calculation of $\bar{n}_{gel}$ is however based on the assumption that the ratio $(\bar{r}^2/M)^{1/2}$ is unmodified by the neutralization of the carboxy telechelic polymer and the occurrence of the gelation. In that respect, we have observed that bulk Ba and K dicarboxylato PBD clearly display a periodic structure with a lamellar character. The lamellar thickness calculated from SAXS data is dependent on the prepolymer chain length, but insensitive to the cation nature, i.e. to the electrostatic interactions between dipoles.[24]   This would mean that the multiplets originally formed at high dilution are able to grow into layered structure without stretching the prepolymer chains significantly.   Furthermore, the lamellar structure is maintained in the presence of toluene up to ca. 50%, at 25°C. It is noteworthy that the observed periodicity is then very close to the $<\bar{r}^2>^{1/2}$ value of the non–neutralized prepolymer under the same conditions of solvent and temperature. For instance, a gel of α,ω–Ba–dicarboxylato PBD ($\bar{M}_n$:4,600) in toluene (50%) at 25°C displays a periodicity of 74 $\pm$ 3 Å, whereas $<\bar{r}^2>^{1/2}$ of the original prepolymer is 77.6 Å in toluene at 25°C, as determined from equation 7.

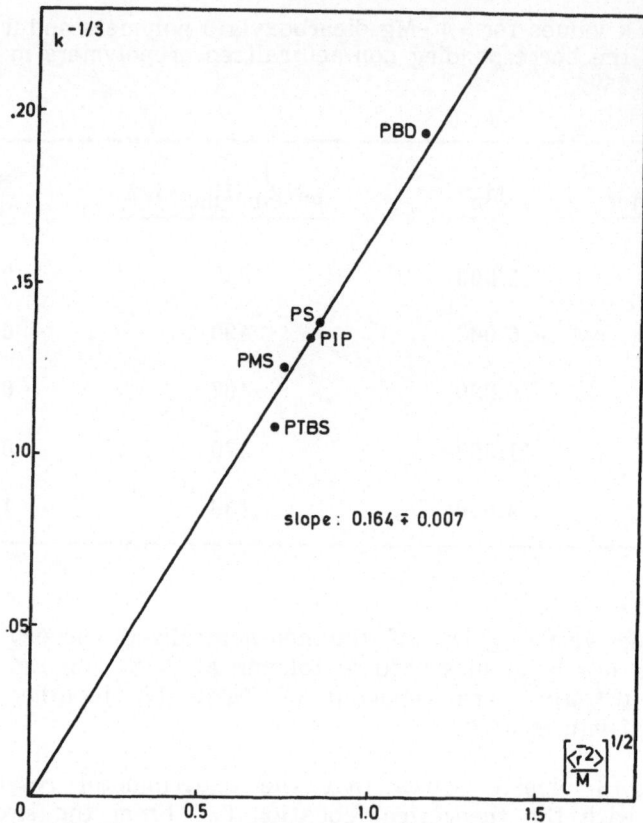

Figure 8. Dependence of the proportionality constant k (eq. 5) on the conformation $(\bar{r}^2/M)^{1/2}$ of the non-neutralized telechelic prepolymer.

VISCOELASTIC PROPERTIES OF SOLUTION OF ALKALINE-EARTH DICARBOXYLATO PBD

Let us discuss the value of $\bar{n}_{gel}$ = 2.3 calculated for a series of α,ω -Mg dicarboxylato polymers in toluene at 25°C. As the functionality of both the metal and the prepolymers is 2, the critical conditions of gelation would correspond to slightly more than one cation per multiplet $(\bar{n}_{gel} > 1)$, at least when the polymer chain-ends are involved into separate multiplets (excluding "ring" formation). Larger values of $\bar{n}_{gel}$ mean that the "ring formation" might really take place but probably also that the multiplets have a finite lifetime resulting in a time dependence of the solution properties. From that point of view, it was attractive to study the viscoelastic properties of the HTP solutions in nonpolar solvents. A series of HTP based on Hycar CTB and alkaline earth cations was investigated by means of the Rheometrics Mechanical Spectrometer RMS-7200 equipped with both the plate-and-plate and the cone-and-plate geometries.

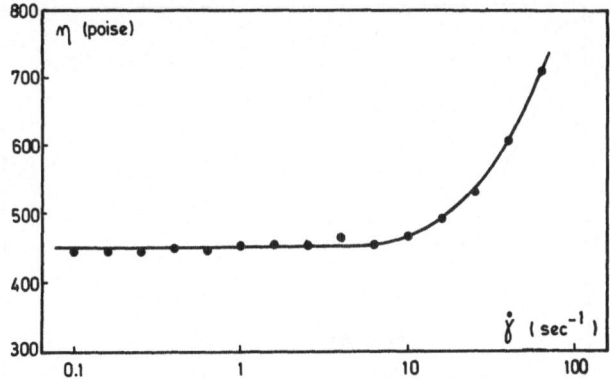

Figure 9.   Shear-thickening behavior of 10 g dL^{-1} solution of α, ω -Mg
            dicarboxylato PBD (Hycar CTB) in decahydronaphthalene).

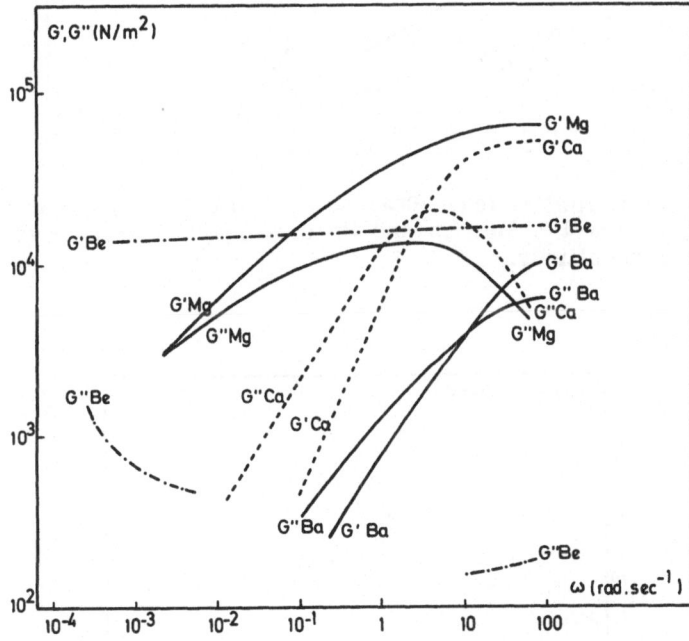

Figure 10. Partial master curves of shear storage and loss moduli for 10
            g dL^{-1} xylene solutions of α, ω -Be, Mg, Ca and Ba
            dicarboxylato PBD (Hycar CTB), respectively.   Reference
            temperature: 297°K.   Six isotherms have been reported
            between 297 and 342°K and obey time-temperature super-
            position.

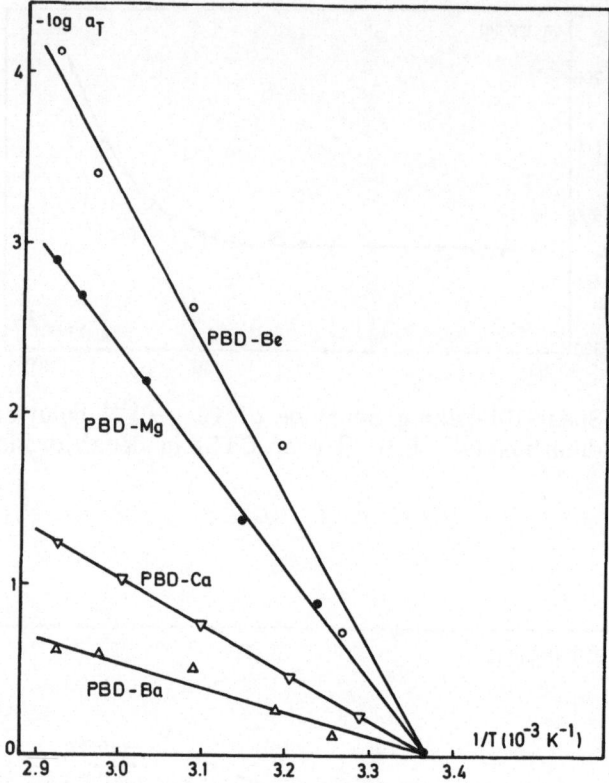

Figure 11.   Shift factors (a_T) versus 1/T for 10 g dL^{-1} xylene solutions
of α,ω –Be, Mg, Ca and Ba dicarboxylato PBD (Hycar CTB),
respectively.

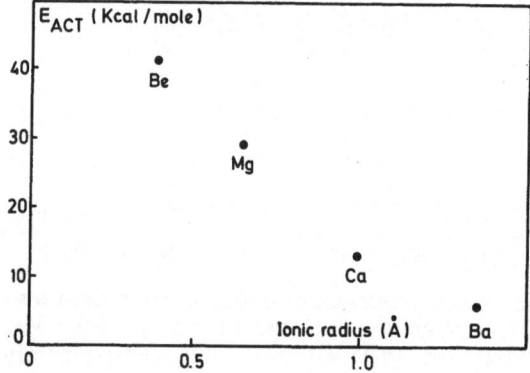

Figure 12.   Dependence of the secondary relaxation activation energy on
the ionic radius for 10 g dL^{-1} xylene solutions of α,ω –
alkaline earth dicarboxylato PBD (Hycar CTB).

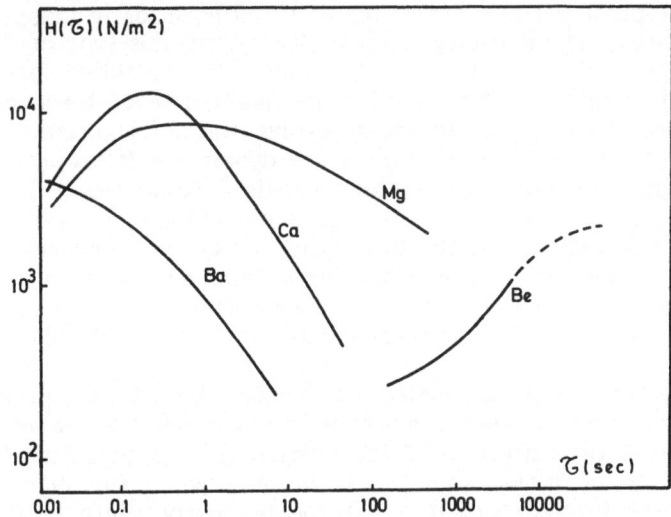

Figure 13.   Relaxation times distribution at 297°K of 10 g dL^{-1} xylene
solutions of α, ω -Be, Mg, Ca and Ba dicarboxylato PBD
(Hycar CTB), respectively.

The steady-flow viscosity of a 10 g.dL^{-1} solution of α , ω -Mg
dicarboxylato PBD in decahydronaphthalene (DHN) is clearly dependent
on the $\overset{\circ}{\gamma}$ shear rate when higher than 6 sec^{-1} (Figure 9). The observed
deviation from the Newtonian behavior confers to the solution a rarely
observed shear thickening (or dilatant) character. The substitution of
DHN (ε= 2.2) by chloroform (ε= 4.4) is responsible for a large decrease
in the steady-flow viscosity (ca. 350 times), which is now quite
independent of the shear rate at least up to 150 sec^{-1}.

Partial master curves of 10 g.dL^{-1} solutions of α,ω-alkaline earth
dicarboxylato PBD in xylene at 297°K are reported in Figure 10, and
result from a good frequency-temperature superposition of the experimental data.[17]   Only the G" master curve of the solution of Be-based
HTP is ill-defined due to the poor accuracy in the determination of the
very small values of G".   The shift factors support an apparent
Arrhenius-type of dependence (Figure 11), from which the activation
energy of the observed secondary ionic relaxation process was calculated
and found to decrease as the radius of the alkaline earth cations
increases (Figure 12).   One also observes that the relaxation spectrum
calculated by the first order approximation of Ninomiya and Ferry[25] is
displaced along the time scale in relation with the cation size (Figure
13).   The dynamic behavior of the 10 g.dL^{-1} solution is obviously
dependent on frequency: a regular transition from a viscous flow (G' <
G") to a gel-like behavior (G' > G") is noted at increasing frequencies.
The point where G' = G" appears at decreasing frequencies as the
alkaline earth cation size decreases and finally the Be-based HTP

exhibits a typical gel-like character as G' is practically independent of frequency and G" significantly smaller. The Arrhenius-type of activation ascertains a deformation process which is controlled by thermal multiplet dissociation. The stability and mean-size of these multiplets are therefore the key parameters governing the deformation mechanism of the HTP solutions. The multiplets are dynamic entities characterized by an average lifetime of the electrostatic interactions. In nonpolar solvents and for the series of alkaline earth cations, it is obvious that the smaller the cation, and the lower the critical gel concentration, the longer the relaxation time and the higher the activation energy. The effects of solvent (DHN compared to chloroform) and cation confirm the electrostatic origin of the general behavior of the HTP solutions.

Multiplets with $\bar{n}$ just higher than 1 should be efficient in promoting the gel formation, but only a more or less viscous solution is observed because the average lifetime of the electrostatic interactions is shorter than the time of measurement. It is only when sufficiently larger multiplets are formed (i.e. $\bar{n} = 2.3$ for Mg carboxylate in toluene at 25°C) that the polymer chains appear to be involved into an infinite network. The dynamic behavior of HTP solutions can also be understood in terms of average lifetime of the multiplets compared to the relaxation times which affect the viscoelastic properties. Similarly, the shear-thickening effects is the consequence of rather long-lived inter-molecular electrostatic interactions which increasingly prevent the individual polymer chains from relaxing when decreasing the time scale of deformation. Under the experimental conditions of Figure 9, the average lifetime of the "cross-links" would be approximately 0.16 sec.

In conclusion, these HTP solutions are distinguished by a thermo-rheological simplicity as well as by a secondary relaxation which is undoubtfully characteristic of the multiplets formed by the metal carboxylate end-groups. These features largely justify the choice of HTP as model compounds for ion-containing polymers. Furthermore, the rheological behavior, and more especially the shear-thickening effect, impart to HTP solutions an important technological interest.

ACKNOWLEDGEMENT

The authors are very much indebted to UNIBRA (Brussels) and the "Services de la Programmation et de la Politique Scientifique" for efficient support and fellowship (to G.B.). They are also grateful to Professor B. Gallot (Centre de Biophysique Moleculaire, Orleans, France) and Professor C. Marco (University of Mons-Belgium) for their collaboration in the investigation of the morphology and the viscoelastic properties, respectively. They acknowledge the gift of Hycar CTB samples by B. F. Goodrich Chem. Co. The skilful technical assistance of Miss J. Goffard and M. G. Renders was greatly appreciated.

The figures are reproduced from "Macromolecules" with the permission of the "American Chemical Society."

REFERENCES

1. W. V. Titow and B. J. Lanham, "Reinforced Thermoplastics," Applied Science Publ. Ltd, 1975.
2. T. L. Smith, Rubb. Chem. Technol., 51, 225 (1978).
3. A. Noshay and J. E. McGrath, "Block Copolymers - Overview and Critical Survey," Academic Press, Inc., 1977.
4. D. R. Paul and S. Newman (Eds.), "Polymer Blends," Academic Press, Inc. vol. 1 and 2, 1978.
5. A. Eisenberg, Contemp. Top. Polym. Sci., 3, 231 (1979).
6. W. J. McKnight and T. R. Earnest, Jr., J. Polym. Sci., Macromol. Rev., 16, 41 (1981).
7. Yu. N. Panov, Europ. Polym. J., 15, 395 (1979).
8. R. D. Lundberg and H. S. Makowski, J. Polym. Sci., Polym. Phys. Ed., 18, 1821 (1980).
9. B. Siadat, R. D. Lundberg and R. W. Lenz, Macromolecules, 14, 773 (1981).
10. N. Ise and T. Okubo, Acc. Chem. Res., 13, 303 (1980).
11. G. Broze, R. Jerome and Ph. Teyssie, Macromolecules, 14, 224 (1981).
12. G. Broze, R. Jerome and Ph. Teyssie, Macromolecules, 15, 920 (1982).
13. J. Economy and J. H. Mason, "Ionic Polymers," L. Holliday, Ed., Applied Science Publishers, Chap. V, London, 1975.
14. J. C. Danjard, C. Niemerich, and M. Pineri, Rev. Gen. Caoutchouc Plast. (R.G.C.P.), 50, 723 (1973).
15. M. Pineri, C. Meyer, A. M. Levelut and M. Lambert, J. Polym. Sci., Polym. Phys. Ed., 12, 115 (1974).
16. E. P. Otocka, M. Y. Hellman and L. L. Blyler, J. Appl. Phys., 40, 4221 (1969).
17. G. Broze, R. Jerome, Ph. Teyssie and C. Marco, Macromolecules, 16, 996 (1983).
18. P. M. Van Der Velden, B. Rijpkema, C. A. Smolders and A. Bantes, Europ. Polym. J., 13, 37 (1977).
19. G. Broze, P. M. Lefebvre, R. Jerome and Ph. Teyssie, Makromol. Chem., 178, 3171 (1977).
20. D. H. Richards and F. J. Burgers, Polymer, 17, 1020 (1976).
21. J. F. Joanny, Polymer, 21, 71 (1980).
22. G. Broze, R. Jerome and Ph. Teyssie, Macromolecules, 15, 1300 (1982).
23. P. J. Flory, "Principles of Polymer Chemistry," Cornell University Press, 1953, 621.
24. G. Broze, R. Jerome, Ph. Teyssie and B. Gallot, J. Polym. Sci., Polym. Letters Ed., 19, 415 (1981).
25. J. D. Ferry, "Viscoelastic Properties of Polymers," John Wiley, New York, 1970.

ON THE MICROENVIRONMENT OF SOLUBLE AND CROSS-LINKED POLYMERS

F. Mikeš[1a], J. Labský[1b], P. Štrop[1c], and J. Králíček[1a]
Department of Polymers
Prague Institute of Chemical Technology
Suchbátarova 1905,166 28 Prague, Czechoslovakia
Institute of Macromolecular Chemistry
Heyrovského nám.2,162 02 Prague, Czechoslovakia and
Institute of Organic Chemistry and Biochemistry
Flemingovo nám.2,166 10 Prague, Czechoslovakia

ABSTRACT

The properties of the microenvironment of soluble synthetic polymers such as polymethacrylamide (PMA), poly(2-hydroxyethyl methacrylate) (PHEMA), poly(2-vinylpyridine) (P-2VP), poly(4-vinylpyridine) (P-4VP), poly(methyl methacrylate) (PMMA), poly(butyl methacrylate) (PBMA), polystyrene (PS), poly(4[5]-vinylimidazole) (PVIm), and poly(N-2-hydroxypropyl methacrylamide) (PHPMA) and cross-lined polymers were studied by the shift and shape of the band in electronic spectra of a solvatochromic "reporter" molecule embedded in polymer chains. Preferential interaction of parts of the polymer molecule with a reporter and the shielding of interactions between solvent molecules and a reporter molecule of a polymer causes a shift and broadening of its solvatochromic band. This shift is mechanistically interpreted as a change in the polarity of the microenvironment of a polymer in solution in comparison with polarity of the solvent used. 4-(4'-Hydroxystyryl)-N-alkylpyridinium-betaine, spiropyran-merocyanine, and 1-dimethylamino-5-sulfonamidonaphthalene (Dansyl) reporters were used. In almost all cases the polarity of the polymer microenvironment was lower than that of the solvent. At the same time, the dependence of the nature of the environment on the distance of the reporters from the polymer chain was studied.

INTRODUCTION

Both synthetic and natural polymers represent microheterogeneous regions in a solution. Therefore, their solutions are intermediate between colloidal and true solutions. In solutions of high molecular weight compounds, particularly in solutions of globular proteins in their native conformation, the spatial distribution of physical properties such as concentration, mass and charge density, polarity, or mobility are different than in solutions containing small molecules and exhibit a higher degree of heterogeneity. As a result of different polarities in the protein microenvironment and in the bulk solution, differences in the pK of an indicator when free or when bound to a protein[2-5] were observed. Data on the local character of a protein (e.g. polarity, mobility, and conformation) were obtained from kinetic data for the reactions with substrates of different hydrophobicites;[6] from spectral data such as the fluorescence of tyrosine, tryptophane, and phenylalanine residues;[7] from solvent perturbation studies;[8] and from PMR spectra.[9]

The local polarity was estimated from band shifts in the electron spectra of solvatosensitive "reporter molecules" sorbed in a specific way[10,11] or bound to proteins.[12-14]

The structure and properties of the solvent in the vicinity of synthetic polymers have been characterized by spectral studies. e.g. fluorescence spectra,[15] the splitting of PMR methylene chloride signals in the presence of polypeptide chains oriented by a strong magnetic field,[16] or the splitting of benzene signals in the system benzene-poly(methyl methacrylate).[17] The different polarity of the micro-environment of a synthetic polymer may cause, as in the case or proteins, considerable differences in pK values of ionizable polymer groups as compared to low-molecular analogues.[5,18,19,20,21] The use of the solvent dependence of the photochromism of spiropyrane derivatives for studying polar and steric effects on polymers was proposed by Vandewyer and Smets[22] who also observed that the behavior of spiropyranes embedded in a polymer chain was much less affected by the solvent polarity than that of model substance.

Properties of the microenvironment of soluble and cross-linked polymers were studied by the shift of bands in the electron spectra of solvatochromic reporter molecules embedded in polymer chains. Generally, the charge-transfer (CT) absorption spectra and emission spectra of a number of compounds were used to correlate solute-solvent interactions with physical and chemical properties of interest. The energy of the band maxima of these chromophores is quite solvent sensitive and is linearly correlated with empirical solvent polarity parameters. The observed shift of the maximum of the solvatochromic reporter embedded in the polymer chains, compared with a low-molecular weight analog in the same solvent, was interpreted in terms of a change in the polarity of the microenvironment of the polymer in solution.

The method proposed by Kosower[23] for the semiempirical deter-mination of the solvent polarity was employed in this study to measure the polarity of the microenvironment of synthetic vinyl polymers. The polarity of the polymer microenvironment and solvent was expressed by the absorption or emission band energy of a solvatochromic compound

$$E = N_A h\nu = \frac{1.197}{\lambda_{max}} \times 10^5 \text{ (kJ/mol)} \sim 2.859 \times \bar{\nu} \times 10^{-3} \text{ (kcal/mol)}$$

where E is the energy in kJ/mol or kcal/mol, h is Planck's constant, ν is the frequency, $\bar{\nu}$ is the wave number (cm^{-1}) and λ is the wavelength of the maximum of solvatochromic reporter molecule.

I. SPECTRAL PROPERTIES OF REPORTERS

An interaction between solvent and solute is often reflected in a change of some physical property of the solvated molecule. Methods of investigation of such changes which are particularly suited for the purpose include some optical methods. A change in these properties accompanying a change in the solvent depends on the type and extent of interactions and has recently been used for this purpose.

The term "polarity" of a solvent has been often used without, however, being unambiguously defined. In this study, "polarity" of the solvent denotes its solvation power, i.e. its ability to solvate the solute. This power is determined by the sum of all interactions between solvent and the solute. Those interactions which lead to defined chemical changes of the dissolved compound are ruled out. That no simple relationship exists between the equilibrium and rate constants and the position of the UV and IR bands, on the one hand, and the physico-chemical properties of solvents, on the other has been proven exper-imentally and is also anticipated on the basis of the complicated character of interactions between solvent and the solute. In particular, the macroscopic dielectric constant is no immediate measure of interactions on the molecular microscopic level.

In recent years, these findings have led to attempts to express the polarity of the solvent empirically. From the effect of solvents on some appropriate standard processes sensitive to solvents (e.g. rate or equilibrium of a standard chemical reaction, light absorption by a solvatochromic standard dye) conclusions are drawn about the polarity of the solvent. The chosen standard processes act as reporters char-acterizing the effect of solvent on the microscopic level.

Using the rate constants of certain reactions in the wide range of solvents, empirical solvent polarity parameters have been derived, representing a universal measure of solvation.

In addition, empirical polarity parameters derived from kinetic measurements have been supplemented by parameters determined from spectral data.

In our study, the properties of the microenvironment of synthetic polymers were investigated by means of various solvatochromic reporters 1-methacryloyloxyethyl-4-carboalkoxypyridinium salts (SCHEME I), 1-alkyl-4-/4-hydroxystyryl/pyridinium-betaine (SCHEME II) and spiropyran-merocyanine type (SCHEME III) all have solvatochromatic bands in the visible absorption;   1-dimethylaminonaphthalene-5-sulphonamido chromophore (SCHEME IV) has a solvatochromic band in the emission spectrum.

A. Pyridinium and Spiropyran-Merocyanine Reporters

MPI[24] is a structural analogue of 1-ethyl-4-carboethoxypyridinium iodide (EPI) (SCHEME I) i.e., of the compound originally proposed by Kosower[23] for semiempirical measurement of the solvent polarity. Very similar wavelengths were found for the maxima of the solvatochromic bands of these compounds.   The polarity scale based on MPI may, therefore, be correlated with a similar scale proposed by Dimroth.[25]

The weakest absorption band at the highest wavelength corresponds to a charge-transfer solvatochromic band.   The molar absorption coefficient of the CT band varies from 20 to 700. The lowest intensities were observed in aprotic dipolar solvents such as DMF and DMSO. The CT band is not sufficiently distinguished in water, whereas in solvents with a lower polarity it is shifted to higher wavelength and is more separated from the other absorption bands.

The solvatochromic monomer SB (SCHEME II) exhibits none of the disadvantages of the compound MPI.   The solvatochromic band which corresponds to the highest wavelength is always separated from the other bands and is intense (the molar absorption coefficient of SB is about 4×10^4); it is also sufficiently sensitive to the solvent polarity.[24]

The polarity of a series of solvents, as measured by SB, was correlated with the Dimroth polarity parameter[25] (Figure 1).   Linear correlation was found over the whole range of applicability of SB, i.e., for solvents with a polarity higher than that of pyridine, E_T (SB) = 45.4 kcal/mol.

The solvatochromic monomer SB facilitates measurements in high polarity solutions, and its high absorption coefficient makes it possible to incorporate less than 0.1 mol % of the SB reporter monomer and so retain the character of the polymer.

The correlation of E_T (SB), the semiempirical polarity parameter based on SB, and dielectric constant is not linear.   Nearly linear correlation was obtained for mixtures of organic solvents with water with dielectric constants higher than 30 - 40.

For measuring the microenvironment polarity of synthetic polymers it is of great importance to know to what extent specific inter-

SCHEME I

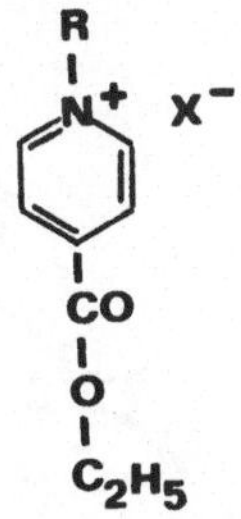

$\underline{\text{EPI}}$ R - ethyl, X^- - iodide
$\underline{\text{MPI}}$ R - methacryloyloxyethyl, X^- - iodide

SCHEME II

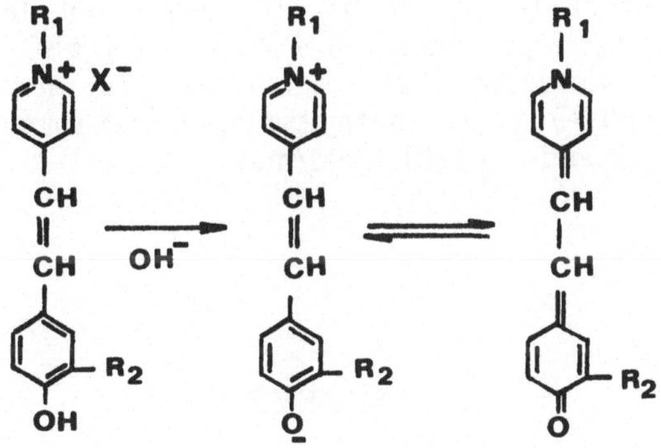

RSX   a   SB   b

$\underline{\text{MSC}}$ R_1 - methacryloyloxyethyl, R_2 - ethoxy, X^- - chloride

$\underline{\text{BSB}}$ R_1 - 4 - bromobutyl, R_2 - ethoxy, X^- - bromide

$\underline{\text{AESB}}$ R_1 - β - aminoethyl, R_2 - H, X^- - bromide

$\underline{\text{C}_1\text{SB}}$, $\underline{\text{C}_4\text{SB}}$, $\underline{\text{C}_{10}\text{SB}}$, $\underline{\text{C}_{16}\text{SB}}$ R_1 - methyl, R_4 - n-butyl,

$\qquad\qquad$ C_{10} - n-decyl or C_{16} - n-cetyl;

$\qquad\qquad$ R_2 - ethoxy, X^- - bromide.

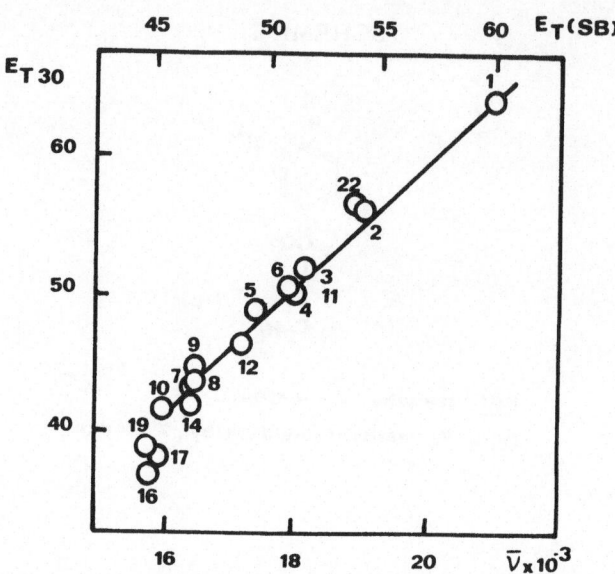

Figure 1. Correlation of a semiempirical polarity scale based on the shift of the solvatochromic band of the compound SB with the Dimroth parameter, E_{T30} . (1) Water, (2) methanol, (3) ethanol, (4) 1-propanol, (5) 2-propanol, (6) 1-butanol, (7) 2-methyl-2-propanol, (8) DMFA, (9) DMSO, (10) pyridine, (11) benzylalcohol, (12) acetonitrile, (14) acetone, (16) dioxane, (17) THF, (19) 2-methylpyridine, (22) formamide. $\bar{\nu}cm^{-1}$, and E_{T30} and E_T(SB) (kcal/mol).

SCHEME III

RNHCH$_2$ ⇌ (hν / Δ) RNHCH$_2$

MSP   R − methacryloyl   **AcSP**   R − acetyl

actions[23,26-28] of the reporter in mixed solvents occur. Specific solvation would naturally lead to discrepancies in correlations between various solvation-sensitive phenomena. However, such correlations are surprisingly good over a wide range of solvent compositions if one considers the differences between the processes or compounds involved. Kosower's Z parameter was correlated with different semiempirical parameters of the polarity of solvents derived from kinetic measurements and determined from spectral data (e.g.[25] E_{T30}, Ω-parameter[29] etc.) Since linear correlations were also found between $E_T(SB)$ and the Z parameter, E_{T30}, as well as the rate of the stilbene trans-cis photoisomerization,[30] one may assume that the specific solvation of the reporters of SB molecules was not the most significant factor. The spread of data which is often found in correlations of solvent polarity dependent phenomena and D,[26,27,29] also observed for $E_T(SB)$, has its origin in the different character of these two parameters.

Irradiation of a spiropyran molecule with light of a suitable wavelength leads to splitting of the bond between the carbon and oxygen atom with production of intensely colored merocyanines (SCHEME III). Recombination of the molecule proceeds after the source of radiation has been removed.

Ethyl acetate/methanol and ethyl acetate/hexane mixtures were the binary solvents used in this investigation. The effect of some one-component solvents was also studied using the position of the absorption maximum of the colored merocyanine form of a low-molecular weight model AcSP. For these solvents, a correlation was found between the Dimroth parameter E_{T30} and the energy $E_T(SP)$ of the absorption maximum of the merocyanine form of the low-molecular compound AcSP. In general, as the polarity of the solvent is increased, the visible absorption maximum of the merocyanines shifts to a shorter wavelength, the molar absorption coefficient decreases, and the half-width of the band increase.[31]

B.   1-Dimethylamino-5-sulfonamidonaphthalene Reporters

Among the most commonly used fluorescent probes for biochemical and biological systems[32] are 1-dimethylamino-5-sulfonamidonaphthalenes (DNS derivatives) (SCHEME IV). The emission spectrum of the model compound 1-dimethylamino-5-(β-isobutyroyl-aminoethyl)sulfonamidonaph-thalene (IB DNS) in absolute methanol is characterized by one band with an emission maximum at 538 nm which corresponds to the excited state of IB DNS, with a nonprotonized dimethylamino group.[33] The emission spectrum was independent on the exciting radiation for wavelengths from 250 to 380 nm.

The main task is the determination of the solvation effect on IB DNS molecules on their fluorescence, and particularly the determination of the effect of a change in the polarity of environment. The energy of the emission maximum IB DNS was measured in a number of dioxane/water and alcohol/water mixtures and in some one-component solvents and was then correlated with the polarity parameters Z and E_{T30} (Figure 2). In no case, however, was a general linear correlation found. An approximately linear correlation can be observed for the individual mixtures. As a consequence, a general interpretation in which the results obtained for the individual mixtures are compared is not completely reliable. The largest deviations from correlation have been observed with aprotic dipolar solvents, similar to other fluorescence polarity indicators. No effect of the viscosity of solvents on E_F(IB DNS) could be detected. The energy of the emission maximum in ethylene-glycol ($\eta \sim 20$ cp) satisfies correlation for homologous alcohols ($\eta \sim 1$-2 cp).

II. POLARITY OF THE MICROENVIRONMENT OF POLYMERS

To determine the polarity of the microenvironment of polymers, polymer labelled with solvatochromic reporters were prepared, either by (1) copolymerization with solvatochromic monomers;[24,34] or (2) polymer analogous reactions (e.g., the reaction of copolymer active esters with primary amino groups of the solvatochromic molecule[35,36] or alkylation reaction of PVIm and cross-linked polymers with solvatochromic molecule.[34,37]. The properties of the microenvironment of polymers were studied by the shift and shape of band in electronic spectra of a solvatochromic reporter molecule embedded in polymers.

A. Soluble polymers

1. Polymers in One-Component Solvents

(a) Pyridinium Type Reporters

The monomer MPI was not found to be suitable for measuring the polarity of the polymer microenvironment. In the region most interesting for measuring either the solvent polarity or the polarity of the polymer microenvironment, the solvatochromic band of compounds MPI and EPI is not separated from a much more intense adjoining band.

In the solvatochromic band region the spectra of compound SB embedded in polymer chains do not differ appreciably from the spectra of free SB in the same solvents. However, we observed a shift of the solvatochromic band a larger half-width of the solvatochromic band for the polymer, and sometimes even at the same wavelength a different resolution of the finer vibrational structure of the band. The band of the solvatochromic compound SB embedded in the polymer is shifted to higher wavelengths, i.e., it indicates a lower polarity of the polymer chain microenvironment as compared to the solvent polarity.

SCHEME IV

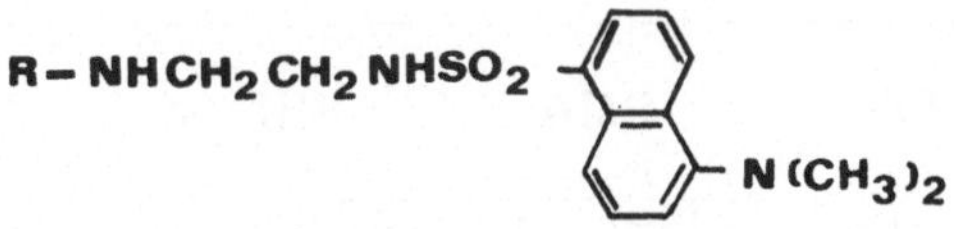

R — NHCH$_2$CH$_2$NHSO$_2$ — N(CH$_3$)$_2$

M DNS   R— methacryloyl   **IB DNS**   R– isobutyroyl

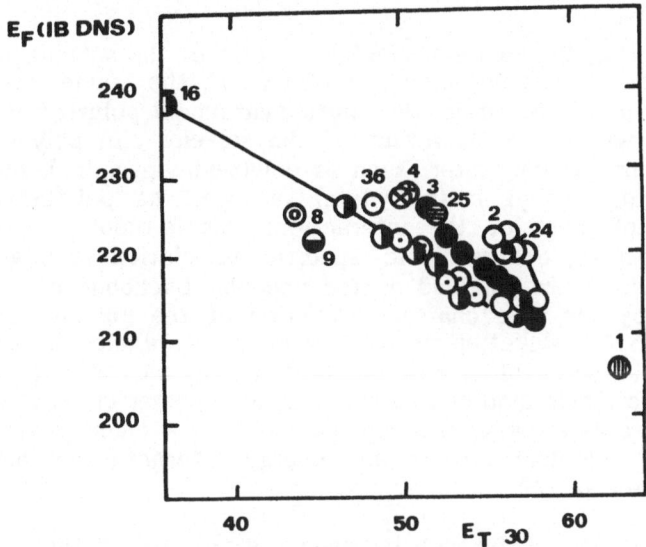

Figure 2. Correlation of a semiempirical polarity scale (E_F (IB DNS)), based on the shift of the solvatochromic emission band of the compound IB DNS, with the Dimroth parameter E_{T30}. For symbols of the solvents see Figure 1; (24) ethylene glycol, (25) ethyl cellosolve. E_F (IN DNS) (kJ/mol) is the energy of the emission maximum of DNS fluorochrome; E_{T30} (kcal/mol).

Table 1

Values of the Energy of the Solvatochromic Band $E_T(SB)$, kcal/mol of SB, and Values of $E_T(SB)$ Imbedded in PMA, PHEMA, PBMA, and PS in Several Solvents (Half-Band Widths, Δ, cm^{-1})

Solvent	SB	PMA		PHEMA		PMMA	PBMA	PS
	$E_T(SB)$	$\overline{E_T(SB)}$	Δ	$\overline{E_T(SB)}$	Δ	$E_T(SB)$	$E_T(SB)$	$E_T(SB)$
Water	60.19	57.63	4920	–	–	–	–	–
Methanol	54.39	–	–	53.35	4120	–	–	–
DMFA	46.98	–	–	–	–	46.43	–	–
Aniline	46.86	–	–	–	–	–	–	46.09
Butanone	46.31	–	–	–	–	46.20	46.13	46.09
Nitromethane	48.60	–	–	–	–	46.89	–	–

In Table I, the semiempirical parameter of the solvent polarity and the polymer microenvironment polarity in the same solvents are compared. In all the cases, the microenvironment polarity of a polymer in solution was lower than that of the solvent. In polymers with a partially nonpolar character, such as poly(4-vinylpyridine), poly(2-vinylpyridine), poly(methyl methacrylate), as well as poly(2-hydroxyethyl methacrylate), part of the interactions (dipole–dipole, dipole–induced dipole, multipole, charge–dipole, specific association such as hydrogen bonding, etc.)[38] are shielded by the nonpolar backbone of the polymer chain and by the side chains. Solvation of the polymer polar group differs from the solvation of the low-molecular analogue also in other respects. In spite of a relative polarity of the polymer units, the orientation of their dipoles to a bound polar reporter or reactive residues is not as free as for a solvent molecule so that a much wider dispersion of orienting electric dipoles and energy interactions[39] may be encountered (see p. 277).

The smallest difference between the polarity of the bulk solution and that of the polymer microenvironment was observed for PHEMA, a larger difference was found for P-2VP, and the largest difference was found for P-4VP. The difference for PBMA was also larger than that for PMMA. This sequence corresponds to that of the expected polarity of the polymers.

Another polymer which was studied from this point of view in one-component solvents was poly(4[5]-vinylimidazole). Two different modified polymers,[34] PVIm-I and PVIm-II, were prepared for the estimation of the polarity of the microenvironment of PVIm (SCHEME V). The polarity of the microenvironment of PVIm-I and PVIm-II was studied in

SCHEME V

POSITION OF N–SUBSTITUTION IS UNKNOWN

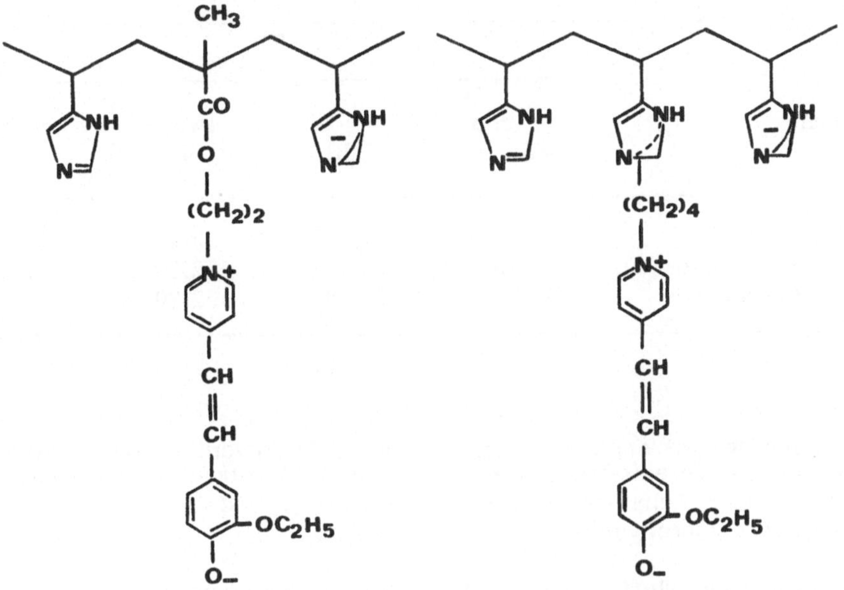

PVIm– I   PVIm – II

methanol, methylcellosolve, and ethylcellosolve. A great difference in E_T values for PVIm-I and PVIm-II was observed in methanol (Table II). The polarity of the microenvironment of PVIm-I in methylcellusolve is equal to the polarity of methylcellosolve itself, while in ethylcellosolve the polarity of the microenvironment of PVIm-I is higher than that of ethylcellosolve. The above results suggest that PVIm in the anionic form is relatively polar.[34] The difference between the energy of the CT absorption band for PVIm-I and PVIm-II in a particular solvent is caused by a different distance of the solvated reporter group from the polymer backbone (see p.   ).

(b)   Spiropyran - Mirocyanine Reporters

The wavelength of the absorption maximum of the colored form of spiropyrane bound on poly(methyl methacrylate) differs from that of the maximum of a low-molecular weight model in the same solvent. Similarly to Irie et al.,[39] we have observed that the absorption maximum of the merocyanine form bound to PMMA in benzene, dioxane, tetra-hydrofurane, ethyl acetate, and chloroform solutions is shifted to shorter wavelengths compared with the maximum of absorption of the low-

Table 2

Values of the Energy of the Solvatochromic Band (E_T, kcal/mol) of MSC and BSB in One-Component Solvents and Values of E_T(SB) in PVIm.

| Solvent | E_T | | | |
| | Model compound | | PVIm-I | PVIm-II |
	MSC	BSB		
Methanol	54.38	54.33	53.18	54.32
Methylcellosolve	52.15	52.21	52.20	52.38
Ethylcellosolve	51.23	51.38	52.20	–

molecular weight analogue AcSP in the same solvent. Bound and free merocyanine were found to behave similarly in tetrachloromethane. In acetone their behavior is almost identical (bound merocyanine 213.0 kJ/mol free merocyanine AcSP 212.8 kJ/mol).

The blue shift indicates that the microenvironment around the merocyanines bonded to poly(methyl methacrylate) is more polar than that of the low molecular analogue in benzene, dioxane, tetrahydrofurane, ethyl acetate, and chloroform. The intramolecular solvation by the ester side groups of the merocyanines causes the blue shift.

The decrease of the polarity of the polymer chain microenvironment results, as for one-component solvents, from the following factors: (a) the apolar contribution of the polymer backbone and its substitutents (i.e., shielding against polar and mobile solvent molecule), (b) the structure of the solvent in the vicinity of the polymer, and (c) reduced "solvation" of dipole reporter SB molecules by groups of the polymer chains since they are less polar and are not quite free to orient their dipoles toward the embedded compound SB.

2.   Polymers in Binary Mixtures

(a)   Pyridinium Type Reporter

The polarity of the microenvironment of polymer chains was studied for the above mentioned polymers also in the binary solvents methanol/water, ethanol/water, 1-propanol/water, 2-propanol/water, pyridine/water, and benzyl alcohol/pyridine.

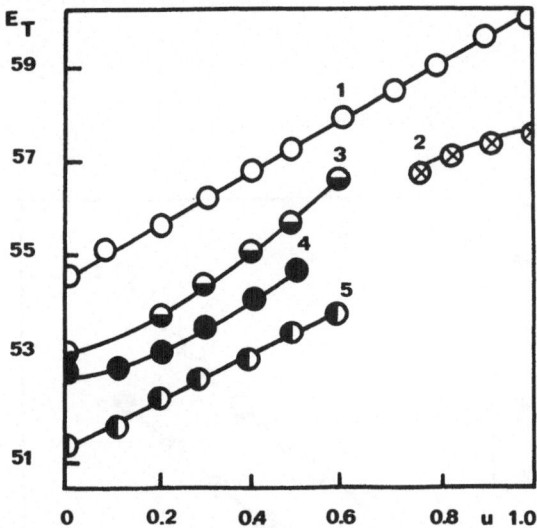

Figure 3. Solvent composition dependence of the polarity (E_T) of the binary solvent methanol/water (1) and the polarity of the microenvironment of PMA (2), PHEMA (3), P-2VP(4), and P-4VP (5). E_T is the energy of the CT absorption band of the compound SB (kcal/mol), u is the volume fraction of water.

Figure 3 shows the dependence of the energy E_T(SB) of the CT solvatochromic band of the model compound SB (characterizing the polarity of a solvent) and of the model compound SB embedded in a polymer chain (characterizing the polarity of the polymer micro-environment) on the binary solvent composition for PMA, PHEMA, P-2VP, and P-4VP in methanol/water mixtures. Similar data are shown in Figure 4 for PHEMA and SB in 1-propanol/water, and in Figure 5 for PHEMA, and SB in dioxane/water and acetone/water. The same dependences were found for PMMA, PBMA, and SB in methanol/toluene. In most cases the absorption maximum of the compound SB embedded in the polymer is shifted to higher wavelengths, i.e., to lower values of E_T, so that the polarity of the polymer chain microenvironment is lower than that of the solvent.

In binary solvents, preferential sorption of components of the solvent on the polymer chain constitutes an additional factor affecting the microenvironment of the polymer. For PHEMA, which was studied intensively in a series of binary solvents, one component of the solvent mixture is preferentially sorbed by the polymer coil.[40] Figure 4 shows the dependence of the limiting viscosity number [η] and the preferential sorption coefficient γ (at γ < 0 water is preferentially sorbed, at γ > 0

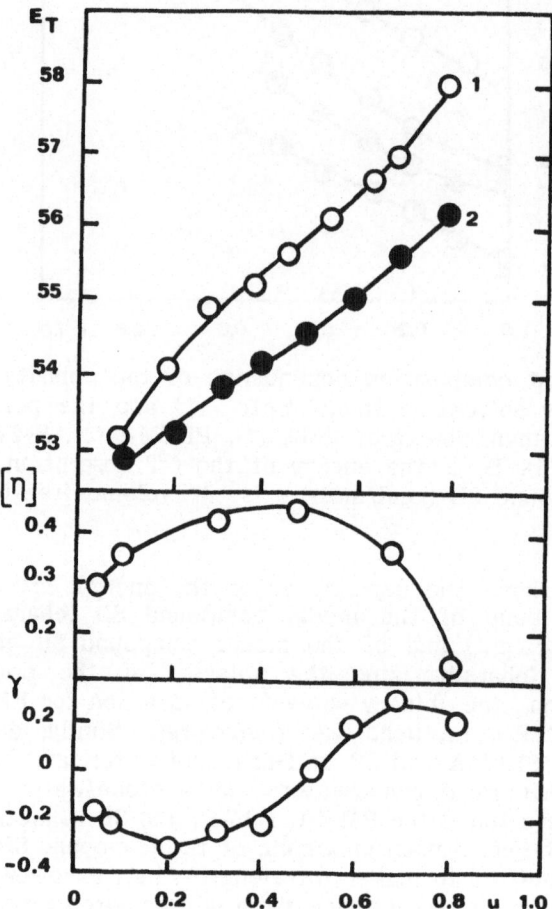

Figure 4. Solvent composition dependence of the polarity (E_T) of the binary solvent 1-propanol/water (1) and the polarity of the PHEMA microenvironment (2), limiting viscosity number [η] (dlg^{-1}), and coefficient of the preferential sorption γ. For a definition of the symbols E_T and u see Figure 3.

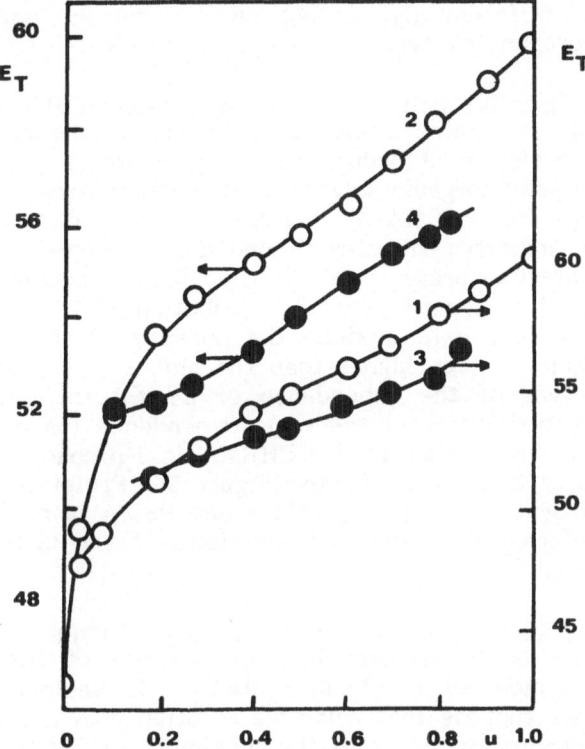

Figure 5. Solvent composition dependence of the polarity (E_T) of the binary solvents dioxane/water (1) and acetone/water (2), and the polarity of the PHEMA microenvironment in these mixtures (3), (4). For a definition of the symbols E_T and u see Figure 3.

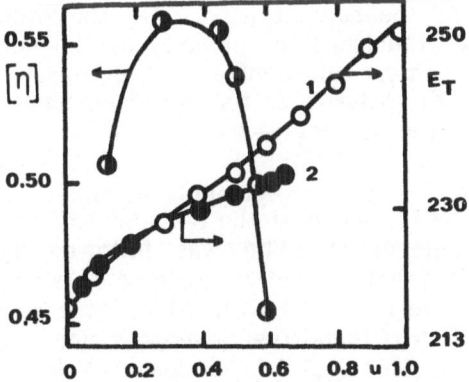

Figure 6. Solvent composition dependence of the intrinsic viscosity ($[\eta]$) of PVIm, polarity (E_T) of ethanol/water mixture (1), and the polarity of microenvironment of PVIm-II (2). For a definition of the symbols E_T (kJ/mol) and u see Figure 3.

1-propanol is preferentially sorbed) on the solvent composition for PHEMA in 1-propanol/water.

With preferential sorption of one component of the binary solvent on the polymer coil, an increase or decrease of the polarity of the polymer microenvironment occurs depending on whether the more polar (water) or less polar (organic solvent) component is sorbed. Preferential sorption occurs for PHEMA in 1-propanol/water, dioxane/water, and acetone/water mixtures (Figures 4 and 5). When the more polar component (water) is preferentially sorbed from mixtures in which its concentration is low, then the apolar contribution of the polymer may be compensated to that extent, since the polarity of the polymer chain microenvironment is even higher than the bulk solvent polarity. As a result, the curves of the dependence of E_T for the polymer on the solvent composition intersect the same dependence for mixed solvents. This phenomenon was observed for PHEMA in 1-propanol/water (Figure 4), dioxane/water, and acetone/water (Figure 5). Preferential sorption is also indicated by the results for PMMA and PBMA in methanol/toluene mixtures. Preferential sorption was previously found in this system by dialysis equilibria.[41]

The dependence of the intrinsic viscosity of PVIm on the composition of ethanol/water mixtures in the presence of lithium chloride (Figure 6) is typical of cosolvent systems. In analogy with similar systems[40,42] we suppose that selective sorption does not take place at the composition corresponding to the maximum in Figure 6 (35 vol % ethanol/water) while at other compositions, the component present at lower concentration is selectively sorbed.

In cosolvent mixtures containing a low fraction of water (for example, ethanol/water, Figure 6) selective sorption of water on PVIm takes place and the polarity of the microenvironment of PVIm-II is higher than the polarity of the solvent used. On the other hand, selective sorption of alcohol diminishes the polarity of the microenvironment of PVIm; at a lower content of alcohol in the binary mixture a higher difference between the polarity of the microenvironment of the polymer and the polarity of the solvent was observed.

PVIm behaves in both one-component and binary mixtures as a polyelectrolyte. The influence of the polymer expansion on the polarity of the microenvironment of PVIm was followed by suppressing the polyelectrolyte effect with an increase in ionic strength. For PVIm-II and the solvatochromic form of BSB in 50 vol % ethanol/H_2O and 50 vol % 1-propanol/H_2O it was found that a suppresssion of the poyelectrolyte effect by addition of KCl (0.0-0.5 M) does not lead to a shift in the maximum of the CT solvatochromic band; i.e., it does not influence the polarity of the microenvironment of the polymer. On this basis one may conclude that the polarity of the microenvironment of PVIm and also of the other polymers is not influenced by their expansion. Thus, the

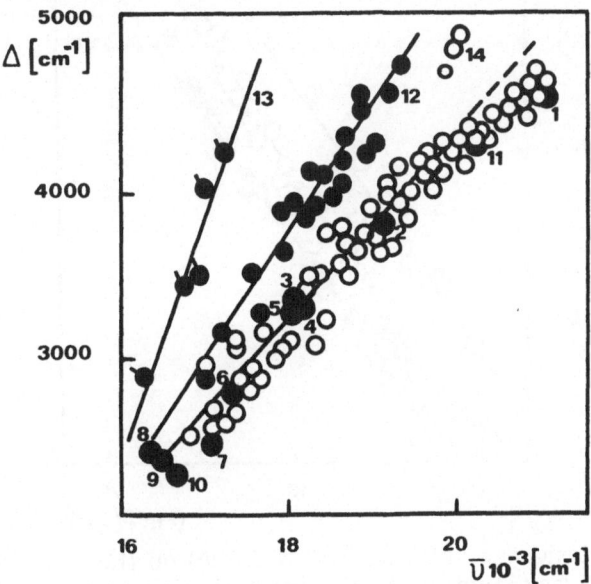

Figure 7.   Dependence of the half-width (Δ) on the wave number (ṽ) of
the maximum of the CT band of the compound SB in water
(1), methanol (2), 2-hydroxyethyl methacrylate (3), ethanol
(4), 1-propanol (5), 1-butanol (5), 2-propanol (6), acetonitrile
(7), 2-methyl-2-propanol (8), aniline (9), dimethyl sulfoxide
(10), and a series of mixed solvents (11) and for the compound
SB embedded in the chain of PHEMA (12), PS (13), and PMA
(14).

detected polarity has probably a local character and does not reflect
polymer coiling and expansion.   This is what would be expected, since
the excluded volume effect concerns the separation of chain segments of
synthetic linear polymers which are very far from each other along the
contour of the chain.   Such segments have a very low probability of
being spatially close to each other.

For all the polymers, and the model compound SB, the half-width
of th CT absorption band was measured in solvents of different
polarities.   The half-width of the CT absorption band of the compound
SB increases with increasing solvent polarity (Figure 7).   At the same
polarity of the polymer microenvironments the half-width of the
absorption band of SB embedded in the polymers decreases in the order

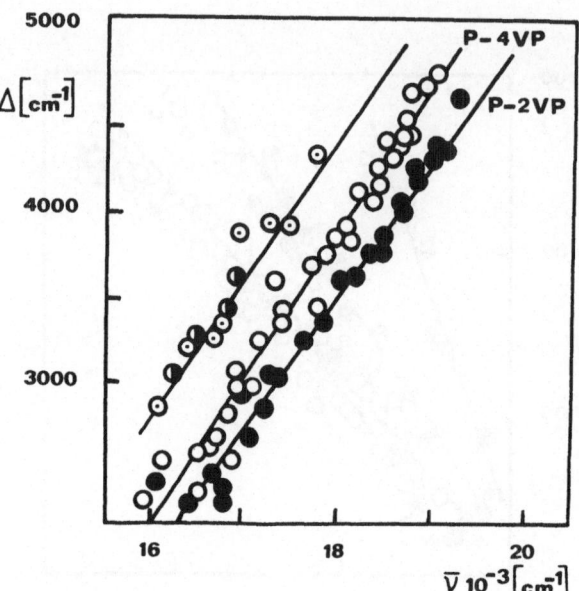

Figure 8.   Dependence of the half-width (Δ) on the wave number (ṽ) of
the maximum of the CT band of the compound SB embedded
in the chain of PMMA (⊙), PBMA (◑), P-2VP, and P-4VP in a
series of solvents.

PS, P-4VP, PMMA, PHEMA, P-2VP (Figures 7 and 8). In all the polymers
a significantly larger half-width was observed than for the free
compound SB at a given polarity. The lower the polarity of the polymer
microenvironment, the larger this difference is generally. In polymers
the orientation of segment dipoles of the chain is accomplished in many
ways which leads to a much wider dielectric dispersion.[39] A larger half-
width of the absorption band indicates that the energy interaction
resulting in the dispersion of vibrational and rotational energy states of
the molecule governing the band width and shape in the electron
spectrum span a wider range for the compound SB embedded in the
polymer than for the free SB in solution. This effect therefore indicates
greater inhomogeneities of the solvation atmosphere of the compound SB
embedded in the polymer. The more polar solvent molecules and the less
polar polymer groups both play a role in the solvation of the embedded
SB compound. The largest half-width was thus observed with the least
polar polymer (PS), probably because in this case the range dipole–dipole
and dipole-induced dipole interactions between the most polar solvent
and the polymer structural unit is most extreme. For more polar
polymers such as, e.g., PMA and PHEMA, a smaller range of interaction
may be expected.

(b)   Spiropyran-Merocyanine Reporter

In ethyl acetate/methanol, with methanol content over   2.5 vol. %,
the wavelength of the solvatochromic maximum of merocyanine bound to

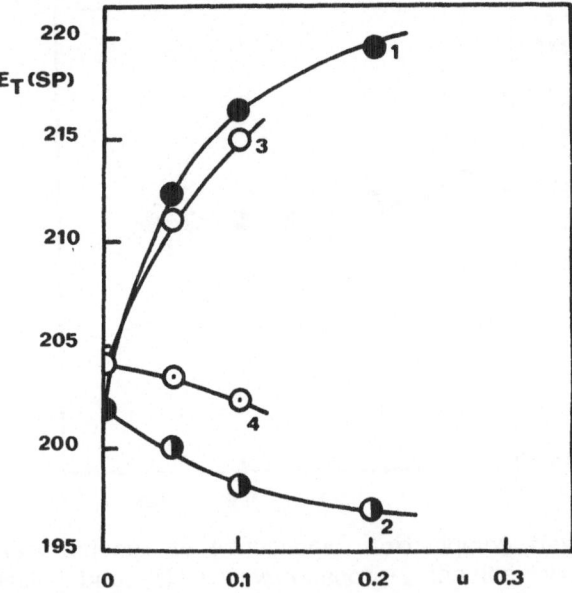

Figure 9. Solvent composition dependence of the polarity (E_T (SP)) of
the binary solvents ethyl acetate/methanol (1) and ethyl
acetate/hexane (2), and the polarity of the microenvironment
of PMMA in these binary solvents (3), (4), respectively. E_T
(SP) is the energy of the absorption band of the merocyanine (kJ/mol), u is the volume fraction of methanol or
hexane.

PMMA is shifted to longer wavelengths compared with the unbound
reporter in the same solvent. In ethyl acetate/hexane the absorption
maxima are shifted in an opposite way[36] (Figure 9). With increasing
polarity of the solvent, a considerable blue shift is observed with the
low-molecular weight analogue; this effect is smaller for the bound
reporter. In the ethyl acetate/methanol the polar solvent causes a
change in the microscopic polarity of the solvent even at low methanol
content.

Lower solvatochromism in the polymer spectrum implies that
merocyanine is solvated by ester groups of the same polymer chain and
are shielded from the bulk solvent. This observation fits in with the
foregoing conclusions on microenvironment polarity as estimated from
the shift of the absorption band of the solvatochromic reporter SB.

(c)   1-Dimethylamino-5-sulfonamidonaphthalene Reporter

The polarity of the microenvironment of poly(2-hydroxyethyl-
methacrylate) in 1-propanol/water solution, expressed by the energy of
the emission maximum of DNS fluorophore incorporated in the polymer
side chain and the empirical solvent polarity parameter is shown in
Figure 10. The polarity of the microenvironment of PHEMA is generally
lower than that of alcohol/water.

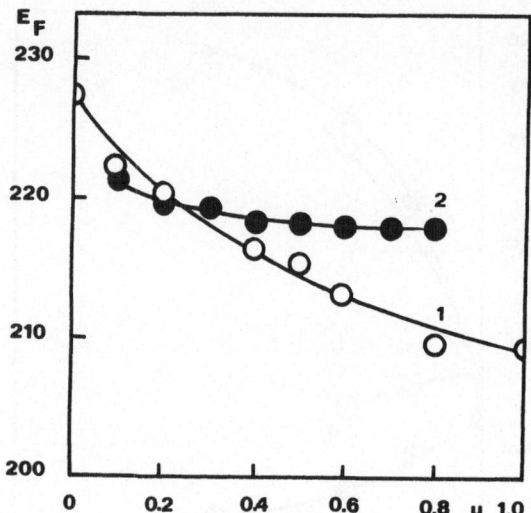

Figure 10. Solvent composition dependence of the polarity (E_F) of the
binary solvent 1-propanol/water (1) and polarity of the
microenvironment of PHEMA (2). For a definition of the
symbols E_F and u see Figures 2 and 3.

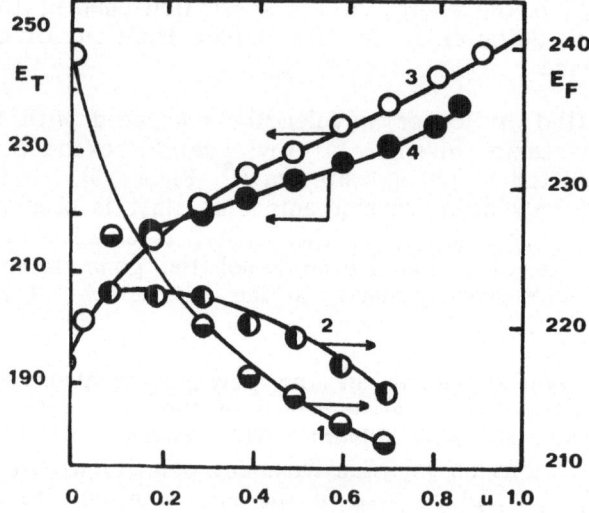

Figure 11. Solvent composition dependence of the polarity (E_T, E_F) of the
mixture dioxane/water (1,3) and the polarity of the microenvironment of PHEMA (2,4). Curves 1,2 DNS reporter;
3,4 SB reporter. For a definition of the symbols E_T (kJ/mol)
and u see Figure 3 and for E_F see Figure 2.

In those binary solvents in which selective sorption of the more polar component on the polymer chain may take place, an increased polarity of microenvironment could be observed, compared with the polarity of the binary mixture used. In all the binary solvents studied here (alcohol/, dioxane/, and acetone/water) a qualitative agreement was found between the polarity of the microenvironment of PHEMA in solution determined from the energy of the CT absorption band of pyridinium–betaine (SB) and values obtained from the energy of the emission transition of PHEMA-bound DNS fluorophore (Figure 11).

3.   The Dependence of the Polarity of the Microenvironment of Polymer on the Distance from the Polymer Backbone

The dependence of the quality of the polymer chain domain on the distance from the polymer backbone was studied using poly(N-(2-hydroxypyropyl methacrylamide)) (SCHEME VI) in water. With increasing distance of the reporter from the polymer backbone, the polarity in the surroundings of the reporter gradually increases, assuming its maximum value in a spacer possessing six methylene groups (Figure 12).   The extension of the spacer between the polymer backbone and the reporter displaces the reporter from the domain of the polymer further into bulk solution. At $n = 6$, the detected polarity is close to that of water. With increasing number of methylene groups, the hydrophobic character of the spacer increases (water is not a good solvent for the spacer) which probably leads to an intramolecular interaction between reporter and polymer or collapse of long aliphatic side chain. Such interpretation is supported by the same dependence of the reciprocal value of relaxation times[43] of the DNS fluorophore in the side chain on the distance between the DNS fluorophore and the polymer chain for the same polymer under the same conditions. Collapse of the long aliphatic side chain causes an increase in the hydrodynamic volume of the DNS fluorophore and thus an increase in the relaxation time. Similarly any interactions of long aliphatic chain possessing DNS reporter would decrease the mobility of the side chain.

B.   Cross-linked Polymers

The same method and reporter (pyridinium type – SB) were employed for estimation of the polarity of the microenvironment of the cross-lined polymers.[37]

We compared various chromatographic materials employed in the separation and immobilization of biopolymers, such as loosely cross-linked gels based on natural polysaccharide polymers (Sephadex, Sepharose, etc.), moderately cross-linked polystyrene ion exchanger (Dowex 1-X2), highly cross-lined hydrophilic macroporous heterogeneous gel with each of 3 - 4 monomeric units cross-linked (Spheron[44] P-300) and porous glass coated with a hydrophilic monolayer.[45]

SCHEME VI

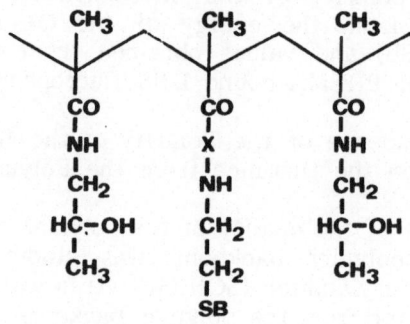

$$X \ldots NH(CH_2)_n CO$$
$$n = 0-7, 10, 11$$

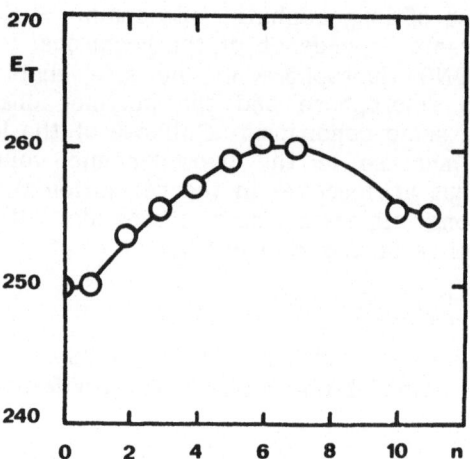

Figure 12. Dependence of the polarity (E_T) of the microenvironment of
PHPMA on the distance (n) between the reporter and the
polymer chain. For a definition of the symbol E_T(kJ/mol) see
Figure 3.

Table 3

Wavelengths of Maxima for the Solvatochromic CT Band (λ_{max}, nm) and the Respective Transition Energies (E_T, kJ/mol) for the BSB Label Bound on Various Chromatographic Materials, for a Free Label with the Substituents C_1, C_4 and C_{10} and for a Label with the Substituents C_4, C_{10} and C_{16} Sorbed on Spheron P-300.

Material	λ_{max} (nm)	E_T (kJ/mol)
Sephadex G-10	510	234.7
Sephadex G-25	502	238.4
Sephadex G-50	500	239.4
Sepharose 4B	499	239.9
Sepharose CL 4B	505	237.0
Sepharose CL 6B	509	235.2
Octyl-Sepharose CL 4B	506	236.6
Octyl-Sepharose CL 4B II	516	232.0
Cellulose	493	242.8
Spheron P-300	520	230.2
Glucose-coated Spheron P-300	533	224.6
SP-Sephadex C-25	486	246.3
CM-Sephadex C-25	490	244.3
QAE-Sephadex A-25	496	241.3
DEAE-Sephadex A-25	502	238.4
DEAE-cellulose	498	240.4
DEAE-starch	494	242.3
DEAE-Spheron P-300	528	226.7
DEAE-Glycophase-coated glass	502	238.4
CM-Glycophase-coated glass	500	239.4
Dowex 1-X2	515	232.4
C_1SB in buffer	467	256.3
C_4SB in buffer	469	255.2
C_{10}SB in buffer	467	256.3
C_4SB after sorption on Spheron P-300	524	228.4
C_{10}SB after sorption on Spheron P-300	534	224.2
C_{16}SB after sorption on Spheron P-300	534	224.2

[a]All measurements were performed in 0.1 M borate buffer (pH 9.15) by recording transmission or reflection spectra.

The wavelengths of the maxima of solvatochromic CT band after binding on the sorbents are given in Table III, which also presents values measured with an unbound label in an aqueous solution containing only buffer. These value (E_T 255.2 and 256.3 kJ/mol) characterize the polarity of the pure solvent. If the maxima of the solvatochromic band of BSB bound on chromatographic materials suspended in this solvent are shifted towards longer wavelengths (lower energies), they indicate a limited solvation of the chromphore near the chromatographic material and a lower local polarity.

In pure buffered aqueous solutions without sorbent, three compounds were measured, containing a methyl(C_1SB), an n-butyl(C_4SB) and an n-decyl(C_{10}SB) substituent on the nitrogen atom of the chromophore. All of these compounds exhibited approximately the same maxima for the solvatochromic CT band. Hence it can be concluded that the size of the substituent per se, even for non-polar aliphatic substituents up to the size of n-decyl, does not perceptibly affect the solvation of the dipole of the chromophore. Shift of the solvatochromic band after binding on chromatographic materials is due only to the different polarities of the medium near the chromatographic material.

It can be seen from Table III that in all materials under investigation the solvation of the bound label is hindered and affected to some extent; hence, the polarity detected in a region near the chromatographic material is lower than the polarity of bulk solution.

Comparison among the chromatographic materials under study (Table III) reveals that the most hydrophilic material is Sepharose 4B (E_T 243.3 kJ/mol), with cellulose (E_T 242.8 kJ/mol) next to it with respect to polarity. An even lower polarity was measured for all types of Sephadex (G-10, G-25 and G-50), which differ in their exclusion limits. Certain differences were observed depending on the degree of cross-linking. The polarity measured for the most highly cross-linked Sephadex G-10 was the lowest (E_T 234.7 kJ/mol). Sephadex G-50, with a higher exlusion limit and a lower degree of crosslinking, was even more polar. Spheron P-300 is the least polar material studied. It is a macroporous rigid copolymer[44] of 2-hydroxyethyl methacrylate and ethylene dimethacrylate. The high degree of cross-linking (20-30%) and the aliphatic character of the basic polymeric backbone are the cause of the lowest measured E_T (226.7 kJ/mol). The decrease in polarity in the vicinity of Spheron P-300 compared with water is larger (20.9 kJ/mol) than that observed with linar poly(2-hydroxyethyl methacrylate), where the difference measured against the bulk solution in most solvents was only 4-8 kJ/mol.

The method of comparison of the polarity of chromatographic materials described above differs from methods described earlier (e.g., ref. 9). The measurement of the local polarity at the surface of the chromatographic materials described here is best suited for more polar materials having a microhomogeneous structure in an aqueous medium

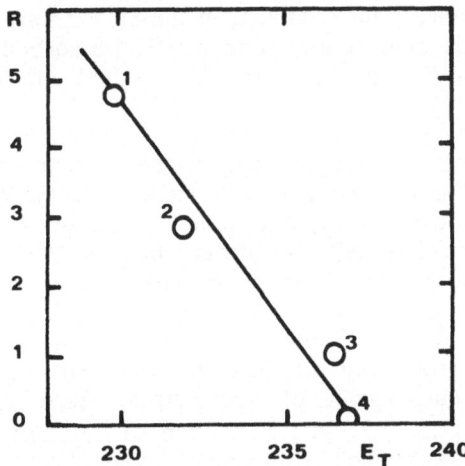

Figure 13.   Dependence of the retention (R) (see text for definition) on
             the polarity E_T (kJ/mol) of the chromatographic material.
             Spheron P-300 (1), octyl-Sepharose CL 4B II (2), octyl-
             Sepharose CL 4B (3), Sepharose CL 4B (4). Chromatographic
             analysis was performed in a 0.5 M NaCl - 0.05 phosphate
             buffer (pH 5.5) at 20°C. For a definition of the symbol E_T
             (kJ/mol) see Figure 3.

and in polar solvents and mixtures. The sorbent must possess reactive
groups, such as hydroxyl, amine or thio groups. This measurement gives
direct information about how the polarity difference between sorbent
and solvent affects the solvation power of the solvent near the sorbent.

     One should bear in mind that in general the local polarity
calculated from the spectrum of the label need not be identical with
that operative in reactions of other molecules on the chromtographic
support. The supports are inhomogeneous, and the measured CT band is
a superposition of the absorption bands of the label bound in various
regions. The determined maximum of the absorption band and the local
polarity calculated there from are a mean values characterizing the
distribution of the label in the support. This distribution is given by the
conditions of the binding reaction, i.e., by the magnitude of physical
interactions of the label with the particular regions and with the
reaction solvent, by the rate of diffusion to reactive groups and by the
kinetics of the binding reaction. An example is the low E_T values (Table
III) calculated from the spectra of the $C_{10}SB$ and $C_{16}SB$ labels adsorbed
on Spheron (Spheron was placed in a solution of the label in methanol,
then thoroughly washed with water and measured in the usual way in
borax buffer, pH 9.15). The label, which in this instance is bound on the
matrix by hydrophobic forces through n–decyl and n–cetyl hydrocarbon
chains, is obviously preferentially localized in less polar domains of the

heterogeneous carrier. In contrast, a label without the hydrophobic chain (C_1SB) did not exhibit any tendency to be sorbed preferentially in non-polar regions, and it could be readily washed out from the supports with water.

The lower local polarity of the microenvironment of the chromatographic material compared with the surrounding bulk medium may be reflected in the retention of less polar compounds, and also in the influence on some chemical equilibria and reactions, through local effective activities which depend on the local dielectric constant. Figure 13 shows the correlation between the local polarity, calculated from the spectra of the bound label, with the retention, R, for hydrophobic chromatography of lysozyme on Spheron and on octyl-Sepharose with various degrees of substitution. Here R is defined as $(V_{el} - V_0)/V_0$, where V_{el} is the elution volume of lysozyme, and V_0 is the value of V_{el} when hydrophobic interaction are absent. The relatively good agreement obtained for these two different types of materials, aerogel and xerogel, shows that the spectra of the bound label predominantly reflect the polarity in the surface regions of the support, which in turn determined the interaction between the latter and macromolecules.

CONCLUSION

The properties of the microenvironment of soluble and cross-linked synthetic polymers were studied using solvatochromic reporters bound to polymers. The polarity of the domain of polymer chains was estimated in one-component and binary solvents and compared with the polarity of solvents used. The polarity was expressed semiempirically by the absorption or emission band energy of a solvatochromic compound. The polarity of the microenvironment of soluble polymers and also the polarity in the vicinity of matrix of cross-linked polymers suspended in aqueous buffer was almost in all cases lower than that of the solvent.

The concept of reduced solvation of the solvatochromic chromophore bound to polymer chain with solvent molecules and preferential intramolecular solvation of the solvatochromic molecule with side subtituent of the polymer backbone was supported by means of time-resolved absorption spectrometry of the merocyanines bonded to poly/methyl methacrylate/.[39]

The polarity of the microenvironment of synthetic polymers in solution and cross-linked polymers was studied in order to characterize semiquantitatively, from one point of view, the influence of the polymer on the reactions of bound functional groups and on interactions of the macromolecules with low- or high-molecular weight compounds.

A linear correlation has been found between the polarity-dependent rate of the photochemical trans-cis isomerization of stilbene residues in the side chain of poly(2-hydroxyethyl methacrylate) and the polarity of the microenvironment of this polymer in solution.[30]

Hydrophobic sorption of compounds with a long aliphatic chain on poly(4[5]-vinylimidazole) in binary mixtures of ethanol/water with water content higher than 30 vol. % has been revealed.[34] The polarity of the microenvironment of this polymer in these mixtures is lower than that of solvents used.

Horie et al.[46] applied two fluorochromes (DNS and 6-p-toluidino-2-naphthalenesulfonate) to estimate the hydrophobicity of the microenvironment of cross-linked polystyrene containing quaternary ammonium groups. They found a correlation between the wavelength of the emission maximum of fluorophores and the rate of decarboxylation of 6-nitrobenzisoxazole-3-carboxylate anion catalyzed by these gels. The hydrophobic microenvironment due to cross-linked matrices was suggested as a cause for this catalytic activity.[47]

The applicability of a semiempirical polarity scale, based on comparison with a solvent of the same polarity, to processes occuring in the region of the polymer chain microenvironment represents only a first approximation. Since the polarity of the microenvironment is determined using a given solvatosensitive process, the applicability of the result to any other process depends on similarities in the various solvation interactions. Nevertheless, characterization of the polymer microenvironment by the method used in this study has been found to be suitable for a semiquantitative interpretation of the reaction rate of polymer sustituents and of the rate of reactions catalyzed by polymer catalysts.

ACKNOWLEDGEMENT

We are indebted to Drs. E. Brynda and Z. Tuzar for their collaboration in this investigation.

REFERENCES

1. a. Prague Institute of Chemical Technology.   b. Institute of
   Macromolecular Chemisty. c. Institute of Organic Chemistry and
   Biochemistry.
2. Klotz, I. M. Science 1958, 128, 815.
3. Klotz, I. M., Ayers, J. J. Am. Chem. Soc. 1957, 79, 4078.
4. Frey, P. A., Kokesh, F. C., Westheimer, F. H. J. Am. Chem. Soc.
   1971.
5. Klotz, I. M., Stryker, V. H. J. Am. Chem. Soc. 1960, 82, 5169.
6. Knowles, J. R. J. Theor. Biol. 1965, 9, 213.
7. Edelman, G.M., McClure, W. O. Acc. Chem. Res. 1968, 1, 65.
   Chen, R. F., Edelhoch, H., Steiner, R. F. In "Physical Principles
   and Techniques of Protein Chemistry," Leach, S. J., Ed., Academic
   Press: New York, N. Y., 1969, Part A, p. 171. Brandt, L. Withold,
   B. Methods Enzymol. 1967, 11, 776. Brand, L., Gohlke, T. R. Ann.
   Rev. Biochem. 1972, 41, 843.
8. Herskovits, T. T. Methode Enzymol. 1967, 11, 748. Krouman, M. J.,
   Robbins, F. "Biological Macromolecules," Fasman, G. D.,
   Timasheff,   S. N., Ed., New York, N. Y., 1967, Vol. 3.
9. O'Sullivan, W. J., Cohn, M. J. Biol. Chem. 1966, 241, 3104 and
   3116.
10. Brenhard, S. A., Lee, B. F., Tashjian, Z. H. J. Mol. Biol. 1966, 18,
    405.
11. Foester, R. J., Coahran, D. R. Fed. Proc. 1963, 22, 245.
12. Hille, M. B., Koshland, D. E. Jr. J. Am. Chem. Soc. 1967, 89, 5945.
13. Horton, H. R., Koshland, D. E. Jr. Methods Enzymol. 1967, 11, 856.
    Kirtley, M. E., Koshland, D. E. Jr. Biochem. Biophys. Res.
    Commun. 1966, 23, 810. Koshland, D. E. Jr., Karkhanis, Y. D.,
    Latharn, H. G. J. Am. Chem. Soc. 1964, 84, 1448.
14. Kallos, J., Avatis, K. Biochemistry 1966, 5, 1979.
15. Witz, G., Van Duuren, B. L., J. Phys. Chem. 1973, 77, 648.
16. Orwoll, R. D., Vold, R. L. J. Am. Chem. Soc. 1971, 93, 5335.
17. Liu, K. J. Trans. N. Y. Acad. Sci. 1971, 33, 333.
18. Štrop, P., Mikeš, F.,Kálal, J.   Paper presented at the   13th
    Prague Microsymposium on Macromolecules, Aug. 1973 J. Polym.
    Sci., Part C, Polym. Symp. 1974, 47, 255.
19. Rakhrianskaia, A. A., Kirsch, Yu. E., Kabanov, V. A. Dokl. Akad.
    Nauk. SSSR 1973, 212, 889.
20. Oster, G. J. Polymer Sci. 1952, 9, 553.
21. Kirsh, Yu. E., Komarova, O. P., Lukovin, G. M. Europ. Polym. J.
    1973, 9, 1405.
22. Vandewyer, P. H., Smets, G. J. Polym. Sci. 1970, Part A-1, 8, 2361.
23. Kosower, E. M. "An Introduction to Physical Chemistry," Wiley:
    New York, N. Y., 1968.
24. Štrop, P., Mikeš, F., Kálal, J. J. Phys. Chem. 1976, 80, 694 and
    702.

25. Dimroth, D., Reichard, C., Siepman, T., Bohlman, F. Annales 1963, 661, 1. Reichardt, Ch., Dimroth, K. Fortschr. Chem. Forsch. 1968, 11, 1.
26. Fainberg, A. H., Winstein, S. J. Am. Chem. Soc. 1956, 78, 2770.
27. McClure, W. O., Edelman, G. M. Biochemistry 1966, 5, 1908.
28. Brandts, J. F., Kaplan, L. J. Biochemistry 1973, 12, 2011.
29. Berson, J. A., Hamlet, Z., Mueller, W. A. J. Am. Chem. Soc. 1962, 84, 297.
30. Mikeš, F., Štrop, P., Kálal, J. Chem. Ind. 1973, 1164.
31. Bertelson, R. C. In. "Photochromism," Brown, G. H., Ed., Wiley: New York, N. Y., 1971, p. 117.
32. Azzi, A. Q. Rev. Biophys. 1975, 8, 237.
33. Strauss, U. P., Vesnaver, G. J. Phys. Chem. 1975, 95, 1558.
34. Mikeš, F., Štrop, P., Tuzar, Z., Labský, J., Kálal, J. Macromolecules 1981, 14, 175.
35. Mikeš, F., Labský, J. will be published.
36. Labský, J.; Koropecký, J., Nešpurek, S., Kálal, J. Eur. Polym. J. 1981, 17, 309. Labský, J., Mikeš, F., Kálal, J. Polym. Photochem. in press.
37. Brynda, E., Štrop, P., Mikeš, F., Kálal, J. J. Chromatogr. 1980, 196, 39.
38. Suppan, P. J. Chem. Soc. 1968, A, 3125.
39. Morawetz, H. Adv. Protein Chem. 1974, 26, 243. Irie, M., Menju, A., Hayashi, K. Macromolecules 1979, 12, 1176.
40. Tuzar, Z., Bohdanecký, M. Collect. Czech. Chem. Commun. 1969, 34, 289.
41. Živný, A., Pouchlý, J., Šolc, K. Collect. Czech. Chem. Commun. 1967, 32, 2753.
42. Tuzar, Z. Kratochvíl, P. Collect. Czech. Chem. Commun. 1967, 32, 3358.
43. Mikeš, F., Labský, J. will be published.
44. Mikeš, O., Štrop, P., Coupek, J. J. Chromatogr. 1978, 153, 23.
45. Chang, H., Gooding, K. M., Regnier, F. E. J. Chromatogr. 1976, 125, 103.
46. Horie, K., Mita, I., Kawabata, J., Nakahama, S., Hirao, A., Yamazaki, N. Polym. J. 1980, 12, 319.
47. Yamazaki, N., Nakahama, S., Hirao, A., Kawabata, J. Polym. J. 1980, 12, 231.

PART V

ORDERED POLYMER—LIGAND COMPLEXES

A. SYSTEMS WITH BIOLOGICAL COMPONENTS

INTERACTION BETWEEN DNA AND DODECYLPYRIDINIUM CATION

Keishiro Shirahama, Tomoya Masaki, and Koji Takashima
Department of Chemistry
Faculty of Science and Engineering
Saga University
Saga 840, Japan

ABSTRACT

Binding of dodecylpyridinium cation to calf thymus DNA was measured by using the ligand-sensitive electrode which utilizes propionyl α-cyclodextrin as a neutral carrier through the electrode membrane. Binding isotherms have a cooperative nature in that binding begins suddenly and is completed within a very narrow range of equilibrium concentration, followed by further massive binding. The cooperative feature was analyzed in terms of the one dimensional Ising model which extracts K, a binding constant of the ligand transferred from the aqueous bulk phase to an isolated binding site on DNA, and u, a cooperativity parameter, or an equilibrium constant for the aggregation of bound ligands. K decreases but u increases on addition of sodium chloride. There is an inversion temperature around 23°C where K has its minimum and u attains a maximum. From the above observation, it is imagined that dodecylpyridinium cation is bound to phosphate groups on DNA as a primary binding site with interalkylchain interaction playing an important role for the binding. The hydrophobicized DNA furnishes further binding sites where the ligands are bound with their alkylchains in contact with each other, while cationic groups of the ligand are exposed to the aqueous phase helping DNA redisperse. The ligand-bound DNA shows a melting temperature much lower than that of a ligand-free DNA.

INTRODUCTION

Surfactants having both strong hydrophilic and hydrophobic groups

299

in a molecule are not only surface-active at two dimensional interfaces,[1] but also bind to linear polymers with and without electric charges. The interaction between surfactants and polymers has been investigated by various methods,[2-4] among which a binding isotherm method directly reflects the interaction and allows thermodynamic interpretation of the phenomena with the aid of statistical mechanics. Binding isotherms are usually characterized by a sudden rise and saturation of binding that occurs within a very narrow range of equilibrium concentration. This feature was interpreted in terms of cooperative nature due to inter-action between alkylchains of bound surfactants.[5] Sodium alkyl sulfates, for example, are bound to polymers with appropriate hydrophobicity such as poly(vinyl pyrrolidone) (PVP),[6-8] and poly(ethylene oxide) (PEO)[5,9,10] providing the anionic surfactants with hydrophobic binding sites. The binding isotherm for sodium dodecyl sulfate (SDS)-PEO system is indeed steep, but it is difficult to explain the steepness by a model in which all the bound surfactants are bound into a single aggregate (e.g., 145 SDS molecules on a PEO molecule with molecular weight 1.5×10^4).[5] Instead, bound surfactant molecules are considered to be divided into subsystems, or clusters containing some 20 - 25 SDS molecules, in accordance with the idea of the thermodynamics of small systems.[11] It is noteworthy that this cluster size is estimated to be smaller than ordinary micellar size, say 90, in the same solvent condition (in 0.1 M NaCl at 30°C). Nuclear and electron spin resonance spectroscopies also suggest smaller cluster sizes.[12-14] Various physical measurements such as electric bire-frigence,[15,16] and free-boundary electrophoresis support a picture that a "string" of macromolecule binds "beads" of surfactant clusters, i.e., a necklace model.[17] Surfactant-polymer complexes solubilize sparingly soluble organic substances just as ordinary micelles do, indicating some micelle-like region in surfactant-polymer complexes.[18-20]

Charged polymers also bind surfactants with the opposite charges. Here again, features of very cooperative interaction are recognized in binding isotherms.[21-23] Sodium decyl sulfate (SDeS) is cooperatively bound on cationic polypeptides such as poly(L-lysine) and poly(L-ornithine) accompanied by conformational changes (coil to β-structure, and α-helix).[21,22] This cooperativity is not solely due to the cooperative conformation changes of the host polypeptides themselves. Binding of SDeS to a cationic copolymer of diallyldimethylamine and sulfur dioxide (PAS)[24] which does not show any cooperative conformation change, is highly cooperative as judged from the binding isotherm.[23] A decyl-pyridinium bromide-poly(vinyl sulfate) system shows a cooperative binding isotherm.[25] Anionic poly(L-glutamate) binds cooperatively decylammonium chloride.[22] It is deduced from these results that signs of charges of surfactant and macroion are not a crucial factor. How electrostatic interaction is contributing to surfactant-polymer binding is seen by comparing the difference of binding isotherms of two systems. Transitional concentration where binding suddenly begins for the SDeS-PAS system appears at $C_f = 0.16$ mM in 100 mM NaCl,[23] while that for the SDeS-PEO system is estimated to be about 13 mM in the same medium.[9]   The 80-fold difference is roughly assigned to twice the electrostatic contribution to binding energy.   On binding SDeS, there is a discharging process in the former system, but a charging process in the latter is unfavorable for binding.

Protein-surfactant interactions have given interesting examples for study,[26-28] but they are too complicated because of the variety of primary and higher order structures of proteins.   Briefly, the binding of surfactants at low concentration to sites with electric charges opposite to the surfactants' is sometimes called "statistical binding."   The surfactant-bound protein is now so hydrophobic that there sometimes appears a precipitate which is, however, redispersed on addition of more surfactant.   At this stage, surfactants are bound with their ionic groups exposed to the aqueous phase, in massive aggregates, somewhat akin to ordinary micelles.   The original backbone of the protein is distorted, as judged from circular dichroic observation, although some local order may increase.   An assessment of the validity of sodium dodecyl sulfate-poly-(acrylamide) gel electrophoresis was carried out based on the accumulated understanding of protein-surfactant interactions.[29,30]

Deoxyribonucleic acid (DNA), on the other hand, may be thought of as a simpler host polymer in terms of primary and higher order structures.[31]   So it is interesting to compare its binding behavior with those of another class of polymers, since the interactions of DNA with various kinds of ligands are biologically and medically important.[31,32]

In a study of interaction between a small molecule and a macromolecule, equilibrium dialysis is a fundamentally important method and it has been applied to many studies.   But in the case of surfactant ligands, it has a drawback: as much as a week may be required to attain an equilibrium.[5,33]   Gel chromatography provides another technique for obtaining binding isotherm, but has been used mostly to determine an amount of binding saturation.[34]   It is supposed that absorption of surfactants into the gel phase may obstruct reaching equilibrium at lower surfactant concentrations.

An electrode selectively responsive to an ionic ligand in equilibrium with ligands bound on a macromolecule can be utilized to construct binding isotherms.[21-23,35]   Dodecylpyridinium halide was chosen in this work as an amphiphilic ligand because an excellent electrode sensitive to it was found.   This amphiphile is regarded as a model compound of many classes of biomedical molecules since it combines hydrophobic, aromatic, and polar (cationic) groups.   The limited number of papers published on DNA-surfactant interaction,[36] further prompted this work.

Interaction between calf thymus DNA and dodecylpyridinium halide (DoPX) was studied through binding isotherms, varying concentrations of added sodium chloride, and temperature.   The results are analyzed in term of a one-dimensional nearest neighbor interaction model.[37]   The thermodynamic parameters obtained were used to develop a physical picture of the DNA-surfactant complex in aqueous solution.

EXPERIMENTAL

Materials. Dodecylpyridinium bromide was synthesized by treating fractionally distilled 1-dodecane bromide in dry pyridine for 12 hr. The crude surfactant was recrystallized twice from acetone followed by decolorization with active charcoal in methanol solution.[38] The resulting white crystal is a monohydrate of DoPBr. The critical micelle concentration (CMC) in aqueous solution as determined by electric conductivity method is 17.4 mM at 30°C in agreement with literature.[39] Calf thymus DNA (sodium salt, SIGMA) was used as received. Residual (nucleotide) concentration was determined by a colloid titration using poly(potassium vinyl sulfate) as a titrant and Toluidine Blue as an indicator.[40] Propionyl-α-cyclodextrin (prop-α-CD) used as a neutral carrier was prepared by esterification of α-cyclodextrin (Tokyo Kasei Co.) with propionic anhydride in dry pyridine at room temperature for 12 hr. The reaction mixture was poured onto ice to obtain a gummy product which was then dissolved in acetone and precipitated in cold water. The dissolution-precipitation was repeated three times. The hydrophobicized α-CD is a white powder.

Surfactant electrode. To a slurry of poly(vinyl chloride) (0.6 g, PVC) and dioctylphthalate (2.37 g, DOP), prop-α-CD (0.03 g) dissolved in tetrahydrofuran (8 ml, THF) was added. The whole mixture was well stirred and heated at 60°C to obtain a clear viscous solution, which was then cast on a flat glass plate. The solvent was gradually evaporated in a desiccator. The resulting gel membrane was cut, and placed on one end of a poly(methacrylate) resin tube (11 cm long and 1 cm diameter). The electrode membrane was further "annealed" at 40°C under vacuum for 12 hr, and was then ready for use. This electrode forms a concentration cell:

$$\text{Calomel} \left| \begin{array}{c} \text{sat} \\ \text{KCl} \end{array} \right| \begin{array}{c} \text{KCl} \\ \text{agar} \end{array} \left| C_S \right| M \left| C_O \right| \begin{array}{c} \text{KCl} \\ \text{agar} \end{array} \left| \begin{array}{c} \text{sat} \\ \text{KCl} \end{array} \right| \text{Calomel}$$

where M is the ion-selective membrane, C_O and C_S are the surfactant concentrations of inner (standard) and outer (to be examined) solutions, respectively. The electromotive force of the above cell was measured on a digital voltmeter with high input impedance (Takeda Riken Co., TR6843). Potentiometric titration was carried out in a cocylindrical cell thermostated by circulating water at a constant temperature.

Melting temperature of DNA. Conformational change of DNA was monitored by optical absorption at 260 nm as a function of temperature on a spectrophotometer (Hitachi, model 139) equipped with a thermostat (Lauda, type K2RD).

RESULTS AND DISCUSSION

The potentiogram for a system of DoPBr in 20 mM NaBr at 25.0°C is shown in Figure 1. The electrode is responsive to DoP cation with the Nernstian slope = 59 mV/(ten-fold change in concentration), and its sensitivity spans a concentration range from 2 µM up to above the CMC (5.1 mM) (shown by open circles) where a break is seen. However, this linearity suddenly deviates when DNA (C_p = 0.870 mM nucleotide) is present in the solution (filled circles). This kind of deviation was also found in various sodium decyl sulfate–cationic polymer systems[21-23] and interpreted as evidence of surfactant binding to polymers. It can be easily understood that ΔC is an amount of bound surfactant and C_f is an equilibrium ligand concentration. The binding isotherm is constructed by plotting X vs. log (C_f) and shown in Figure 2, where X = $\Delta C/C_p$, a degree of binding. It is clear that binding occurs in two steps. The first binding begins suddenly and soon attains X = 1, where the isotherm shows a slightly flattening tendency, followed by a steep increase of the second binding. The fact that the first binding seems saturated at X = 1 suggests that anionic phosphate groups in DNA are the primary binding sites, as is well known for many kind of cationic ligands.[31,32,36] Interaction between hydrocarbon tails of adjacently bound surfactants may add another free energy gain. It was sometimes found that the solution became turbid around X = 0.8 - 1.2. This turbidity may be responsible for slightly reduced binding before the steep second binding begins as shown later (cf. isotherm at 45°C in Figure 4, for example).

Highly hydrophobicized DNA in aqueous solution might not be stable enough at higher temperature. In the second binding process it is deduced that the DNA which has been hydrophobicized as a result of the first binding offers other binding sites for some massive aggregation of surfactants akin to ordinary micelle formation. These aspects can be seen in different modes in the effect of added electrolyte as seen in Figure 3, where binding isotherms in various NaCl concentrations are shown. Binding isotherms shift to the right, i.e., to a higher C_f, at the first step binding, while there is a tendency that the order is reversed in the second binding, on increasing the added NaCl concentration. The added electrolyte shields the electrostatic potential which is one component of binding forces in the first step. Thus the added electrolyte lowers surfactant binding affinity, leading to the shift of the binding isotherm to a higher equilibrium concentration. In the second binding step, in contrast, surfactant binding accumulates more positive electrostatic charge which is now repulsive against further bindings. Shielding of electrostatic potential by adding electrolyte at this stage promotes the binding of cationic ligand, causing the binding isotherm to shift to the lower equilibrium concentration. The latter mechanism is

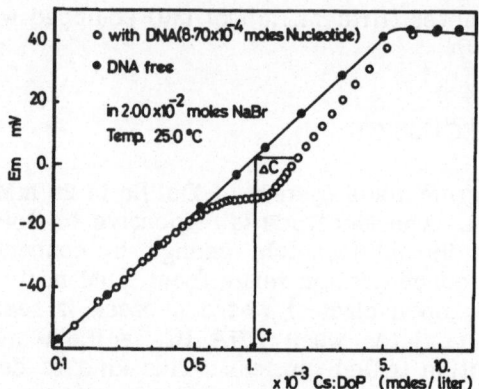

Figure 1. E_m vs. $\log(C_S)$ in 20 mM NaBr at 25.0°C. C_S; dode-
            cylpyridinium bromide (DoPBr) concentration, ΔC; amount of
            bound DoP, and C_f; equilibrium concentration.

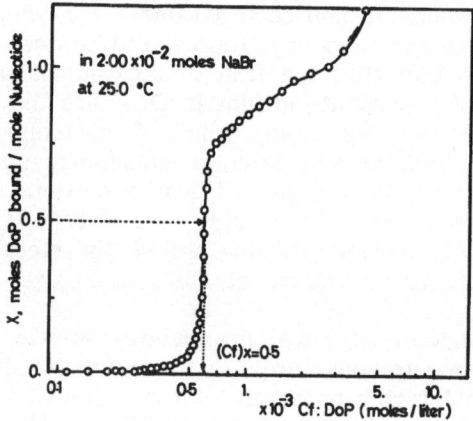

Figure 2.   Binding isotherm of DoPBr–DNA system in 20 mM NaBr at
            25.0°C. $X = \Delta C/C_p$, C_p; DNA concentration in mM
            nucleotide.

quite similar to ordinary micelle formation which is enhanced by added
electrolyte.[41]   The biphasic binding mode is contrasted to the multi-
phasic binding observed in many surfactant-protein systems,[26,27] although there are similarities in that ionic groups on macroions are the
primary binding sites, and bound ligands facilitate further massive
binding.

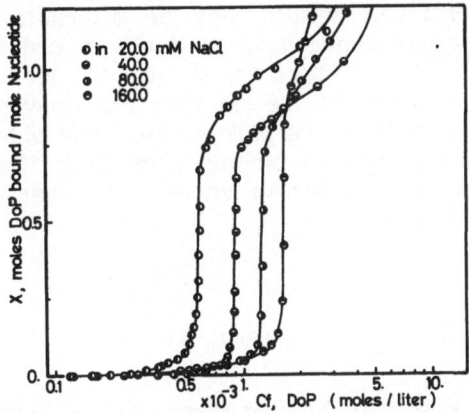

Figure 3.　Effect of added NaCl on binding isotherms at 25.0°C.

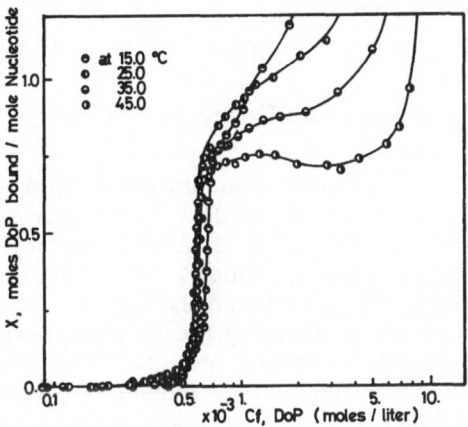

Figure 4.　Effect of temperature on binding isotherms at 20 mM NaCl.

The effect of temperature is shown in Figure 4, where binding isotherms at 4 temperatures are plotted. At a glance, it is clear that there is very little temperature dependence even with an inversion temperature. Insensitivity to temperature is thermodynamically related to a small value of binding enthalpy. This might have to do with a delicate balance or cancellation of enthalpy changes related to electrostatic, (de)hydration, conformation, and hydrophobic interactions.

The above-mentioned results may be alternatively described in terms of the one dimensional Ising model[37] in order to extract the feature of the interactions. There are several theories that treat DNA-ligand interactions, but one of the simplest was conveniently chosen to express cooperative nature of the phenomena. The phosphate groups on DNA are viewed as an array of binding sites along the poly(nucleotide) helices, each site being occupied or vacant. The partition function, z, for such system is written as,

$$z = (1,1) \begin{pmatrix} 1 & 1 \\ s/u & s \end{pmatrix}^{n} \begin{pmatrix} 1 \\ 0 \end{pmatrix} \tag{1}$$

where $s = KuC_f$ is a sort of binding "pressure," K being an equilibrium constant of the ligand for a transfer process from the aqueous bulk phase to an isolated binding site (one without neighboring ligand) on DNA, u an equilibrium constant for bound ligands to aggregate, or a cooperativity parameter, and n number of binding sites. Equation (1) yields many useful relations among which are:

$$k = 1/(uC_f(0.5)) \tag{2}$$

and $$[d\ln(X)/d\ln(C_f)]_{X=0.5} = \sqrt{u}/4 \tag{3}$$

where $C_f(0.5)$ is an equilibrium concentration giving X = 0.5. The parameters, K and u extracted from data in Figures 3 and 4 are listed in Tables 1 and 2. As for the effect of added electrolyte, K decreases, because of the shielding effect, while u is increased. The apparent increase in cooperativity is interpreted as follows: Binding of organic cation to DNA is a process discharging the anionic phosphate groups, and there remain fewer anionic phosphate groups to be bound. This means that the electrostatic potential, or the binding affinity is reduced with the advancement of binding. This is a kind of negative cooperativity

TABLE 1.   Effect of Added NaCl on Binding Parameters at 25.0°C.

C_a/mM	u	K/M^{-1}
20	2500	0.67
20 (NaBr)	2400	0.70
40	4100	0.27
80	6700	0.12
160	12300	0.05

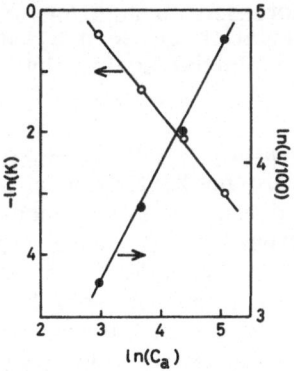

Figure 5. $-\ln(K)$ and $\ln(u/100)$ vs. $\ln(C_a)$. C_a; added NaCl concentration in mM, and K in M^{-1}.

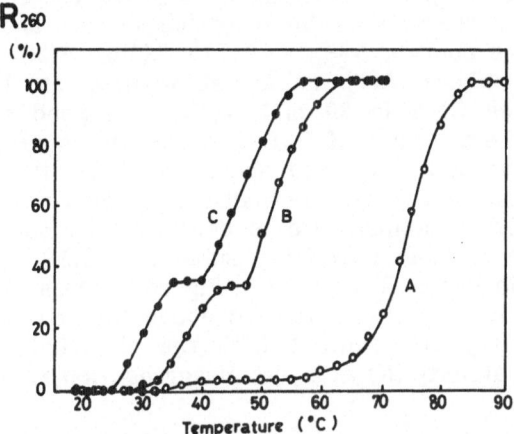

Figure 6.   "Melting" curves for DNA in 20 mM NaCl. $C_p = 0.870$ mM nucleotide, DoPCl concentrations; A, 0, B, 0.9, and C, 1.0 mM, respectively.

TABLE 2.   Effect of Temperature on Binding Parameters in 20 mM NaCl.

Temp./^{o}C	u	K/M^{-1}
15.0	630	2.39
25.0	2500	0.67
30.0	1300	1.30
45.0	820	2.02

which, in this case, is decreased on addition of extraneous electrolyte. The partly compensating relation between u and K is seen in Figure 5, where ln(u) and -ln(K) are plotted against ln(C_a), C_a being the added electrolyte concentration.

The effect of temperature is a little more complex. Inspection of Table 2 reveals extrema around 23°C, above which K increases but u decreases. Contrarily K decreases and u increases below the critical temperature. Similar break points were reported for systems of tetradecyltrimethylammonium bromide-, and hexadecylpyridinium bromide-DNA.[36] These observations are very hard to understand, but a tentative explanation will be given. Below the critical temperature, binding to an isolated site is an exothermic process driven by discharging the phosphate groups, and hydrophobic interaction between alkylchains makes the aggregation process endothermic. These are the same effects as observed in the temperature dependence below the critical temperature. In Figure 6, normalized "melting" curves are shown by plotting relative absorbance change, R_{260} = [A(T) - A(20)]/[A(90 - A(20)] against temperature, T, where A(T) is an absorbance at T°C. Melting temperature of the DNA in 20 mM NaCl is in good agreement with literature.[42] In the presence of DoPCl in the same medium, the melting temperature is very much lowered and some minor transition emerges, suggesting a conformation change induced by bound surfactant. For example, the melting temperature is decreased by some 25°C in the presence of 1 mM of DoPCl (which assures X = 0.5 in reference to a binding isotherm in Figure 3) as seen in Figure 6. As a result of thermal denaturation enhanced by the surfactant above the critical temperature, exposed base moiety can contact with the alkylchains of surfactant gaining additional affinity (K) at a loss of aggregation affinity among the alkylchains (u).

In conclusion, the cationic surfactant is cooperatively bound on DNA, and inter-alkylchain interaction plays a substantially important role. As a result of marked cooperativity, bound surfactants are considered to aggregate into groups. The average aggregate size, $\bar{l}$, at X = 0.5, is expressed in terms of the Ising model[21,37] as,

$$\bar{l} = \sqrt{u} + 1 \qquad (4)$$

i.e., some 30 to 100 DoP are clustered on the average at X = 0.5, in reference to Tables 1 and 2.

Binding sites are localized on charged groups in the case of macroions, but delocalized in uncharged polymers so that surfactant clusters move along the macromolecular chains. In any case, bound aggregates are viewed as hydrophobic domains dispersed in aqueous media by host macromolecules. Cooperativity is a common feature in surfactant binding to any class of macromolecules such as nucleic acids, proteins, polypeptides, and synthetic polymers with and without electric charges, and is deeply rooted in the nature of surfactant action. It is by aggregation, whether by ordinary micellization, cluster formation on macromolecule, or hemi-micellization on two-dimensional interface, that surfactants may effectively escape from their uncomfortable aqueous environment. This is the origin of the surface activity exhibited by surfactants.

REFERENCES

1. J. F. Scamehorn, R. S. Schechter, and W. A. Wade, J. Colloid Interface Sci., 85, 463 (1982).
2. I. D. Robb, in "Anionic Surfactants," E. H. Lucassen-Reynders, Ed., Marcel Dekker Inc., New York, 1981, p. 109.
3. H. Lange, Chemiker-Zeitung, 96, 192 (1972).
4. M. M. Breuer and I. D. Robb, Chemistry and Industry, 1, 530 (1974).
5. K. Shirahama, Colloid Polymer Sci., 252, 978 (1974).
6. H. Arai, M. Murata, and K. Shinoda, J. Colloid Interface Sci., 37, 223 (1972).
7. M. L. Fishman and F. R. Eirich, J. Phys. Chem., 75, 3135 (1971).
8. H. Lange, Kolloid Z. Z. Polymere, 243, 101 (1971).
9. K. Shirahama and N. Ide, J. Colloid Interface Sci., 54, 450 (1976).
10. M. N. Jones, J. Colloid Interface Sci., 23, 36 (1967).
11. T. L. Hill, "Thermodynamics of Small Systems" Vol. 1, Bejamin Pub., New York, 1963, p. 142.
12. B. Cabane, J. Phys. Chem., 81, 1639 (1977).
13. M. L. Smith and N. Muller, J. Colloid Interface Sci., 52, 507 (1975).
14. K. Shirahama, M. Tohdo, and M. Murahashi, J. Colloid Interface Sci., 86, 282 (1982).
15. P. J. Rud and B. R. Jennings, J. Colloid Interface Sci., 48, 302 (1974).
16. A. K. Wright, M. R. Thompson, and R. L. Miller, Biochemistry, 14, 3224 (1975).
17. K. Shirahama, K. Tsujii, and T. Takagi, J. Biochem., (Tokyo), 75, 309 (1974).
18. J. Steinhardt, J. R. Scott, and K. S. Birdi, Biochemistry, 16, 718 (1977).
19. F. Tokiwa and K. Tsujii, Bull. Chem. Soc. Jpn., 46, 2684 (1973).
20. S. Saito, T. Taniguchi, and K. Kimura, J. Colloid Interface Sci., 37, 154 (1971).
21. I. Satake and J. T. Yang, Biopolymers, 15, 2263 (1976).
22. I. Satake, K. Gondo, and H. Kimizuka, Bull. Chem. Soc. Jpn., 52, 361 (1979).
23. K. Shirahama, H.Yuasa, and S. Sugimoto, Bull. Chem. Soc. Jpn., 54, 375 (1981).
24. S. Harada and M. Katayama, Makromol. Chem., 90, 177 (1966).
25. K. Shirahama and M. Tashiro, Bull. Chem. Soc. Jpn., 57, 377 (1984).

26.   J. Steinhardt and J. A. Reynolds, "Multiple Equilibria in Proteins,"
      Academic Press, New York, 1969, p. 234.
27.   C. Tanford, Adv. Protein Chem., $\underline{23}$, 121 (1968).
28.   M. N. Jones, "Biological Interfaces," Elsevier Sci. Publ, Ams-
      terdam, 1975, p. 101.
29.   T. Takagi, K. Tsujii, and K. Shirahama, J. Biochem. (Tokyo), $\underline{77}$,
      939 (1975).
30.   J. Reynolds and C. Tanford, Proc. Natl. Acad. Sci., $\underline{66}$, 1002 (1970).
31.   M. T. Record, Jr., S. J. Mazur, P. Melancon, J. -H. Roe, S. L.
      Shaner, and L. Unger, Ann. Rev. Biochem., $\underline{50}$, 997 (1981).
32.   A. J. Hopfinger, "Intermolecular Interactions and Biomolecular
      Organization," John Wiley & Sons, New York, 1977.
33.   R. Pitt-Rivers and F. A. Impionbato, Biochem. J., $\underline{109}$, 825 (1968).
34.   S. Makino, J. A. Reynolds, and C. Tanford, J. Biol. Chem., $\underline{248}$,
      4926 (1973).
35.   B. J. Birch, D. E. Clarke, R. S. Lee, and J. Oakes, Analytica Chim.
      Acta, $\underline{70}$, 417 (1974).
36.   R. Chatterjee and D. K. Chattoraj, Biolpolymers, $\underline{18}$, 147 (1979). In
      the course of editing this book, a following paper appeared: K.
      Hayakawa, J. P. Santerre, and J. C. T. Kwak, Biophys. Chem., $\underline{17}$,
      175 (1983).
37.   G. Schwarz, Eur. J. Biochem., $\underline{12}$, 442 (1970).
38.   W. P. Ford, R.H. Ottewill, and H. C. Parreira, J. Colloid Interface
      Sci., $\underline{21}$, 522 (l966).
39.   P. Mukerjee and K. J. Mysels, "Critical Micelle Concentrations of
      Aqueous Surfactant Solutions," U. S. Government Print. Office,
      Washington, D. C., 1971.
40.   R. Senju, "Koroido Tekiteiho," Nankodo, Tokyo, 1969.
41.   K. Shinoda, T. Nakagawa, B. Tamamushi, and T. Isemura, "Colloidal
      Surfactants," Academic Press, New York, 1963, p. 58.
42.   C. Schildkraut and S. Lifson, Biopolymers, $\underline{3}$, 195 (1965).

ORDERED CONFORMATION OF POLY(L-LYSINE) AND ITS HOMO-
LOGS IN ANIONIC SURFACTANT SOLUTIONS[1]

Jen Tsi Yang and Shigeo Kubota
Department of Biochemistry and Biophysics and
Cardiovascular Research Institute
University of California, San Francisco
San Francisco, California 94143

ABSTRACT

The conformation of poly(L-lysine) and its homologs with side group $R = -(CH_2)_m NH_2$ in anionic surfactant solutions has been studied by circular dichroism. In neutral $NaDodSO_4$ solution at 25° only $(Lys)_n$ (m = 4) adopts a β-form, whereas the m = 2, 3, 5, 6 and 7 homologs all form a right-handed helix as that found in proteins. Most unusual is the m = 1 homolog, which has a left-handed helix in $NaDodSO_4$ solution but a right-handed helix in alkaline methanol [pH(apparent)$\geq$8]. In alkaline $NaDodSO_4$ solution $(Lys)_n$ undergoes a β-to-helix transition, which is temperature dependent. Raising the temperature shifts the transition pH downward from about 12 at 10° to 10.5 at 55°. The conformation of $(Lys)_n$ also depends on the chain length of alkyl sulfates. It is the β-form in sodium decyl, tetradecyl, and hexadecyl sulfate solution just as in $NaDodSO_4$ solution, but it becomes helical in octyl sulfate solution. The conformation of $(Lys)_n$ is also the β-form in sodium dodecyl sulfonate and tridecanoate solution at 45° and pH 9.3. However, in sodium dodecyl phosphate solution at 45° $(Lys)_n$ shows a pH-induced β-to-helix transition with a midpoint at pH 9.2 and it converts the helix back to the β-form at pH above 11. The binding of $NaDecSO_4$ to $(Lys)_n$, $(Orn)_n$ (m = 3) and $(D,L-Orn)_n$ has been determined by potentiometric titration. The binding is highly cooperative regardless of the conformation of the three polypeptides; it suggests a micelle-like clustering of the surfactant ions onto the polypeptide chain. The titration curves can be analyzed by adapting the Zimm-Bragg theory for the helix-coil transition of polypeptides to the binding isotherms between cationic polypeptides and anionic surfactants.

311

INTRODUCTION

In 1951 Pauling and Corey proposed the α-helix and β-pleated sheets for proteins and polypeptides. Within a few years Doty, Blout and their coworkers confirmed that synthetic poly(γ-benzyl-L-glutamate) exists in the helical form in 'poor' solvents and in the unordered form (often referred to as 'random coil') in 'good' solvents.[2] On the other hand, oligomers of γ-benzyl-L-glutamate form intermolecular β-aggregates.[3] Thus began an exciting era in organic and physical chemistry of polypeptides in solutions. The advance in this field was further stimulated by the renaissance of optical rotatory power in the middle 1950s. It is a powerful technique in the identification of various conformations of polypeptides and proteins in solution.[4]

Three major classes of biopolymers — proteins, nucleic acids and polysaccharides -- are all optically active or chiroptical. However, the current interest is not in studying the configurations of asymmetrical carbons but in probing the structural elements of macromolecules. The idea is not new; in fact, it was foreseen by Louis Pasteur more than a century ago:[5]

> Imagine a winding stair, the steps of which shall be cubes, or any other object with a superposable image. Destroy the stair, and the dissymmetry will have disappeared. The dissymmetry of the stair was the result only of the mode of putting together its elementary steps.

[Pasteur clearly expounded his views on molecular dissymmetry in the second of two lectures in 1860, although Augustin Fresnel (1788-1827) pre-dated Pasteur by anticipating the optical rotatory power of a helicoidal arrangement of the molecules from his theory of rotatory polarization.] An α-helix can be twisted into two nonsuperimposable forms (a right-handed helix of an L-polypeptide is the mirror image of a left-handed helix of its D-isomer). The helix of a polypeptide or a protein is chiroptical if only one kind of handedness of the helix prevails. Despite some early skepticism and also the belief that an L-polypeptide chain should have equal probability of winding into a right- and left-handed helix, optical rotatory dispersion (ORD) of L-polypeptides and proteins, which are made up of L-amino acids, shows that the helix is one-handed.[4] So far only right-handed helices are found in crystalline proteins from X-ray diffraction studies. In addition, many proteins contain β-pleated sheets (or β-form) and β-turns. These structural elements are also chiroptical.

Today circular dichroism (CD*) has replaced ORD as one of the most sensitive techniques in determining the conformations of proteins and polypeptides; each conformation has its characteristic CD bands (Figure 1).[6] Briefly speaking, CD measures the difference in absorbances of the left- and right-circularly polarized light by a chiroptical

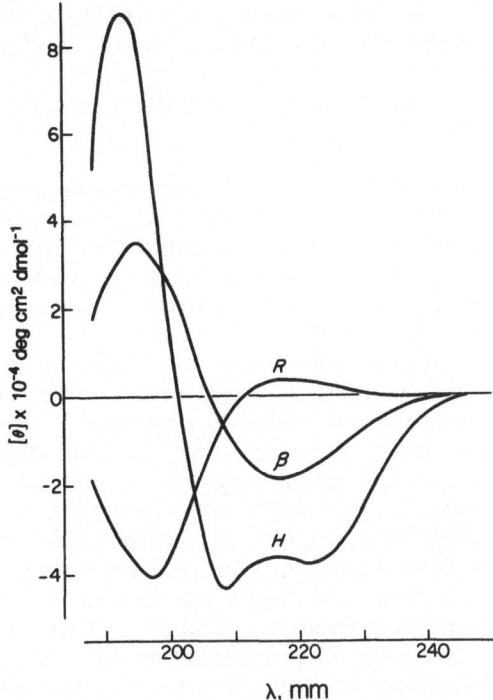

Figure 1.   Circular dichroism of poly(L-lysine) ($\underline{M}_r$ = 193,000) in water at 25°.   Curves: R, unordered form at neutral pH; H, α-helix at pH 10.8; β, β-form at pH 11.1 after 15 min heating of the helix at 52° and cooling back to 25°. Lys conc'n: 0.56 mM (residue).

substance, whereas ORD measures the difference in refractive indices of the left- and right-circularly polarized light (circular birefrigence).[7] An α-helix shows three strong CD bands in the wavelength range of 185 and 240 nm: a negative band at 222 nm (n-π* transition), another negative one at 208-210 nm and a stronger positive one near 191 nm (π-π* transitions). The β-form usually has a negative CD band around 216-218 nm and a stronger positive one near 195 nm. The unordered form has a strong negative band near 198 nm, a weak positive one near 220 nm and an even weaker negative one near 235 nm. Of the many types of the β-turn the most common one appears to have a positive CD band around 205 nm and a negative one below 190 nm plus a weak negative one near 225 nm (not shown in Figure 1).

Because the CD of the helix predominates over that of other conformations, the appearance of a double minimum at 222 and 208-210 nm and a maximum near 191 nm in the CD spectrum of a polypeptide or protein indicates the presence of a helical conformation; the larger the CD magnitude, the higher the helicity in the macromolecule.

Computer programs have been developed to analyze the CD spectrum of proteins and thereby the amounts of various conformations in the molecule, although the method is still empirical; the calculated helicity is often good, but the estimates of other conformations remain uncertain.

The CD data of polypeptides and proteins are often expressed as either molar (or mean residue) differential absorption coefficient, $\varepsilon_L - \varepsilon_R$, in cm^{-1} M^{-1} or molar (or mean residue) ellipticity, $[\theta]$, in deg cm^2 $dmol^{-1}$.[7] [Historically, J. B. Biot in 1836 introduced the definition of the specific rotation, $[\alpha] = \alpha/\ell c$), where α is the rotation in degrees, ℓ the light path in decimeters and c the density of a liquid or the concentration of a solute in grams per milliliter. Biot expressed the length in decimeters instead of centimeters "in order that the significant figures may not be uselessly preceded by two zeros." For similar reasons the molar rotation, $[M]$, is now defined as $[M] = [\alpha]M/100$, where M is the molecular weight (or molar mass). By analogy, the specific ellipticity, $[\psi]$, and molar ellipticity, $[\theta]$, are defined as $[\psi] = \psi/(\ell c)$ and $[\theta] = [\psi]M/100$, respectively, where ψ is the ellipticity in degrees. Today we find it awkward to use decimeters as the unit for the length, not to mention that Biot's concern does not exist for the optical activity of macromolecules. Likewise, there is no reason to divide 100 into the molecular weight (molar mass). But the introduction of a new set of definitions or units, however attractive they may be, would merely compound confusion amony the users. The magnitude of the extrema of a 100% helix is of the order of 10^4 in deg $cm^2 dmol^{-1}$, but it is understood that experimentally the mean residue ellipticity has only three significant figures followed by two zeros.

This review will briefly discuss the interactions of poly(L-lysine) and its homologs with anionic surfactants that were studied in our laboratory. The general formula of this series of polypeptides is:

$$H-(HN-CH-CO)_n-OH$$
$$(CH_2)_m$$
$$NH_2$$

with m = 1 for poly(L-α,β-diaminopropionic acid)
     2   poly(L-α,γ-diaminobutyric acid)
     3   poly(L-ornithine) or $(Orn)_n$
     4   poly(L-lysine) or $(Lys)_n$
     5   poly(L-α,ω-diaminoheptanoic acid)
     6   poly(L-α,ω-diaminooctanoic acid)
     7   poly(L-α,ω-diaminonanoic acid)

---

*Abbreviation used: CD, circular dichroism; $(Lys)_n$, poly (L-lysine); $(Orn)_n$, poly (L-ornithine); $NaDodSO_4$, sodium dodecyl sulfate; $NaDecSO_4$, sodium decyl sulfate; $NaOctSO_4$, sodium octyl sulfate; $NaTetradSO_4$, sodium tetradecyl sulfate; $NaHexadSO_4$, sodium hexadecyl sulfate.

High molecular-weight polypeptides were synthesized by the standard N-carboxy-α-amino acid anhydride (NCA) method. The ω-carbobenzoxy groups of the polypeptide in dioxane solution were removed with hydrogen bromide gas. The water-soluble polymers were then dialyzed repeatedly against dilute HCl and finally water. The concentrations of the polypeptides were determined by microKjeldahl analyses. All CD spectra were measured on a Jasco J-10 or, more recently, J-500A spectropolarimeter (for its calibration, see Ref. 8).

CONFORMATION OF $(Lys)_n$ AND ITS HOMOLOGS IN AQUEOUS SOLUTION

The very simplicity of synthetic polypeptides makes it attractive to characterize their various conformations in solutions under different experimental conditions. In particular, the helix-coil transition of polypeptides has been extensively studied both experimentally and theoretically.[9] Although synthetic polymers are unlike real proteins, they have often been used as model compounds for the ordered conformations of proteins. Among water-soluble polypeptides poly(L-glutamic acid) and poly(L-lysine) have been studied in most detail. Of particular interest is $(Lys)_n$ which can adopt three conformations merely by varying the pH and temperature of its aqueous solution (cf. Figure 1). It is a 'random coil' in neutral solution because the polypeptide behaves as a polyelectrolyte with its ω-amino side groups protonated as $-NH_3^+$ ions. Raising the pH to above 10 begins to dissociate the protons and induce a coil-to-helix transition, which is reversible.[10] The helical form can be converted to a β-form upon mild heating at elevated temperature for several minutes.[6] Cooling the β-form at, say, 4^0 for several days restores the helical form. The β-form is probably anti-parallel; it can be either intra- or intermolecular, depending on the concentration of the polypeptide used.[11]

The $(Lys)_n$ homologs are also unordered at neutral pH, as reflected by the appearance of a strong negative CD band near 200 nm, except that potentiometric titrations of the m = 6 and 7 homologs seemed to indicate that they were partially helical even when their side groups were fully protonated (data not shown).[12] Higher homologs with m = 5, 6 and 7 showed a sharp pH-induced helix-coil transition with midpoints at pH 9.6, 9.0 and 8.7 in 0.05 M KCl at 25^0, respectively, as compared with 10.1 for $(Lys)_n$ under similar conditions (the m = 7 homolog was highly aggregated above pH 9) (data not shown). Thus, the larger the number of methylene groups in the side chain, the more stable is the helical conformation of these homologs. The helical form of these higher homologs is so stable that it is not easily converted to the β form at elevated temperatures. For instance, at 80^0 the CD spectra gradually changed from a double minimum at 222 and 208-210 nm to a broad minimum at 218-221 nm with a reduced magnitude, suggesting that the helix and β-form may coexist.[12] Lower homologs with m = 1, 2 and 3 cannot have a stable helical conformation. Deprotonated $(Orn)_n$ has only about 20 to 25% helix,[13] whereas the m = 2 homolog may have a trace of helix[14,15] and m = 1 homolog none at all even at pH above 12.[15]

Qualitatively, we can summarize the results of (Lys)$_n$ and its homologs in aqueous solution as follows: Poly(L-alanine) with a -CH$_3$ group in the side chain is completely helical in aqueous solution [this polypeptide can be solubilized by synthesizing a block polymer with, say, poly(D,L-glutamic acid)].[16] Replacing one hydrogen atom in the methyl group by an amino group (m = 1 homolog) immediately destabilizes the helical conformation even when the polypeptide is uncharged. However, this instability can be overcome by lengthening the -CH$_2$- groups to four for un-ionized (Lys)$_n$, which is fully helical. The pH-induced helix-coil transition of higher homologs of (Lys)$_n$ shifts toward lower pHs with increasing -CH$_2$ groups (these higher homologs tend to aggregate and precipitate in aqueous solution). On the other hand, the position of the -CH$_2$- group can also affect the conformational stability; for instance, the helical conformation of uncharged poly(N$^\delta$-methyl-L-ornithine) is less stable than its isomer (Lys)$_n$.[17] For comparison, in the (Glu)$_n$ series with a side group -(CH$_2$)$_m$COOH, un-ionized poly(L-aspartic acid) (m = 1) does not adopt a helical conformation,[18] whereas (Glu)$_n$ with m = 2 is helical at pH below 4,[19] so is poly(L-α-aminoadipic acid) (m = 3).[20]

CONFORMATION OF (Lys)$_n$ AND ITS HOMOLOGS IN ALKYL SULFATE SOLUTIONS

A. Effect of NaDodSO$_4$. The conformation of (Lys)$_n$ and its homologs in neutral NaDodSO$_4$ solution is intriguing. The addition of NaDodSO$_4$ converts the unordered form of these cationic polypeptides into an ordered form, as reflected by the changes in their CD spectra (Figures 2 and 3). With increasing surfactant concentrations (the polypeptide concentration being kept constant), (Orn)$_n$ undergoes a coil-to-helix transition and (Lys)$_n$ a coil-to-β transition (Figure 2).[21] The NaDodSO$_4$ concentrations used were well below the critical micelle concentration, except the two curves at the highest NaDodSO$_4$ concentrations. Thus, it is the ratio of surfactant to polypeptide, not the critical micelle concentration of NaDodSO$_4$, that determines the conformation of the polypeptide-surfactant complexes. At the ratio of one surfactant ion per polypeptide residue the complexes were insoluble (see potentiometric titration below), but the complexes redissolved at higher NaDodSO$_4$ concentrations and the CD spectra of both polypeptides were unaffected by the presence of excess surfactant micelles, suggesting a saturation of surfactant ions bound to the polypeptide molecules.

More surprisingly, (Lys)$_n$ distinguishes itself by being the only polypeptide among its homologs that adopts a β-form in NaDodSO$_4$ solution,[22] whereas m = 5, 6 and 7 homologs in addition to m = 3 homolog all form a right-handed helix under similar conditions (Figure 3).[12] However, the magnitudes of the two negative CD bands were

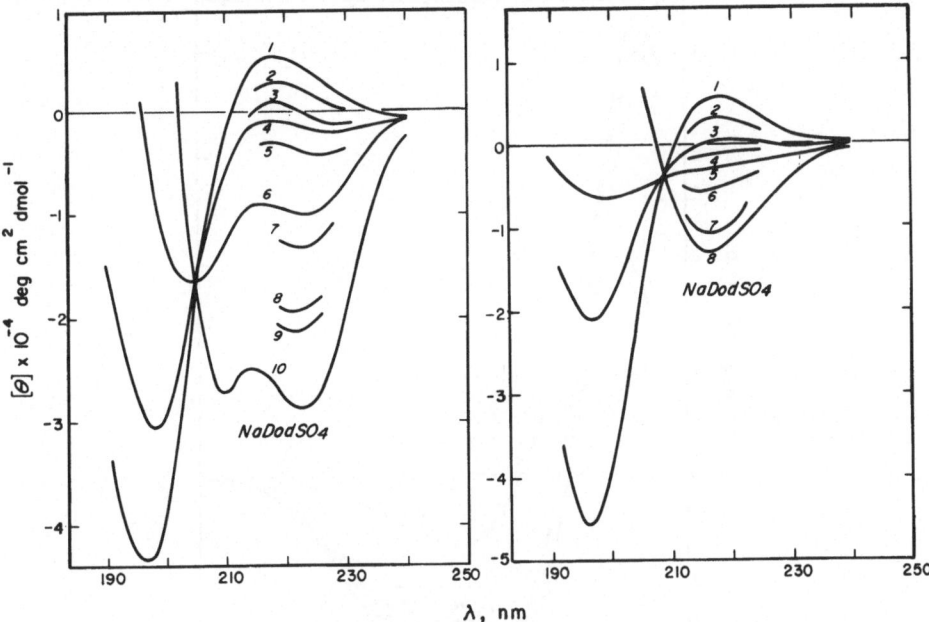

Figure 2.   CD of (Orn)$_n$ and (Lys)$_n$ in sodium dodecyl sulfate solutions at
25°.   (Left) (Orn)$_n$: 0.108 mM (monomer) ; NaDodSO$_4$: 1,
none; 2 to 10, 0.0245, 0.0340, 0.0490, 0.0596, 0.102, 0.119,
0.153, 0.170 and 25.5 mM, respectively.   (Right) (Lys)$_n$: 0.117
mM (monomer); NaDodSO$_4$: 1, none; 2 to 8, 0.0255, 0.0511,
0.0702, 0.0894, 0.114, 0.170 and 25.5 mM, respectively.
[Redrawn from Ref. (21) with permission by Academic Press.]

smaller than those at high pH without the surfactant (cf. Figure 1). The
m = 2 homolog is also helical in neutral NaDodSO$_4$ solution at room
temperature (Figure 4, right).[15]   It can undergo a thermally-induced
helix-to-β transition; raising the temperature converts the CD double
minimum to a single minimum with a concomitant reduction in
magnitude. This transition for m = 2 homolog is gradual and irreversible
at temperatures above 30°. The conformation of this homolog in other
alkyl sulfate solutions is essentially the same as that in NaDodSO$_4$
solution (Figure 4, left) (see, however, Section C. below).

Most unusual is the m = 1 homolog, which adopts a left-handed
helix in neutral NaDodSO$_4$ solution, but a right-handed helix in 80%
methanol at pH (apparent) 8 or higher (Figure 5).[15] The CD spectrum in
the surfactant solution shows a positive maximum at 222 nm, a shoulder
near 215 nm and a stronger negative band near 195 nm. This is almost
a mirror image of the spectrum for a right-handed helix (cf. Figure 5,
left and right). [Strictly speaking, the left- and right-handed helices of
an L-(or D-)polypeptides are not exactly the mirror image of each other.]
As far as we know, this is the only water-soluble L-polypeptide that can
form a left-handed helix; in fact, no such helices have been found in

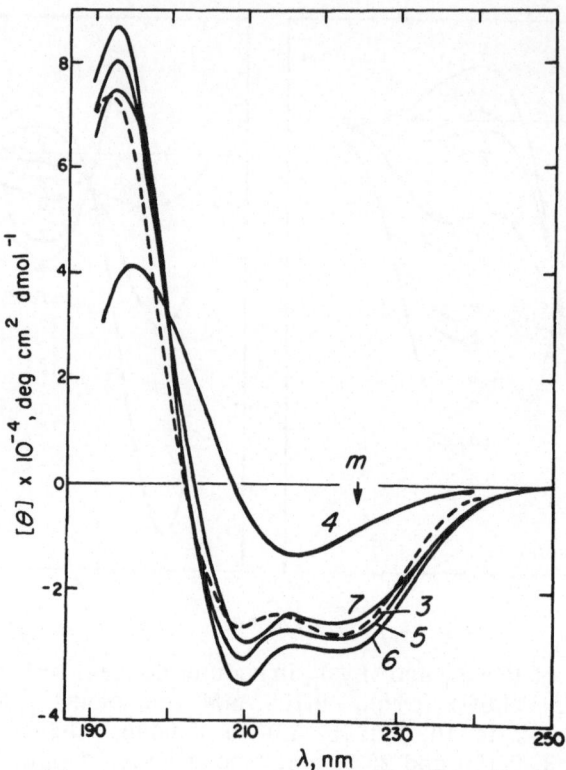

Figure 3.   Molar residue ellipticities of (Lys)$_n$ and its homologs in 25 mM
NaDodSO$_4$ at 25°.   Side chain R = -(CH$_2$)$_m$NH$_2$·HCl.
Polypeptide concentration: 0.1 mM (monomer).   [Redrawn
from Ref. (12) with permission by John Wiley & Sons.]

proteins. Why a polypeptide can have one kind of helix in a surfactant
solution and another kind in alkaline methanol is difficult to explain.
Equally puzzling is that only (Lys)$_n$ among its homologs has a β-form in
neutral NaDodSO$_4$ solution at room temperature and the m = 2 homolog
undergoes a helix-to-β transition in the surfactant solution at elevated
temperatures. Thermodynamically, the difference in Gibbs free energy
between the formation of the left- and right-handed helices and that of
a helix and a β-form may not be too great.   However, a theoretical
explanation for these findings is still lacking.

B.   Phase Diagram.   While (Lys)$_n$ in NaDodSO$_4$ solution at pH
below 10 has a negative CD band at 219 nm, which is characteristic of
a β-form, it can undergo a thermally-induced β-to-helix transition at
higher pHs; that is, the single CD minimum is converted to a double
minimum at 222 and 209 nm.   The transition is cooperative and
reversible; raising the temperature shifts the transition pH downward

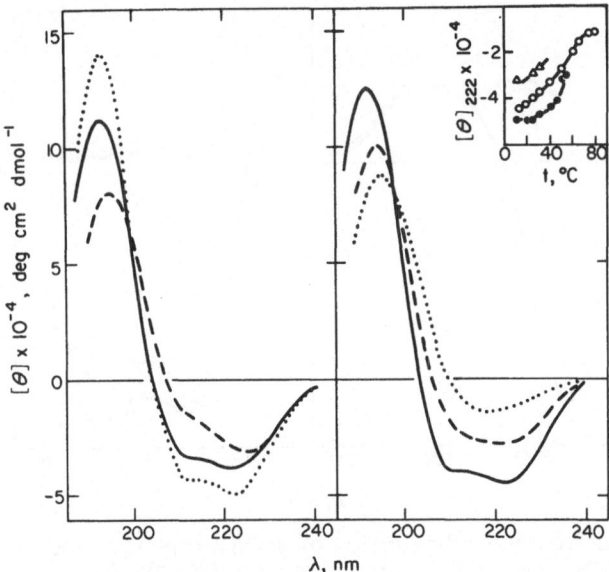

Figure 4.   Molar residue ellipticities of poly(L-α,γ-diaminobutyric acid) in 50 mM alkyl sulfates at neutral pH.   Polypeptide concentration: 0.5 mM (monomer).   Left: NaDodSO$_4$ (-); NaDecSO$_4$ ($\cdots$); and NaOctSO$_4$ (---) (all at 25°).   Right: in NaDodSO$_4$ solution at 14° (-), 51° (---) and 80° ($\cdots$).   Inset: temperature dependence in NaDodSO$_4$ (0), NaDecSO$_4$ ($\bullet$) and NaOctSO$_4$ (Δ).   [Taken from Ref. (15) with permission by John Wiley & Sons.]

(Figure 6, left).[23]   On the high-pH side, the absolute CD magnitude decreases with increasing temperature, probably because of a partial breaking-up of the helical conformation of (Lys)$_n$.   That (Lys)$_n$ in NaDodSO$_4$ solution can adopt the helical conformation at high temperature contrasts markedly with (Lys)$_n$ in aqueous solution without the surfactant.   For instance, (Lys)$_n$-NaDodSO$_4$ complexes at pH 11 are helical even after 1-h heating at 70°, whereas uncharged (Lys)$_n$ in aqueous solution without the surfactant readily undergoes a helix-β transition after mild heating at 50° (cf. Figure 1).

The results in Figure 6 (left) enables us to construct a conformational phase diagram (Figure 6, right).[23]   The shaded area represents the β-helix transition region.   The transition pH shifts downward from about 12 near 10° to below 11 at 55°. Raising the temperature of the β-form above 55° precipitated the polypeptide.   We attribute this β-helix transition to the deprotonation of the ω-NH$_3^+$ side groups, which in turn dissociates the polypeptide-surfactant complexes.   High temperature increases the apparent dissociation constant of lysine side groups, thus shifting the transition pH downward.   However, (Lys)$_n$ in NaDodSO$_4$ solution remains helical at pH 12.9 and 45°, suggesting that the

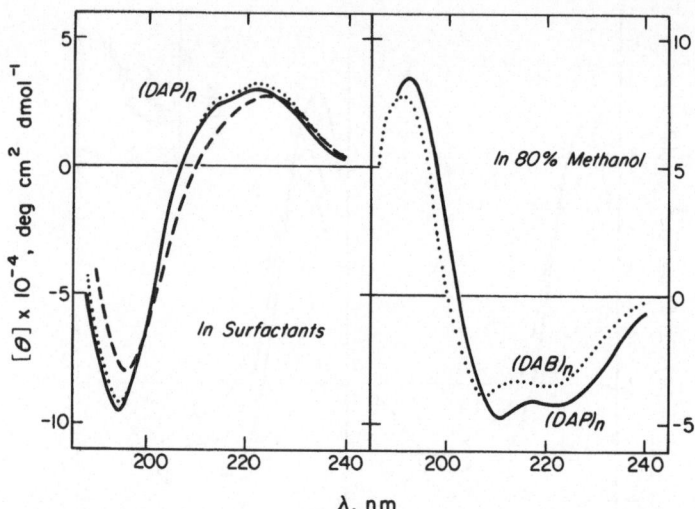

Figure 5.   Molar residue ellipticities of poly(L-α,γ-diaminopropionic acid)
[(DAP)$_n$] at 25°.   Left: 0.5 mM (DAP)$_n$ (residue) in 50 mM
NaDodSO$_4$ (-), NaDecSO$_4$ (···) and NaOctSO$_4$ (---) at neutral
pH.   Right: 1.4 mM polypeptide (residue) in methanol at pH
(apparent) 11.6.   Poly(L-α,γ-diaminobutyric acid) [(DAB)$_n$] is
included   for   comparison.   [Taken   from   Ref.   (15)   with
permission by John Wiley & Sons.]

polypeptide–surfactant complexes may not completely dissociate even at
pH 13 [above pH 12.9 (Lys)$_n$ precipitated even in the presence of excess
NaDodSO$_4$].   Previously, Feldshtein et al. and Zezin et al. also reported
a reversible helix-β transition of (Lys)$_n$ in NaDodSO$_4$ solution over a pH
range of 11.6 and 11.9 (temperature unspecified),[24] which is in accord with
our findings.

        C.   Effect of other alkyl sulfates.   The ordered conformation of
(Lys)$_n$ and its homologs depends not only on the number of methylene
groups in the side chain of the polypeptides but also on the chain length
of the nonpolar tail of the surfactant.   In neutral solutions of sodium
alkyl sulfates, $CH_3(CH_2)_xSO_4Na$ with x = 7, 9, 13 and 15 (x = 11 for
NaDodSO$_4$), (Lys)$_n$ retains its β-form in decyl (X = 9), tetradecyl (x = 13)
and hexadecyl (x = 15) sulfate solutions, but it becomes partially helical
in octyl (x = 7) sulfate solution (Figure 7).[21]   The amount of the β-form
among   the   four   alkyl   sulfate   solutions   appears   to   increase   with
increasing chain length (Table I).   In contrast, (Orn)$_n$ is helical in all five
alkyl sulfate solutions;   the amount of the helicity also seems to increase
with increasing chain length.   Recently, Takeda et al.[25] studied the

Table I

Molar Residue Ellipticities of $(Lys)_n$ and $(Orn)_n$ in Alkyl Sulfate Solutions[a]

Surfactant[b]	$\underline{t}$	λ_{max}	$[\theta]$	λ_{min}	$-[\theta]$	λ_{min}	$-[\theta]$
Poly(L-ornithine):							
C_{16}	45° [c]	191.5	81000	209	31000	222	32600
C_{14}	35° [c]	191.5	77000	209	30600	222	32900
C_{12}	25°	191.5	72000	209	25800	222	28600
C_{10}	25°	198	67000	210s[d]	4200	226	26100
C_8	25°	195	58000	210s[d]	9700	225	24400
Poly(L-lysine):							
C_{16}	45° [c]	195	45000	217	15400		
C_{14}	35° [c]	195	41000	217	14400		
C_{12}	25°	195	40000	217	13200		
C_{10}	25°	197	32000	219	11000		
C_8	25°	192.5	55000	210s[d]	15900	225	28300

[a]Polypeptide concentration: 0.05 to 0.5 M; the units of λ's are nm and $[\theta]$'s deg cm^2 $dmol^{-1}$.

[b]$CH_3(CH_2)_xSO_4Na$: C_{16}, hexadecyl; C_{14}, tetradecyl; C_{12}, dodecyl; C_{10}, decyl; C_8, octyl. Surfactant concentrations: 0.01 to 0.1 M for C_{16}, C_{14} and C_{12}; 0.1 to 1 M for C_{10} and C_8.

[c]C_{16} and C_{14} are sparingly soluble in water at room temperature.

[d]s = shoulder.

change in conformation of $(Lys)_n$ in $NaOctSO_4$ solution with the surfactant concentration. $(Lys)_n$[0.28 mM (monomer)] was partially helical in the range of 4 to 6 mM $NaOctSO_4$, but the helix was gradually converted to a β-form above 6 mM $NaOctSO_4$ (no measurements above 8 mM were reported). We have used $(Lys)_n$ [0.05 to 0.5 mM (monomer)] in 0.6 to 1 M $NaOctSO_4$. Thus, our molar ratio of $NaOctSO_4$ to Lys was over 5000 and that of Takeda et al. about 20. Because of this difference in experimental conditions, any possible discrepancy between our results and theirs must be resolved by future studies.

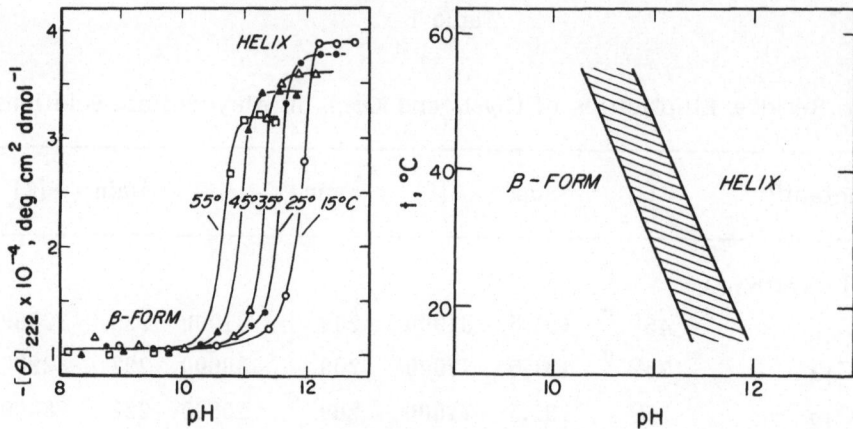

Figure 6. (Left) The helix-β transition of (Lys)$_n$ in NaDodSO$_4$ solution.
(Right) The conformational phase diagram of (Lys)$_n$ in
NaDodSO$_4$ solution based on the data in the left figure.
Shaded area indicates the β-helix transition region. Concen-
trations: (Lys)$_n$, 0.46 mM (monomer); NaDodSO$_4$, 16 mM.
[Redrawn from Ref. (23) with permission by John Wiley &
Sons.]

        D.   Effect of surfactants other than alkyl sulfates.   (Lys)$_n$
assumed a β-form in sodium dodecyl sulfonate (NaDodSO$_3$) and sodium
tridecanoate (NaDodCOO) solutions at 45° pH 9.3 (data not shown),[26]
just as it does in NaDodSO$_4$ solution.   (These two surfactants were
sparingly soluble at room temperature and NaDodCOO was only soluble
above pH 9.).   In 25 mM sodium dodecyl phosphate at 45° (Lys)$_n$ is in
a β-form at neutral pH and undergoes a pH-induced β-helix transition
with a midpoint at pH 9.2, as compared with 10.9 in NaDodSO$_4$ solution
at 45° (Figure 8).[26]   Dodecyl phosphoric acid has two dissociation
constants; its apparent pKs were about 3 and 8 if they were the same
as those of a phosphoric acid in 0.1 M NaCl.[27]   Thus, dodecyl phosphate
carries one negative charge in acidic solution and the second proton of
the phosphate begins to ionize above pH 8.   Probably the electrostatic
repulsion between bound dodecyl phosphate ions at alkaline pHs will
destablize the β-form, but not the helix, of the polypeptide-surfactant
complex because of longer separations between adjacent charged side
groups in the helix than in the β-form.   Raising the pH of the dodecyl
phosphate solution to above 11 causes another conformational transition
of (Lys)$_n$ probably from the helix to a β-form, which is complete within
a few hours and has a midpoint at pH 11.4. For comparison, ionic (Orn)$_n$
at 45° and pH 9.3 adopted a helical conforamtion in the presence of
sodium dodecyl sulfonate or phosphate just as it was in NaDodSO$_4$
solution.[26]   But this polypeptide remained unordered in sodium tridecanoate solution.

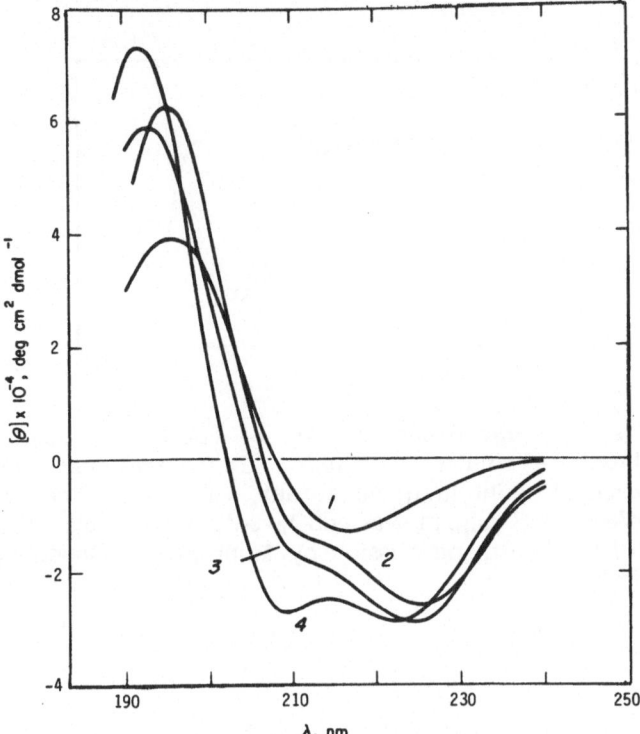

Figure 7.   CD of $(Lys)_n$ and $(Orn)_n$ in $NaDodSO_4$ and $NaOctSO_4$ solutions at 25°.   Curves: 1 and 4, 0.1 mM $(Lys)_n$ and $(Orn)_n$ in 26 mM $NaDodSO_4$, respectively; 2 and 3, about 0.06 mM $(Orn)_n$ and $(Lys)_n$ in 700 mM $NaOctSO_4$, respectively.   Polypeptide concentrations are based on monomers (residues).   [Redrawn from Ref. (21) with permission by Academic Press.]

E.   Binding isotherms.   So far we have described the ordered conformation of $(Lys)_n$ and its homologs in excess surfactant solutions. The mode of interaction between the polypeptide and surfactant can be studied by potentiometric titration.   We choose $(Lys)_n$, $(Orn)_n$ and the racemic $(D,L-Orn)_n$, which are in the β-, helical and unordered form, respectively, in $NaDodSO_4$ solution.   We used $NaDecSO_4$ instead of $NaDodSO_4$ because the latter's strong affinity toward the polypeptides made it difficult to accurately measure the free surfactant concentration.

A titration cell with a liquid membrane that is selectively permeable to surfactant ions was constructed:[28]

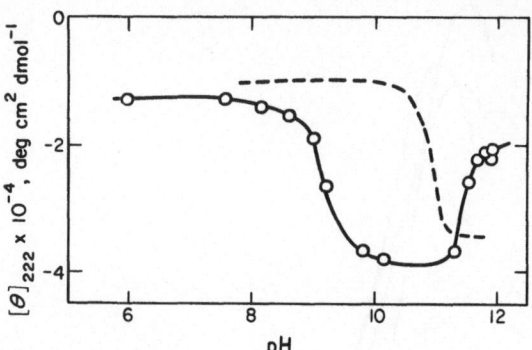

Figure 8. The pH dependence of CD of (Lys)$_n$ in sodium dodecyl
phosphate solution at 45°.   Concentrations: polypeptide
(residue), 0.06 mM; surfactant, 100 mM.   The broken line
refers to (Lys)$_n$ in NaDodSO$_4$ solution at 45°. [Taken from
Ref. (26) with permission by Munksgaard, Copenhagen, Denmark.]

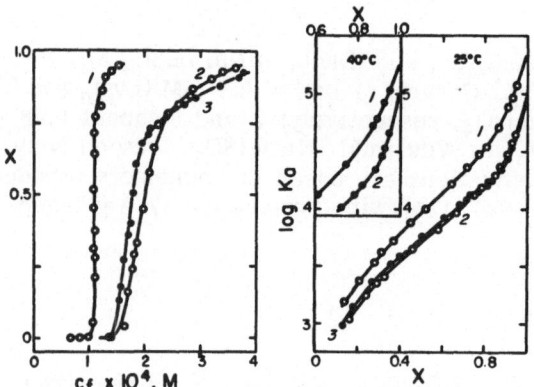

Figure 9. (Left)   The binding isotherms of (Lys)$_n$ (1), (Orn)$_n$ (2) and
(D,L-Orn)$_n$ (3) with NaDecSO$_4$ at 25°. Symbols: x, degree of
binding; c_f, free NaDecSO$_4$ concentration.   Polypeptide
concentration = 0.21 - 0.25 mM (monomer).   (Right)   The
apparent association constant, K_a, of (Lys)$_n$ (1), (Orn)$_n$ (2) and
(D,L-Orn)$_n$ (3) as a function of the degree of binding.
[Redrawn from Ref. (28) with permission by John Wiley &
Sons.]

> Calomel | 1 M NH_4NO_3 agar bridge | reference solution ($NaDecSO_4$ + NaCl) | liquid membrane | sample solution (polypeptide + $NaDecSO_4$ + NaCl) | 1 M NH_4NO_3 agar bridge | Calomel

Here the liquid membrane was made up of dodecyl ammonium decyl sulfate, $[Dod(CH_3)N_3][DecSO_4]$ in nitrobenzene.

Assuming that the polypeptides in a salt solution do not appreciably affect the activity coefficient of free surfactant concentration, a decrease in surfactant activity in the presence of the polypeptide as reflected by the measured electromotive force of the cell represents a decrease in free surfactant concentration because of the binding of the surfactant to the polypeptide. The degree of binding, x, is simply

$$x = (c_t - c_f)/c_p \tag{1}$$

where c_t and c_f are the total and free $NaDecSO_4$ concentration and c_p is the polypeptide concentration on a monomer basis. The binding isotherms rise sharply over a narrow range of c_f for $(Lys)_n$ and somewhat less sharply for the two polyornithines (Figure 9, left).[28] The binding is cooperative regardless of the polypeptide conformation. This is further illustrated in Figure 9 (right), where the apparent association constant, K_a:

$$K_a = x/c_f(1 - x) \tag{2}$$

increases rather than decreases with increasing x. It is the opposite of hydrogen ion titrations of $(Lys)_n$[29] or $(Glu)_n$,[30] for which K_a for protonation decreases with increasing x. Thus, the contribution of electrostatic free energy alone cannot account for the observed curves in Figure 9. Rather, the hydrophobic interactions between bound surfactant molecules overshadow the coulombic repulsions between the nearest-neighbor charged side groups of the polypeptide chain and thus increase successive bindings of the surfactant ions onto the polypeptide chain. We visualize that the bound surfactant ions contact each other and form clusters rather than uniformly distribute themselves on the surface of the polypeptide molecule. Such interactions between nonpolar alkyl chains of the surfactant should present no problem for the unordered form and also the helix of the polypeptide. The situation is somewhat unclear for the β-form: because the side groups alternately extend out on either side of the β-pleated sheets, the distance between residues i and i + 2 is too far apart for Lys side groups to contact each other. However, the alkyl chains of adjacent bound surfactant ions in the same polypeptide chain or between neighboring strands of the same β-sheet could bend toward each other. The atomic model of $(Lys)_n$ further indicates that $DodSO_4^-$-Lys^+ side groups of one β-sheet can interdigitate quite nicely with those of neighboring β-sheets. Recently, we observed that sequential polypeptides such as $(Lys-Ala)_n$

and (Lys-Leu)$_n$ adopted a β-form in excess NaDodSO$_4$ solution, whereas (Lys-Lys-Ala-Ala)$_n$ and (Lys-Lys-Leu-Leu)$_n$ actually existed in a helical conformation under similar conditions, even though the latter two polypeptides also have equimolar Lys and Ala or Leu residues.[31] On the basis of atomic models, it is not possible to pack (Lys$_2$-Ala$_2$)$_n$ and (Lys$_2$-Leu$_2$)$_n$ into a regular β-form with Ala or Leu against Ala or Leu and DodSO$_4^-$-Lys$^+$ side group against DodSO$_4^-$-Lys$^+$ side group. Thus, these two polypeptides in NaDodSO$_4$ solution converted to the helical conformation instead.

The titration curves of the three polypeptides can be analyzed by adapting the Zimm-Bragg theory for the helix-coil transition[32] to the cooperative binding of the surfactant to the polypeptide, provided that the thermodynamic contribution of any conformational change accompanying the binding is comparatively small and that the electrical potential on the surface of the polypeptide molecule can be regarded to be constant. The theory defines two parameters: the equilibrium constant s and an initiator σ. If the polypeptide sequence is schematically written as

$$0\ 0\ 0\ 0\ 1\ 1\ 1\ 1\ 0\ 0\ 0\ 1\ 1\ 1\ 0\ 0\ .\ .\ .\ .\ .$$

then

$$\sigma s = [01]/[00]$$
$$= \underline{K}c_f \tag{3}$$

and

$$s = [11]/[10]$$
$$= \underline{K}c_f/\sigma \tag{4}$$

where the digit 0 represents an unoccupied site and 1 a surfactant-bound site, the bracket denotes the concentration, and $\underline{K}$ is the binding equilibrium constant of a surfactant molecule to a site with another unoccupied nearest neighbor.

The degree of binding, x, can be determined from

$$x = d \ln \lambda_0/d \ln s \tag{5}$$

where λ_0 represents the larger of two eigenvalues of the statistical weight matrix and is given by

$$\lambda_0 = \{1 + s + [(1 - s)^2 + 4\sigma s]^{1/2}\}/2$$

It follows that

$$x = 1/2 + (s - 1)/2[(1 - s)^2 + 4\sigma s]^{1/2} \tag{6}$$

According to the theory the more cooperative the binding, the smaller is the σ value.

It can be shown that[28]

$$(d \ x/d \ \ln \ \underline{c}_f)_{x=1/2} = 1/4\sigma^{1/2} \tag{7}$$

From the results in Figure 9 we found that $1/\sigma = 77$ for $(Orn)_n$, 161 for $(D,L\text{-}Orn)_n$ and $\geq 10^4$ for $(Lys)_n$.

Next, the number of bound surfactant ions is given by $d \ \ln \ \lambda_0^n/d \ \ln \ \sigma$. The average bound surfactant cluster, $\bar{m}$, can be calculated from[28]

$$\bar{m} = 2x[(1 - \sigma)/\sigma]/\{[4x(1 - x)(1 - \sigma)/\sigma + 1]^{1/2} - 1\} \tag{8}$$

The calculated $\bar{m}$ values for the three polypeptides at several x values (Table II) clearly suggest that the binding is more cooperative for $(Lys)_n$ than for the polyornithines. The average cluster size for the β-form of $(Lys)_n$ is sufficiently large to promote the ordered conformation even at low degrees of binding, whereas four or five bound surfactant ions are necessary to start one helical turn of $(Orn)_n$. The CD results (data not shown) indicated that the onset of the helical conformation of $(Orn)_n$ began at about $x = 0.3$; that is, after seven or eight surfactant ions were bound to the polypeptide, which can form two helical turns.

We have neglected the effect of conformational changes in our treatment; thus, it may only be approximately applied to the L-polypeptides which undergo a coil-to-helix or coil-to-β transition. However, the close similarity in the binding isotherms and titration curves between $(D,L\text{-}Orn)_n$ and $(L\text{-}Orn)_n$ (Fig. 9) seems to indicate that the contribution of conformational free energy may not significantly complicate our analysis. When the bound surfactant ions cluster onto the polypeptide chain, the electrical potential on the polymer surface will remain virtually constant unless the size of successive vacant sites becomes extremely small. Indeed, we found that the agreement between the experimental and calculated values of the binding isotherms and titration curves of the three polypeptides was good to about $x = 0.7$, beyond which the calculated curves began to deviate from the observed ones, implying that the assumption of a constant surface potential was no longer valid as x approached unity.

CONFORMATION OF $(Lys)_n$ IN LIPID SOLUTIONS.

To conclude this review we will breifly mention the conformation of $(Lys)_n$ in lipid solutions. These lipids are associated with membranes and lipoproteins. We will illustrate the effect of two phosphoglycerides, which consist of a glycerol backbone, two fatty acid chains and a phosphorylated alcohol (Figure 10). These lipids are insoluble in water but

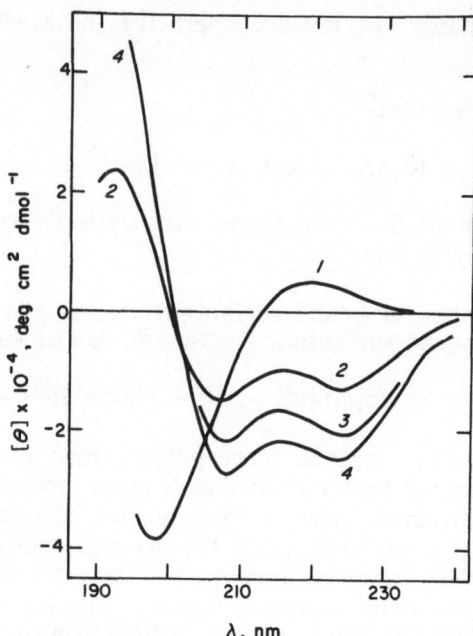

Figure 10. CD of (Lys)$_n$ in neutral phosphatidylserine (PhSer) solutions at 25°. Polypeptide concentration: 0.06 mM (monomer). Curves: 1, in water; 2, 3 and 4, in 0.0025%, 0.025% and 0.025% PhSer and 0.0075%, 0.2% and 0.1% hexadecylpoly-(oxyethylene) ether, respectively. Molar (residue) ratios of polypeptide/PhSer/nonionic surfactant: 2, 1:0.5:1.5; 3, 1:5:40; 4, 1:5:20. [Taken from Ref. (33) with permission by Munksgaard, Copenhagen, Denmark.]

they can form mixed micelles with a nonionic surfactant such as hexadecylpoly(oxyethylene) ether, which are stable in water. Solubilized phosphatidylserine with a positive and negative charge on serine promotes a helical conformation in (Lys)$_n$ at neutral pH (the nonionic surfactant is inert toward the polymer).[33] This contrasts with the induced β-form of (Lys)$_n$ in NaDodSO$_4$ solution. Perhaps the two fatty acid chains are too bulky to be accommodated in a β-form but not in a helix. Not surprisingly, solubilized phosphatidylcholine with a positive charge on choline cannot bind to (Lys)$_n$ and induce an ordered conformation in this polypeptide. Thus, the first step of the complex formation between (Lys)$_n$ and the lipid probably also involves an electrostatic interaction between the ω–NH$_3^+$ group of lysine and the –COO$^-$ group of the lipid; the induced helix is then surrounded by the mixed micelles.

Table II

Average Cluster Size ($\bar{m}$) of Bound Decyl Sulfate Ions on

$(Lys)_n$ and $(Orn)_n$ at Different Degrees of Binding (x)

x	$(Lys)_n$	$(Orn)_n$	$(D,L-Orn)_n$
0.1	33.9	3.5	4.8
0.2	50.6	5.0	7.0
0.3	66.2	6.5	9.0
0.4	82.5	8.0	11.2
0.5	101	9.8	13.7

To summarize, the complex formation between cationic polypeptides and anionic surfactants is initiated through coulombic interaction and followed by a cooperative binding of successive surfactant ions. The nonpolar tails of clustered surfactant molecules can provide a hydrophobic environment that stabilizes an ordered conformation — a helix or a β-form, which depends not only on the number of methylene groups of lysine and its homologs but also on the chain length of the surfactants. Excess surfactant molecules may cluster onto the bound surfactant ions in a "double layer" fashion with their polar heads exposed to the aqueous medium and nonpolar tails shielded from the polar environment. For the β-pleated sheets a bilayer can also be formed between bound surfactant molecules which are sandwiched between two polypeptide chains.

ACKNOWLEDGMENTS

We thank Drs. I Satake, K. Shirahama, Y. W. Tseng and K. Ikeda for their collaborative work during their stay in this laboratory, which makes this review possible. This work was supported by U.S. Public Health Service Grant GM-10880-23.

REFERENCES AND NOTES

1.   A preliminary account of this work was published in Polymer
      Preprints (Am. Chem. Soc.) 1982, 23, 109.
2.   Doty, P.; Holtzer, A. M.; Bradbury, J. H.; Blout, E. R. J. Am.
      Chem. Soc. 1954, 76, 4493. Doty, P.; Bradbury, J. H.; Holtzer, A.
      M. Ibid. 1956, 78, 947.
3.   Blout, E. R.; Asadourian, A. J. Am. Chem. Soc. 1956, 78, 955.
4.   Cf. Yang, J. T.; Doty, P. J. Am. Chem. Soc. 1957, 79, 761.
      Moffitt, W.; Yang, J. T. Proc. Natl. Acad. Sci. U.S.A. 1956, 42,
      596.
5.   Two lectures on the "Researches on the molecular dissymmetry of
      natural organic products," Presented by L. Pasteur to the Chem-
      ical Society of Paris on Jan. 20 and Feb. 3, 1860. Translated from
      "Lecons de chimie professees en 1860," by W. S. W. Ruschenberger
      and reprinted in the Am. J. Pharm. 1862, 34 (Ser. 3, v. 10), 1, 97.
6.   Cf. Greenfield, N.; Fasman, G. D. Biochemistry 1969, 8, 4108.
      Gratzer, W. B.; Cowburn, D. A. Nature (London) 1969, 222, 426.
7.   Cf. Yang, J. T.; Chen, G. C.; Jirgensons, B. in "Handbook of
      Biochemistry and Molecular Biology, Proteins, vol. III," 3rd Ed.
      (Fasman, G. D., Ed.), CRC Press, Cleveland, Ohio, 1976, pp. 3.
8.   Chen, G. C.; Yang, J. T. Analy. Lett. 1977, 10, 1195.
9.   Poland, D.; Scheraga, H. A. "Theory of Helix-Coil Tranisitons in
      Biopolymers," Academic Press, New York, 1970.
10.   Applequist, J.; Doty, P. in "Polyaminoacids, Polypeptides and
      Proteins" (Stahmann, M. A., Ed.), University of Wisconsin, Madison,
      Wisconsin, 1962, pp. 168.
11.   Wooley, S.-Y. C.; Holzwarth, G. Biochemistry 1970, 9, 3604.
12.   Tseng, Y. W.; Yang, J. T. Biopolymers 1977, 16, 921.
13.   Grourke, M. J.; Gibbs, J. H. Biopolymers 1967, 5, 586. Chaudhuri,
      S. R.; Yang, J. T. Biochemistry 1968, 7, 1379.
14.   Hatano, M.; Yoneyana, M. J. Am. Chem. Soc. 1970, 92, 1392.
15.   Kubota, S.; Yang, J. T. Biopolymers 1979, 18, 743.
16.   Gratzer, W. B.; Doty, P. J. Am. Chem. Soc. 1963, 85, 1193.
17.   Yamamoto, H.; Yang, J. T. Biopolymers 1974, 13, 1093.
18.   Brahms, J.; Spach, J. Nature (London) 1963, 200, 72.
19.   Doty, P.; Wada, A.; Yang, J. T.; Blout, E. R. J. Polymer Sci.
      1957, 23, 851.
20.   Kubota, S.; Gaskin, F.; Yang, J. T. J. Am. Chem. Soc. 1972, 94,
      4328.
21.   Satake, I.; Yang, J. T. Biochem. Biophys. Res. Commun. 1973, 54,
      930.
22.   Sarkar, P. K.; Doty, P. Proc. Natl. Acad. Sci. U.S.A. 1966, 55, 981.
23.   Satake, I.; Yang, J. T. Biopolymers 1975, 14, 1841.
24.   Feldshtein, M. M.; Zezin, A. B.; Gragerova, J. J. Biokhimia 1972,
      37, 305. Zezin, A. B.; Felshtein, M. M.; Merzlov, V. P.; Maletina,
      J. J. Molekul. Biol. 1973, 7, 174.
25.   Takeda, K.; Iba, A.; Shirahama, K. Bull. Chem. Soc. Japan 1981,
      54, 1793.

26. Ikeda, K.; Yang, J. T. Intl. J. Peptide Protein Res. 1980, 16, 225.
27. Abramson, M. B.; Katzman, R.; Wilson, C. E.; Gregor, H. P. J. Biol. Chem. 1964, 239, 4066.
28. Satake, I.; Yang, J. T. Biopolymers 1976, 15, 2263.
29. Grourke, M. J.; Gibbs, J. H. Biopolymers 1971, 10, 795.
30. Nagasawa, M.; Holtzer, A. M. J. Am. Chem. Soc. 1964, 86, 538.
31. Kubota, S.; Ikeda, K.; Yang, J. T. Biopolymers 1983, 22, 2219; ibid., 2237.
32. Zimm, B. H.; Bragg,, J. K. J. Chem. Phys. 1959, 31, 526.
33. Shirahama, K.; Yang, J. T. Intl. J. Peptide Protein Res. 1979, 13, 341.

STRUCTURAL COMPLEXES OF CATIONIC POLYSOAP-PHOSPHOLIPID

Lisbeth Ter-Minassian-Saraga
Physico-Chimie des Surfaces et des Membranes
Equipe de Recherche du CNRS
associée à l'Université Paris V
UER Biomédicale
45 rue des Saints-Pères
75270 Paris cedex 06
FRANCE

ABSTRACT

The interaction between an acidic phospholipid, the natural (wheat) phosphatidylinositolmonophosphate PI and a linear cationic polysoap the poly(2-methyl-5-vinyl-hexylpyridinium bromide) PVPC6 has been studied with mixed spread monolayers and with hydrated (40%, w/w) mixed bilayers. The "electrostatic" interaction between PI and PVPC6 involves monolayer condensation and affects the bilayers hydration. In addition, the free energy of the bilayers structural water is modulated by this interaction.

Analysis of the mixed PI + PVPC6 monolayer composition demonstrated that the "electrostatic" interaction is in fact the result of an ion exchange reaction between the monolayer constituents. This reaction modulates the monolayer and the multibilayer properties considered. It is stressed that the multibilayer hydration is relevant to biological membrane stability, adhesion or fusion.

INTRODUCTION

The present review is based on published material.[1-4] The underlying studies dealt with the behavior of one cationic linear polysoap, the poly(2-methyl-5-vinyl-hexylpyridinium bromide) PVPC6 incorporated into hydrated phospholipid two-dimensional planar systems. They are either monolayers[1-3] spread at the air-water interface or

333

fully hydrated multibilayers.[4]   Two phospholipids were used: the natural (wheat) NaPhosphatidyl-inositol-monophosphate PI and the synthetic 1,2-dipalmitoyl-3-sn-phosphatidyl-choline DPPC.   The positive cooperativity of PI association with PVPC6 in monolayers and the analytical expression for the binding isotherm were reported in 1978.[2] In contrast to PI, DPPC is bound by PVPC6 in a non-cooperative mode.[4] Whatever the mode of phospholipid binding by PVPC6 we show[4] that the hydration of the lipids and of PVPC6 is not an additive property when association occurs.   As we assume that PVPC6 is a model for an integral membrane protein, we believe that the hydration studies results may be relevant to biological membrane changes of hydration following the modification of its lipid and protein composition and state.

MATERIALS AND METHODS

Materials

The average molecular mass of poly(vinyl-pyridine) is 290.000 Daltons. It is 88% quaternized by hexylbromide. The molecular mass of PI is 824 Daltons. That of DPPC is 734 Daltons.

The monolayers are spread from a solution in a chloroform/methanol mixed solvent on an aqueous KCl 10^{-3} M + HCl 10^{-5} M substrate. Eventually KSCN is added to it up to 10^{-4} M.

The experiments were carried out at 23°-25°C.

Methods

a)   Monolayer studies

Surface pressure.   Various amounts of molecules were deposited onto the aqueous substrate surface contained either in a small dish ($\sim$70 sq.cm cross section)[1] or inside a large Langmuir through made by Lauda FRG.[2,3]   The average (per monomer or per polar head) area a [sq.nm/polar head] was deduced from the film area A and the number of spread PI molecules and PVPC6 monomers under various conditions.

In the case of small areas A, the film surface pressure Π [mNm^{-1}] was determined with an accuracy of 0.2 mNm^{-1} from the difference γ_0 - γ between the surface tensions of the pure and of the film covered substrate respectively. The tension was measured by the Wilhelmy technique using a roughened Pt slide and a Sanborn force transducer made by Hewlett-Packard.[1]   The accuracy of the surface pressure manometer Lauda is about 0.2 mNm^{-1} too. The variation of the surface pressure Π with average area per molecule a (not shown) was established for the mixed films PVPC6 + PI for which x_{PIt} is 0, 0.1, 0.2, 0.3, 0.4, 0.5, 0.6, 0.7, 0.8, 0.9, 1 and where x_{PIt} is the mol fraction of spread PI inside the films.[1] The excess area per polar head a^E at a given surface pressure Π in a mixed film of given x_{PIt} was deduced. By definition it is equal to:

$$a^E = a - x_{PIt}\, a_{PI}^o - (1 - x_{PIt})\, a_+^o \qquad (1)$$

where

a_{PI}^o, a_+^o are the molecular and monomer areas for the pure PI and PVPC6 monolayers respectively. Negative values of a^E imply either net attractive forces or association between PI and PVPC6 molecules.

Surface potential. The (connected) two-half cells and our differential method for measuring the surface potential of monolayers, described in,[5] were used for this study.

The two-half cells contain the aqueous substrate. The mixed monolayers are spread onto the surface of one of the two-half cells. Two "identical" [241]Am 0.7 mCi ionizing electrodes, purchased from the Radiochemical Center Amersham (U.K.), are placed above the half cells close to the substrate surface. The difference in electric potential of the two electrodes is measured with a high impedence electrometer. By definition[5] this difference is equal to the surface potential ΔV of the monolayer located on the surface on one of the two-half cells. The surface potential ΔV is measured as a function of the spread monolayer molecular area a for each one of the mixed film compositions as above (see surface pressure studies).

Ionic composition of mixed ionized monolayers. Both PVPC6 and PI form ionized monolayers insoluble in the aqueous substrates under our (10^{-3} KCl) experimental conditions. Then the ions Cl^- and K^+ constitute the counter ions of PVPC6 and PI respectively.

When the substrate of a pure PVPC6 monolayer contains both Cl^- (10^{-3} N) and ^{14}C labelled SCN^- (10^{-4} N) anions the measurement of the monolayer β-activity, using a thin window β-ray detector[6] placed above the substrate, indicates the high selectivity of the M^+ monomers of PVPC6 for SCN^- ions. Subsequently[1,2] we use the ^{14}C labelled SCN^- ions as indicators of the unassociated (with PI) or free M^+ monomers of the mixed PVPC6 + PI monolayers. We consider that the following reaction takes place inside the monolayer:

$$PI\ K + poly\text{-}(M^+\ SCN^-) = PI^-\ M^+ + K\ S^{14}CN\ (soluble). \qquad (2)$$

It is found that, owing to the solubility of $K\ S^{14}CN$, the surface radioactivity decreases when the molecular ratio PI^-/M^+ increases i.e. the net monolayer (positive) charge density decreases.

The composition of the mixed monolayers is expressed as a

function of the mol fractions x_i of its three constituents: the $PI^- M^+$ ion-pairs and the "free" PI^- and M^+ groups.

The surface densities of the spread M^+ cationic monomers and of the adsorbed SCN^- ions are designated by δ_+, δ_{SCN^-}. Also $\delta_{SCN^-} \gg \delta_{Cl^-}$. They are expressed as molecules (or ions)/unit area.

We define a degree Θ of PI^- binding by M^+ according to [2]:

$$\Theta = \frac{\delta_+ - \delta_{SCN}}{\delta_+} . \tag{3}$$

If δ_{PIt} and δ_{PI} are the surface densities of spread and of free PI molecules and if a 1/1 monomer/PI complex or ion-pair is formed, we have:

$$\delta_{PIt} - \delta_{PI} = \delta_+ - \delta_{SCN}. \tag{4}$$

Knowing that the total surface density of the mixed monolayer is equal to $\delta_{PIt} + \delta_+$, from Equation [4] the mol fractions x_{PI} of "free" PI, x_+ of free M^+ and x_{PIM} of ion pairs in the monolayer are obtained. The original mol fraction x_{PIt} of the spread PI is defined as:

$$x_{PIt} = \frac{\delta_{PIt}}{\delta_{PIt} + \delta_+} . \tag{5}$$

Using [3], [4] and [5] we deduce the mol fractions x_{PIM}, x_{PI}, x_+ as functions of the original surface densities δ_{PIt} and δ_+.

Resistance to collapse of mixed PVPC6 + PI monolayers.[3] Compression-expansion cycles of mixed films are carried out under conditions comparable to those involved in the surface pressure and surface potential studies (see above).

b)   Multibilayer studies

The mixed hydrated multibilayers PVPC6 + PI are prepared from the solutions used for the spread film studies.[4] After evaporation of the organic solvent known amounts of water were added and equilibrium swelling is allowed for while evaporation of the water is avoided. The various mixed bilayers obtained were analyzed using low (t < 0°C) differential scanning calorimetry DSC as described in.[4,7] A Du Pont de Nemours thermal analyzer 990-910 equipped with a mechanical cooling accessory was used. Heating and cooling thermograms were obtained between -65°C and 70-80°C at 2°C min⁻¹ and 5°C min⁻¹ scanning rates.

The thermograms shown in Figure 1 are established for various PI + PVPC6 mixtures. As explained in,[7,8] the endothermic (exothermic) peaks correspond to melting of frozen water inside the sample (reference) cup. The opposition of the heat inputs inside the sample and reference cups produces the apparently anomalous exothermic peaks of the thermograms. The fact that these two peaks are found inside separate temperature intervals implies[7,8] that the bilayers perturb the thermal behavior of part of the water inside the sample cups.

The areas of the peaks in the thermograms are measured with a planimeter and the measured heats of melting are deduced. They are Q^s and Q^r for sample and reference respectively. Usually $Q^s < Q^r$. Dividing Q^r by the bulk molar enthalpy L_0 of pure ice melting, the mass m^s of the structural of bilayer perturbed water is obtained.[7,8] This mass m^s includes the strongly perturbed non-freezing nf water molecules which do not contribute to the heat Q^s of melting of the frozen water inside the sample cup. The nf water explains only partly the observed difference $\Delta Q = Q^r - Q^s$. In[4,7] we assume that the bilayers lower the structural water free energy and involve a local molar enthalpy of "ice" melting L. We use an original approach to deduce this (local) value of L from the shape of the thermogram in the temperature interval close and below 0°C.

For PI, a highly natural unsaturated phospholipid, the chain melting temperature occurs well below 0°C as shown by Figure 1. The area of these very low temperature distinct peaks provide the molar enthalpy of chain melting ΔH_{PI} for PI.

RESULTS AND DISCUSSION

a)   Spread films

The relevant results for PVPC6 + PI films are shown by Figure 2, Figure 3, Figure 4. Figure 2 represents the mol fractions of the three components M and PI free and associated MPI pairs inside equilibrated mixed films of various initial composition $0 < (x_{PIt}) < 1$. Around an initial composition of 1/1 M/PI the association is maximum. The effect of adding PI to PVPC6 produces condensation or contraction of the mixed film as shown in Figure 3. The excess area a^E is negative. Condensation is maximum for same 1/1 PI/M mixed film composition.

Finally, the interaction demonstrated by Figure 2 and Figure 3 is evident also in Figure 4 which shows that the surface potential of the PI + PVPC6 mixed film does not conform to an additivity rule.

The results of Figure 2 have been used in[2] to establish the binding isotherm of PI by PVPC6. This isotherm has the expression:

$$\Theta = \frac{x_{PI}\, Ke^{-2\Theta\alpha'}}{1 + x_{PI}\, Ke^{-2\Theta\alpha'}} \tag{6}$$

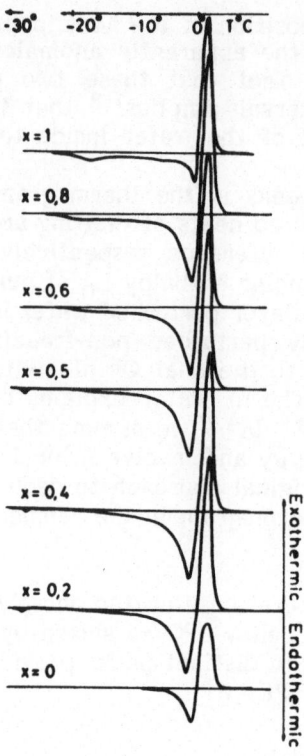

Figure 1.   Thermograms for PVP6 + PI hydrated multibilayers.
            40% (w/w) nominal hydration; x_{PIt}: mol fraction of PI in
            the dry mixture.   Scanning rate 5° min^{-1}.   Phospholipid
            "chain melting" corresponds to the lowest temperature ($\sim$ -
            20^oC) endothermic peaks.

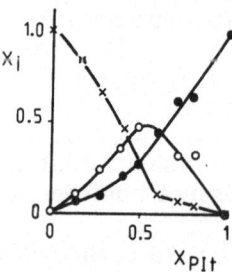

Figure 2.   Composition of the spread mixed PVPC6 + PI monolayers at
            the surface pressure 20 mNm^{-1}.
            x_{PIt}: original mol fraction of (spread) PI; x_i: mol fractions;
            x: free PVPC6 monomers M^+; •: free PI^- anions; o: M^+PI^-
            pair.   Substrate: KCl 10^{-3} M + HCl 10^{-5} M.

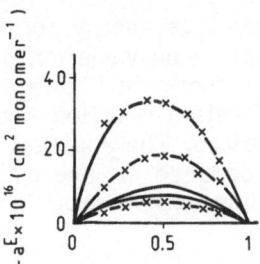

Figure 3.   Excess area a^E per polar head in the mixed PVPC6 + PI monolayers at given surface pressures. From the top: 1, 2, 4, 8, 16–20 mNm^{-1}. Substrate, x_{PIt} (see Figure 2).

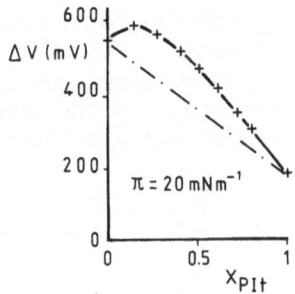

Figure 4.   Surface potential of various spread mixed PVPC6 + PI monolayers at the surface pressure Π = 20 mNm^{-1}. x_{PIt} and substrate (see Figure 2).

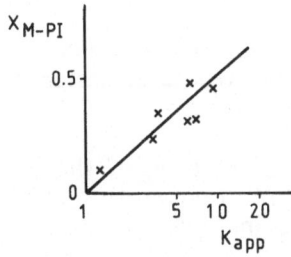

Figure 5.   Test of equation (7). x_{MPI}: mol fraction of associated M^+PI^- pairs. K_{app}: apparent binding constant.

where α' is the exchange free energy for mixing PI and PVPC6 free monomers and K a constant. The variation of Θ with the nominal mol fraction x_{PIt} in the film is shown in Figure 6a. This S shaped isotherm corresponds to the neutralization reaction Equation [2]. As shown in[1,2] the soluble counter-ions desorb. The apparent equilibrium constant K_{app} of the reaction[2] has been deduced in[2] and expressed as a function of x_i, the mol fractions of the film constituents. At given salt composition of the substrate we have:

$$K_{app} = K' \exp \{ - 2x_{MPI}\alpha' \} = \frac{x_{MPI}}{x_{PI}\, x_M} . \qquad (7)$$

Using the results of Figure 2, Equation [7] was tested as shown in Figure 5. We obtain $K' \simeq 1$ and $\alpha' \simeq -2.3$. This value of α' implies the positive cooperativity for the association described by Equation [2] and Figure 6a.

b)   Multibilayers

Figure 1 reproduces from[4] the thermograms for the system PI + PVPC6 including 40% (w/w) water. The shapes and areas of the low temperature peaks have been analyzed using the original approach described in.[4,7,8] Thus the shape of the peak corresponding to the melting of the frozen structural water is given by the expression:

$$\varepsilon(T) \simeq \dot{T}_p \frac{m_S L}{18\alpha} (- \frac{\hat{T}_{min}}{2})^{1/\alpha} (- \hat{T})^{- \frac{\alpha+1}{\alpha}} \qquad (8)$$

where $\varepsilon(T)$ is the deflection from the base line at a given temperature T and where $\dot{T}_p$, m_S, L, α, $\hat{T}_{min}$, $\hat{T}$ are the scanning rate, the mass of structural water, the molar enthalpy of melting of this water, a parameter relevant to the water-bilayer interaction, the location temperature of the endothermic peak maximum and a given shift from 0^oC. From the area of the same peak the heat of melting Q^S of a percentage of structural water is obtained. It is equal to:

$$Q^S = \frac{m_S L}{18} \times (\% \text{ of freezing and melting structural water}). \qquad (9)$$

In our approach, m_S, L are determined independently[4,7,8] and the percentage of non-freezing structural water nf is deduced from [9] and the thermograms in Figure 1.

The molar enthalpy of melting of the lipid chains of PI, ΔH_{PI} is obtained from the lowest temperature endothermic peaks in the thermograms as explained in.[4,7,8]

In Figure 6a we compare results for ΔH in bilayers with the PI binding isotherm or Θ in monolayers. Both experiments indicate positive cooperativity for PI-PVPC6 association.

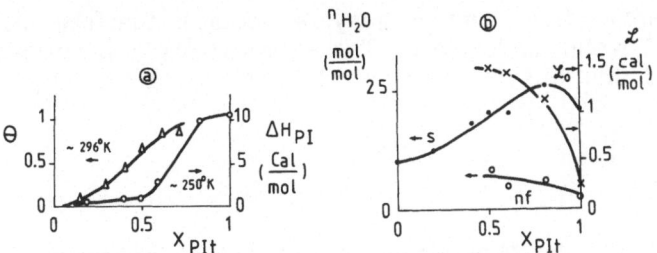

Figure 6.   Cooperativity of the PI and PVPC6 association in mixed
multibilayers and spread monolayers.   Effect of this
association on multibilayer structural water.

For x_{PIt} and composition of the systems, see Figure 2 and
Figure 1.

(a) Θ : degree of PI binding by PVPC6 in monolayers: ΔH_{PI}:
molar enthalpy of PI "chain melting" (DSC studies).

(b)   s: structural water, nf: non-freezing water; L, L_0
molar enthalpy of melting for the bilayers frozen water and
for the bulk ice; n_{H_2O}: moles of H_2O/monomer constituent
in the PI + PVPC6 dry mixtures.

In Figure 6b we plot the values obtained for the mass m_S of
structural water and for the non-freezing structural water nf (calculated per monomer unit) in multibilayers as well as the molar enthalpy
of melting L of the frozen structural water. For pure PI bilayers L is
smaller than L_0 for bulk water. When the net charge of the mixed film
decreases (x_{PIt} → 0.5) L increases and tends to the bulk value L_0.
Therefore when stronger interactions betwen the bilayer constituents
are established the neighboring water molecules are less perturbed.
Under the same conditions in monolayers of mixed films PI + PVPC6 the
strong interaction between the film constituents involves film condensation (Figure 3). One may speculate that monolayer condensation by
strong electrostatic interactions, on one hand, and the modification of
structural water properties following the same interactions inside mixed
bilayers, on the other hand, are related phenomena.

CONCLUSION

The association of the acidic phospholipid PI with the cationic
linear polysoap PVPC6 is cooperative in both mixed spread monolayers
and mixed swollen multibilayers.[9] This association has a condensation
effect on the spread mixed monolayers on one hand and affects the
properties of the interbilayer structural water of swollen mixed
multibilayers on the other hand. Namely the increase in the degree of

PI-PVPC6 association in mixed bilayers increases the free energy resp. enthalpy of water molecules. These are strongly perturbed by the phospholipid multibilayers and possibly also by the spread phospholipid monolayers.

REFERENCES

1. Homma, I.; Ter-Minassian-Saraga, L. "Electrical Phenomena at the Biological Membrane Level"; Elsevier Sci. Publ.: Amsterdam, pp. 273-286, 1977.
2. Ter-Minassian-Saraga, L. J. Colloid Interface Sci. 1979, 70, 245-264.
3. Ter-Minassian-Saraga, L. J. Colloid Interface Sci. 1981, 79, 222-236.
4. Ter-Minassian-Saraga, L.; Madelmont, G. J. Colloid Interface Sci. 1981, 81, 369-384.
5. Plaisance, M.; Ter-Minassian-Saraga, L. J. Colloid Interface Sci. 1972, 38, 489-495.
6. Plaisance, M.; Ter-Minassian-Saraga, L. J. Colloid Interface Sci. 1976, 56, 33-41 and 1977, 59, 113-122.
7. Ter-Minassian-Saraga, L. Cryo-Letters, 1981, 2, 161-166.
8. Ter-Minassian-Saraga, L.; Madelmont, G. J. Colloid Interface Sci. 1982, 85, 375-388.
9. Note added in proof. The same conclusion has been arrived at by Sixl, F.; Galla, H-J. Biochim. Biophys. Acta 1982, 693, 466-478, who used the approach described in[2] for studying the interaction between an acidic phospholipid and a low molecular weight, postively charged protein polymixin. Therefore PVPC6 is a reasonable model for the study of protein-phospholipid interaction.

HIGH-SENSITIVITY DIFFERENTIAL SCANNING CALORIMETRY OF POLYMER-PHOSPHOLIPID MIXTURES

David A. Tirrell, Anne B. Turek, Kenji Seki, and
Doreen Y. Takigawa
Department of Chemistry
Carnegie-Mellon University
Pittsburgh, PA 15213

ABSTRACT

The interactions of synthetic polymers with cell surfaces dominate many processes currently in use in biomedical science and technology, yet these interactions are very poorly understood. High-sensitivity differential scanning calorimetry reveals important changes in the structural organization of phospholipid bilayers which are treated with water-soluble synthetic polymers. These effects are apparent even at polymer concentrations of 0.1% or less. Polymer-phospholipid interactions may be modulated by variations in polymer charge, charge density, and chain microstructure. These experiments constitute a first step toward the development of a logical basis for the design of synthetic polymers with well-defined biological properties.

The interactions of synthetic polymers with the surfaces of plant and animal cells are of major importance in many areas of biomedical science and technology (cf. Table I). Insoluble polymeric surfaces are currently in widespread use as organ and tissue implants in humans[1], and are finding increasing application in the separation of mixed cell populations[2] and in fundamental experiments probing biomembrane structure[3]. Soluble polymers may also be used in cell separation processes,[4] as well as in cell fusion,[5] and in studies of membrane processes such as endocytosis.[6] Rather more speculative applications of synthetic polymers in biotechnology include use as drugs,[7] food additives,[8] agricultural chemicals,[9] and cosmetics.

Table I

Synthetic Polymers in Biomedical Science and Technology

Insoluble Polymers	Soluble Polymers
Organ and tissue implants	Cell fractionation
Cell fractionation	Cell fusion
Probes of biomembrane structure	Endocytosis
	Drugs
	Food additives
	Agricultural chemicals
	Cosmetics

If these projected uses are to be realized, a body of fundamental knowledge concerning polymer-cell surface interactions must be developed. The flexibility available to synthetic polymer chemists interested in preparing materials with useful biological properties is immense, but at this point we just do not know what kinds of polymers to make. There are a number of very general empirical rules, but their power is limited.

Our work on polymer-cell surface interactions addresses three points: (i) the site of interaction, (ii) the nature and magnitude of the forces involved, and (iii) the extent of perturbation of biomembrane structure and function which results from the interaction. We have recently described a new photolabeling technique which we feel will provide information concerning the binding sites of synthetic polymers on cell surfaces.[10,11] In this chapter, we summarize the results of calorimetric experiments which reveal significant changes in the structure of model biomembranes prepared in the presence of synthetic polymeric solutes.[12-15] We find such structural changes to depend very strongly on polymer structure, suggesting that synthetic polymer chemists may in fact be able to manipulate biomembrane properties within wide limits.

HIGH SENSITIVITY DIFFERENTIAL SCANNING CALORIMETRY

The details of our experimental procedures are provided elsewhere,[12-15] but it is useful here to describe the calorimetric experiment in a general way. The instrument used throughout this work is a Microcal, Inc., MC-1 Scanning Microcalorimeter, an adiabatic instrument capable of resolving temperature-dependent changes in heat capacity of 2×10^{-5} cal/deg/cm^3. The high sensitivity of this instrument allows the use of dilute aqueous samples and low scanning rates (ca. 0.15 - 0.20°C/min), so that the system is held close to equilibrium throughout the run. Mabrey and Sturtevant have reviewed the applications of high-sensitivity DSC in the study of biomembranes and related model systems.[16]

A model system which is most useful is a dispersion of a pure phospholipid in water or in aqueous buffer, prepared with vigorous mechanical agitation. Under these conditions, phospholipids form "multilamellar suspensions," sometimes called "liposomes," which consist of concentric spherical lipid bilayers separated from one another by excess water. The assumption is that one can gain some understanding of the lipid bilayer component of natural biomembranes through the study of these simpler molecular systems.

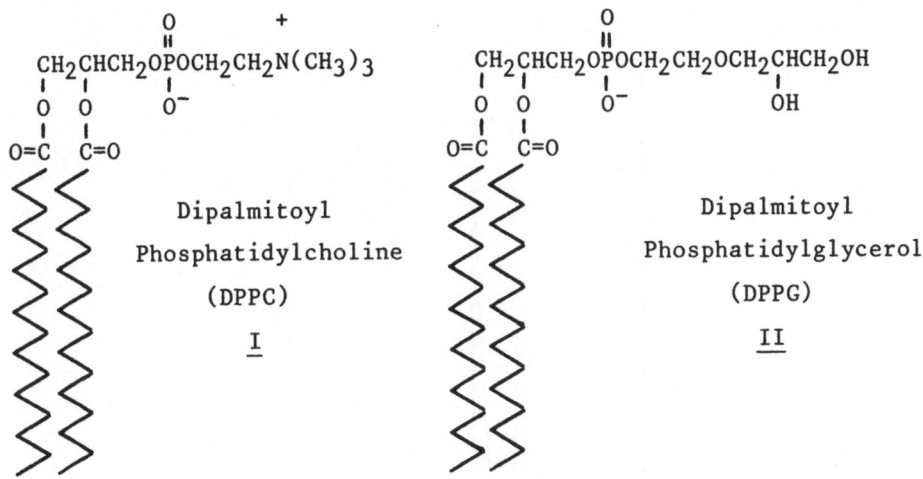

The phosphatidylcholines (I) and the phosphatidylglycerols (II) represent typical phospholipid structures, with a polar phosphodiester "headgroup" attached via a diacylglycerol unit to two long hydrocarbon "tails." At low temperatures, the tails of saturated phospholipids with identical acyl chains adopt a conformation in which all (or very nearly all) of the C-C bonds are trans. As the temperature is raised, one observes a phase transition resulting from a cooperative disordering of the hydrocarbon tails -- a process which requires 8-10 kcal/mole for two C_{16} chains.[16,17] This order-disorder transition is studied most simply and directly by high-sensitivity DSC.

We will show in the following section of this chapter that water-soluble synthetic polymers can be used to modify the phase transition behavior of model biomembranes. In these experiments, 1 mg of dry lipid is dispersed with vigorous shaking in 1 ml of buffer containing 1 mg of the polymer of interest. The transition behavior of the polymer-phospholipid mixture is in all cases compared to that of a control sample of lipid hydrated in the absence of polymer.

In designing synthetic polymers with useful biological properties, one can control a number of structural variables. Among these, one might expect electrostatic charge to be of paramount importance. The following discussion is organized in terms of polymer charge type, i.e., polycations, polyanions, uncharged polymers, and zwitterionic polymers are discussed in turn.

HIGH-SENSITIVITY DIFFERENTIAL SCANNING CALORIMETRY OF POLYMER-PHOSPHOLIPID MIXTURES. POLYCATIONS.

The polycations which we have studied most thoroughly are the ionenes of structure III.[18] These polymers are extremely useful in

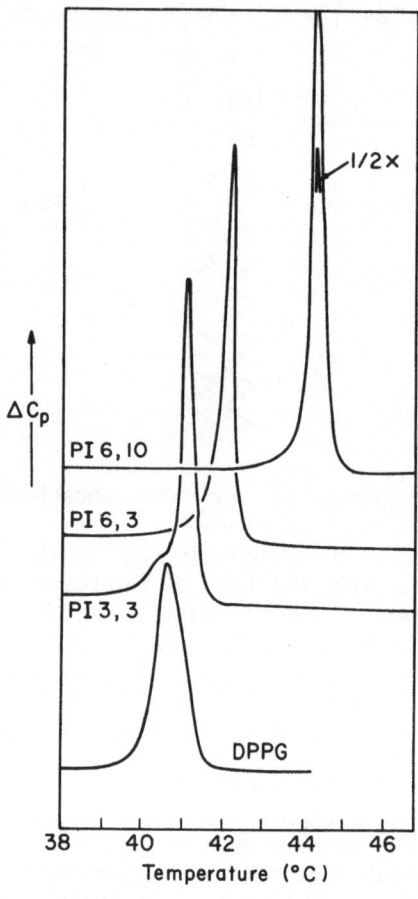

Figure 1.   Melting endotherms for DPPG (1 mg/ml) suspended in 50 mM Tris-HCl buffer, pH 7.4. Bottom curve: DPPG alone. Upper curves: Ionenes as labeled, at a concentration of 1 mg/ml.

$$\begin{array}{ccc} CH_3 & Br^- & CH_3 & Br^- \\ | & & | & \\ \text{[}-N^+-(CH_2)_x-N-(CH_2)_y\text{]} & & & \text{Ionene-x,y} \\ | & & | & \\ CH_3 & & CH_3 & \end{array}$$

III

studies of polymer-biomembrane interaction, because of the simplicity of their structures, and because their charge density is readily varied in a systematic way. In addition, though, the ionenes display interesting biological properties in their own right, including antimicrobial and antifungal activity, and Rembaum has speculated that the inhibition of bacterial growth by ionenes may result from adsorption of the polymers at the bacterial cell surface, causing changes in membrane transport characteristics.[18] We show below that ionenes do in fact perturb the organization of pure dipalmitoyl phosphatidylglycerol suspensions, and the presence of phosphatidylglycerols in bacterial cell membranes suggests that these results may be relevant to the mechanism of the ionenes' antibacterial actions. Our primary concern, however, is the fundamental nature of the polymer-phospholipid interaction.

Figure 1 shows the main phase transition region for dipalmitoyl phosphatidylglycerol (DPPG) suspended in 50 mM Tris-HCl buffer, pH 7.4, in the absence of polymer and in the presence of ionenes 3,3; 6,3; and 6,10. The main phase transition (T_m) in DPPG occurs at 40.7°C under these conditions. Earlier work on DPPG, using lower-sensitivity calorimeters and higher scan rates, gave similar values for T_m.[19,20] The transition enthalpy, obtained by integrating the melting peak, is found to be 8.8 kcal/mole, in good agreement with the results of Findlay and Barton,[20] but about 1 kcal/mole larger than the value reported by Jacobson and Papahadjopoulos.[19]*

The presence of ionene polymers in the suspending medium causes visible aggregation of the lipid suspension, an increase in the transition temperature and transition enthalpy, and a slight narrowing of the transition. As is apparent from Figure 1, the magnitude of each of these effects increases upon changing the polymer from ionene-3,3 to ionene-6,3 to ionene-6,10, i.e., with decreasing positive charge density on the polymer chain. This is perhaps surprising in view of the fact that the lipid surface is negatively charged due to the presence of the uncompensated phosphate anion in DPPG. We will comment on this result below.

---

*In our most recent experiments, we have observed sharper transitions associated with larger enthalpy changes (10.0 kcal/mole). Since the source and lot number of the lipids have remained the same, the origins of the differences in melting behavior are unknown. However, the ionene-induced changes in melting behavior are the same, irrespective of the details of the control melting endotherm.

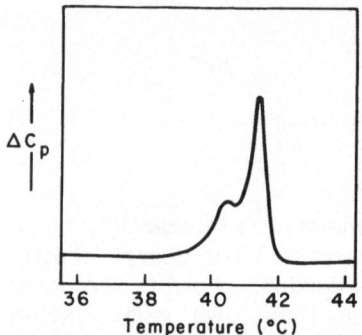

Figure 2.   Melting endotherm for DPPG (1 mg/ml) suspended in 50 mM
             Tris-HCl buffer, pH 7.4, containing 0.15 mg/ml of ionene-
             6,3. The ratio of polymer bound ammonium sites to lipid
             phosphate sites is 0.6:1.

The addition of polycations to phospholipid suspensions might be
expected to produce changes in lipid organization similar to those
observed upon addition of divalent metal cations. To an extent this is
true. Jacobson and Papahadjopoulos[19] report that titration of DPPG
suspensions with increasing concentrations of $CaCl_2$ shifts the melting
transition to higher temperatures, and Sacre et al.[21] describe similar
observations concerning dilauroyl phosphatidylglycerol (DLPG).   The
zwitterionic lipid dipalmitoyl phosphatidylcholine (DPPC) is essentially
unaffected by low concentrations of Ca^{2+}.[19] As shown in Figure 1, the
ionenes also shift the DPPG melting endotherm, though the effect is
smaller than that of Ca^{2+}. We also find that the transition behavior of
DPPC is unaffected by the addition of ionene polymers at these
concentrations.

A third and perhaps most interesting similarity in the behavior
of divalent cations and polycations, is shown in Figure 2: at less than
stoichiometric concentrations of ionene-6,3, we observe two separate
melting transitions in DPPG samples -- an unshifted peak at 40.7°C, and
a second peak at 41.7°C. The most straightforward interpretation of
this result is that the polymer causes separation of the lipid into two
phases -- a polymer-treated phase melting at 41.7°C and an unmodified
lipid phase melting at 40.7°C. Whether these two phases coexist within
a single bilayer (or aggregate of bilayers) is unknown. Papahadjopoulos
and coworkers have observed multiple endotherms in DPPG suspensions
treated with poly-L-lysine,[22] and Sacre et al. describe similar multiple
transitions in Ca^{2+}/DLPG mixtures containing a deficiency of Ca^{2+}.[21]

Calorimetry has also provided insight into the importance of polymer molecular weight in modulating ionene-DPPG interactions. For ionene-6,3, variation of the molecular weight from 3500 to 16,000 produced no observable change in the calorimetric behavior of mixtures of the polymer with DPPG (i.e., the transition temperature is increased to 42.2°C in each case). On the other hand, compound IV, a dimer model of ionene-6,3, leaves the DPPG melting endotherm unchanged. Thus, while the structural alteration caused by ionene binding is not strongly dependent on molecular weight, it appears that some minimum number of ammonium sites must be enchained before any perturbation is observed. Experiments in progress will define this lower limit.

$$(CH_3)_3 \overset{+}{N} \overset{Br^-}{\underset{}{}} (CH_2)_6 \overset{+}{N} \overset{Br^-}{\underset{}{}} (CH_3)_3$$

IV

The observed increases in the melting temperature of DPPG in the presence of ionene polymers suggest that the polymers stabilize the low temperature (ordered) lipid phase. Support for this idea comes from Raman spectroscopy,[15] which shows that adsorption of ionene-6,3 restricts the gradual introduction of gauche rotamers and the gradual loosening of the hydrocarbon chain packing which are observed on warming untreated DPPG suspensions below T_m. We suggest that the ionenes adsorb on the charged headgroup region of DPPG, with significant ion-pairing of polymer-bound ammonium sites with the acidic phosphate sites of the lipid surface. The more pronounced effects of ionene-6,10 may then reflect the greater flexibility of the polymer chain (a consequence of its lower charge density) and a more effective matching of ionic sites at the surface.

Although it is not our objective to elucidate the mechanism of antibacterial action of ionene polymers, comparison of our calorimetric results with the antibacterial data of Rembaum[18] reveals an interesting parallel. In Table II are listed the minimum concentrations of ionene polymers which are required for inhibition of bacterial growth. The antibacterial activity of the ionene polymers increases with decreasing charge density, and the six-carbon dimer model is inactive. Calorimetry shows that perturbation of phosphatidylglycerol structure by ionenes also increases with decreasing polymer charge density, and that the dimer model leaves the lipid structure unchanged. The significance of this correlation is unknown.

Table II

Antibacterial Activity of Ionenes and Model Compounds[18]

	Minimum Inhibitory Concentration (ppm)	
Ionene	S. Aureus	E. Coli
3,3	> 128	> 128
6,6	16	16
6,10	4	4
$(CH_3)_3N(CH_2)_6N(CH_3)_3$	Inactive	

POLYANIONS

While the experiments just described address the effects of polymer charge density and molecular weight on polymer-phospholipid interactions, our work on polyanions concerns a third structural variable: chain microstructure.

Poly(α-ethylacrylic acid)s of variable tacticity can be prepared by anionic polymerization of methyl α-ethylacrylate, followed by exhaustive hydrolysis of the pendant ester groups. (Scheme I)[14]   Table III gives conditions for preparation of three such samples.

Scheme I

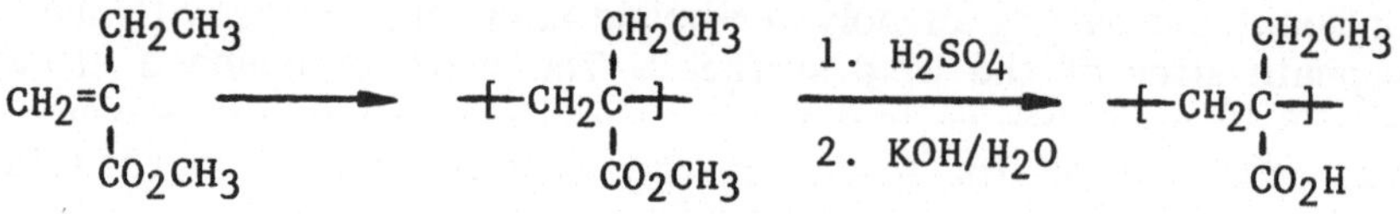

Table III

Poly(α-ethylacrylic acid)s of Variable Microstructure[a]

Sample	Initiator	Solvent	Temperature (°C)	Triad Tacticity(%)		
				I	H	S
1	$(i\text{-}Bu)_2AlNPh_2$	Toluene	-40	1	11	88
2	n-BuLi	Toluene	-78	49	18	33
3	n-BuLi	Toluene	0	91	7	2

[a]Conditions for preparation of poly(methyl α-ethylacrylate) precursors.

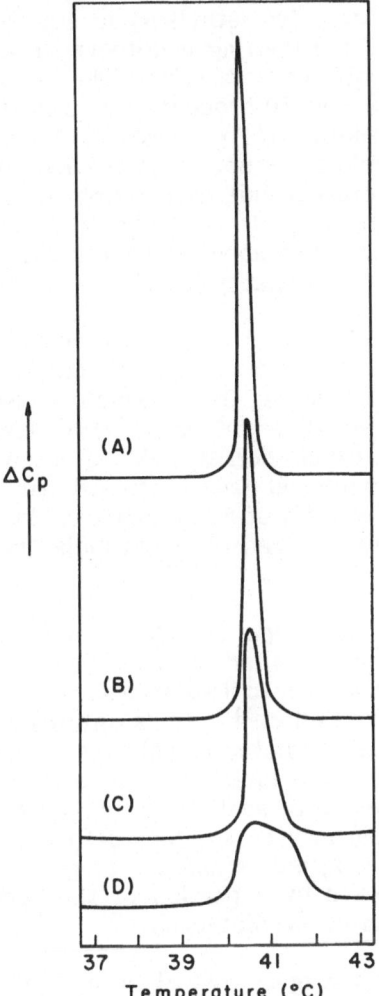

Figure 3.   Melting endotherms for DPPC (1 mg/ml) suspended in 50
mM Tris–HCl buffer, pH 7.4. Curve A: DPPC alone. Curve
B: DPPC plus Sample 1. Curve C: DPPC plus Sample 3.
Curve D: DPPC plus Sample 2. Polymer concentration 1
mg/ml in curves B–D.

Figure 3 shows the melting behavior of mixtures of these polymers
with the zwitterionic lipid DPPC, a synthetic analogue of the choline
lipids which predominate in the outer monolayer of the lipid bilayer of
mammalian cell membranes. As shown in curve A, DPPC alone melts
sharply at 40.4°C, in good agreement with the results of previous

workers.[16]   Addition of poly(α-ethylacrylic acid)s to DPPC suspensions causes a broadening of the melting endotherm in all cases (curves B-D), but the effect of polymer microstructure is quite apparent.   The syndiotactic polymer (curve B) broadens the transition only very slightly (the transition half-width $\Delta T_{1/2}$ increases from $0.21^{\circ}C$ to $0.30^{\circ}C$), suggesting only very minor disruption of the hydrocarbon chain packing. In terms of the cooperative unit description of transition width,[16] the size of the cooperative unit is reduced from approximately 420 molecules in the pure lipid to about 290 molecules in the sample treated with syndiotactic poly(α-ethylacrylic acid).

The isotactic polymer is somewhat more disruptive ($\Delta T_{1/2}$ = $0.48^{\circ}C$, cooperative unit = 180 molecules), but it is the stereoirregular polymer — Sample 2, curve D — which causes the most severe broadening of the melting endotherm.   The transition half-width is increased to $1.10^{\circ}C$, which corresponds to a cooperative unit of 80 molecules (i.e., a symmetrical transition with equivalent ΔH_m and T_m would have a cooperative unit of 80 molecules). In each case, the effect of the polymer on the pre-transition parallels the effect on the main transition.

Hatada et al. have reported that poly(methyl α-ethylacrylate) prepared under conditions similar to those used for the preparation of Sample 2 can be separated into two fractions, one highly isotactic and the other highly syndiotactic.[24]   It is interesting, then, that a 1:1 mixture of isotactic and syndiotactic polymers (Samples 1 and 3) causes almost no broadening of the melting endotherm of DPPC, even though the average microstructure of such a mixture (46% iso-, 9% hetero, and 45% syndiotactic triads) is very similar to that of Sample 2.   We have not fractionated Sample 2, but it appears at present that the disruption of lipid packing is caused by a portion of the sample which is neither highly isotactic nor highly syndiotactic.

These observations suggest that chain microstructure can influence in a significant way the interaction of synthetic polymers with lipid surfaces.   An understanding of these results must await a more thorough study of the conformational properties of poly(α-ethylacrylic acid)s in aqueous solution.

UNCHARGED POLYMERS

We have examined the interactions of seven uncharged polymers — poly(N-vinylpyrrolidone) (PVP), poly(acrylamide), poly(N,N-dimethyl-acrylamide), poly[N-(2-hydroxypropyl)methacrylamide] (PHPMA), poly-(vinyl alcohol), poly(ethylene oxide), and dextran — with multilamellar suspensions of DPPC or DPPG, under conditions similar to those used in the work described above.   In none of these experiments have we observed significant changes in the phase transition behavior of the phospholipid suspensions. These results suggest that electrostatic forces are likely to dominate polymer-lipid interactions, and that uncharged polymers will in general be only weakly adsorbed at cell surfaces. This is consistent with the observations of Lloyd and Kopecek and their coworkers,[25] who have found the rate of endocytic uptake of PVP and PHPMA by rat visceral yolk sac tissue to be equal to the rate of uptake of the incubation medium. Significant adsorption of the polymer on the tissue surface would require a higher rate of accumulation of the polymeric solute. Of course, the calorimetric technique does not measure adsorption directly, and a correlation between our results and cell surface adsorption remains to be established.

ZWITTERIONIC POLYMERS

Our experiments with zwitterionic polymers are in a very preliminary stage, and have concerned only a single polymer: poly(α-amino-acrylic acid). Suspensions of DPPC or DPPG are unaffected by this polymer at pH 7.4 or at pH 10. These experiments are continuing.

CONCLUSIONS

High-sensitivity differential scanning calorimetry of polymer-phospholipid mixtures reveals: (i) a stabilization by polycations of the low temperature ordered phase of acidic phospholipids, (ii) a dependence of this stabilization on polymer charge density, (iii) an insensitivity of this effect to polymer molecular weight in the range of 3500-16,000, (iv) a requirement for enchainment of more than two cationic sites before ordered phase stabilization is observed, (v) a separation of acidic lipid phases in the presence of less than stoichiometric amounts of polycations, (vi) an insensitivity of zwitterionic lipids to the addition of polycations, (vii) a parallel — perhaps coincidental — between perturbation of lipid bilayer organization and the antibacterial potency of synthetic polycations, (viii) the importance of chain microstructure in controlling the interaction of polyanions with zwitterionic lipid bilayers, and (ix) a lack of perturbation of bilayer structure by uncharged polymeric solutes.

We believe that continued exploration of polymer-lipid and polymer-biomembrane interactions by this technique will contribute to the development of new polymers with useful and predictable biological properties.

REFERENCES

1.   S. D. Bruck, Blood Compatible Synthetic Polymers, Charles C. Thomas, Springfield, 1974.
2.   K. Katoaka, M. Maeda, T. Nishimura, Y. Nitadori, T. Tsuruta, T. Akaike and Y. Sakurai, J. Biomed. Mat. Res., 14, 817 (1980).
3.   D. I. Kalish, C. M. Cohen, B. S. Jacobson and D. Branton, Biochim. Biophys. Acta, 506(1), 97 (1978).
4.   D. Fisher, Biochem. J., 196, 1 (1981).
5.   J. A. Lucy, in G. Poste and G. L. Nicolson, eds., Membrane Fusion, Elsevier, 1978, p. 267.
6.   M. K. Pratten, R. Duncan and J. B. Lloyd, in C. J. Ockleford and A. Whyte, eds., Coated Vesicles, Cambridge University Press, 1980, p. 179.
7.   L. G. Donaruma, R. M. Ottenbrite and O. Vogl, eds., Anionic Polymeric Drugs, Wiley, 1980.
8.   N. M. Weinshenker, in L. G. Donaruma and O. Vogl, eds., Polymeric Drugs, Academic Press, 1978, p. 17.
9.   G. G. Allan, J. W. Beer, M. J. Cousin, W. J. McConnell, P. C. Powell and A. Yahiaoui, ref. 8, p. 185.
10.   D. A. Tirrell, Proc. IUPAC 28th Macromolecular Symp., 372 (1982).
11.   V. Ramaswami and D. A. Tirrell, submitted for publication.
12.   D. A. Tirrell, Polym. Prepr., Am. Chem. Soc., Div. Polym. Chem. 22(2), 393 (1981).
13.   A. B. Turek and D. A. Tirrell, Polym. Prepr., Am. Chem. Soc., Div. Polym. Chem. 23(1), 56 (1982).
14.   K. Seki and D. A. Tirrell, Macromolecules, in press.
15.   D. A. Tirrell, A. B. Turek, D. A. Wilkinson and S. D. Merajver, submitted for publication.
16.   S. Mabrey and J. M. Sturtevant, in E. D. Korn, ed., Methods in Membrane Biology Vol. 9, Plenum, 1978, p. 237.
17.   J. F. Nagle, Ann. Rev. Phys. Chem., 31, 157 (1980).
18.   A. Rembaum, Appl. Polym. Symp., 22, 299 (1973).
19.   K. Jacobson and D. Papahadjopoulos, Biochemistry, 14, 152 (1975).
20.   E. J. Finlay and P. G. Barton, Biochemistry, 17, 2400 (1978).
21.   M. M. Sacre, W. Hoffmann, M. Turner, J. F. Tocanne and D. Chapman, Chem. Phys. Lipids, 69, 69 (1979).
22.   D. Papahadjopoulos, M. Moscarello, E. H. Eylar and T. Isac, Biochim. Biophys. Acta, 401, 317 (1975).
23.   The most widely quoted value for the melting point of DPPC in aqueous multilamellar suspension is $41.4^{\circ}C$ (e.g., ref. 16). There is not at the present time a standard procedure for calibration of the absolute temperature scale of the MC-1 calorimeter, so we are reporting uncorrected melting temperatures. This in no way affects the conclusions based on this work.
24.   K. Hatada, S. Kokan, T. Niinomi, K. Miyaji and H. Yuki, J. Polym. Sci., Polym. Chem. Ed., 13, 2117 (1975).
25.   J. Kopecek, personal communication, July 1982.

B. SYNTHETIC SYSTEMS

FLUORESCENCE PROBE STUDIES OF THE AGGREGATION STATE OF SURFACTANTS IN AQUEOUS POLYMER SOLUTIONS

R. Zana, J. Lang, P. Lianos
Centre de Recherches sur les Macromolecules,
CNRS 6, Rue Boussingault
67083 Strasbourg-Cedex, France

ABSTRACT

The interaction between polymers (polyethyleneoxide, POE, and polyvinylpyrrolidone, PVP) and ionic surfactants (sodium dodecylsulfate, SDS, and tetradecyltrimethylammonium bromide, TTAB) has been studied in aqueous solution by means of conductivity and fluorescence probing (fluorescence decay of micelle-solubilized pyrene, fluorescence spectra of micelle-solubilized pyrene and dipyrenylpropane). The difference of behavior noted, with both POE and PVP, between SDS (strong interactions) and TTAB (very weak, if any interactions), clearly indicate that the polymer-surfactant interaction occurs at the micelle surface and strongly depends on the nature of the surfactant head group. The local concentration of the polymer repeat unit in the polymer coil, rather than the stoichiometric polymer concentration, determines the changes of critical micellization concentration (CMC) and micelle ionization degree. The fluorescence probing methods show that the SDS micelles bound to POE are smaller than the micelles formed in the absence of POE : the larger the polymer concentration or the smaller the surfactant concentration, the smaller the micelle size. The fluorescence emission spectrum of the micelle solubilitized pyrene also indicate a more open water-penetrated structure of the POE-bound SDS micelles. The results for the PVP-SDS system are somewhat similar to those for the POE-SDS. However some results suggest a possible specific interaction between the fluorescence probe and PVP, which makes difficult the study of the interaction between this polymer and surfactants.

357

INTRODUCTION

The study of the interaction between polymers and surfactants is interesting from both the fundamental and applied point of views.

From the fundamental point of view, the additional of surfactants to polymer solutions may induce a conformational change of the polymer[1]. Indeed, the binding of ionic surfactants to polymer chains, irrespectively of the aggregation state of the bound surfactant, results in the formation of a polymer–surfactant complex similar to a polyelectrolyte. The electrostatic repulsions between head groups of bound surfactants then bring about an extension of the polymer chain. This effect is partly responsible for the unfolding of globular proteins. On the other hand, the addition of polymers to micellar solutions of surfactants may modify the micellar properties (critical micellization concentration and micelle ionization degree and aggregation number), if interactions occur between the polymer and surfactant micelles[2,3]. Non-polymeric additives, such as alcohols, amines, carboxylic acid, etc ... do bring about changes of micellar properties, particularly significant when the additive is partially solubilized by the micelles[4].

From the applied point of view, the understanding of polymer–surfactant interactions may help in the formulation of the surfactant–polymer systems used in the process of enhanced oil recovery by surfactant flooding. In this process, which is a very promising one to improve the output of oil wells[5], a surfactant solution is pushed through the pores of the oil reservoir by applying pressure on a viscous polymer solution injected after the surfactant solution. Owing to the increased mixing of the two solutions as they move in the oil reservoir, the surfactant–polymer interactions become more and more important, and may strongly affect the feasibility of the process.

Most of the studies dealing with polymer–surfactant interactions have focused on the thermodynamics of these interactions and/or the stoichiometry and nature of the surfactant–polymer complexes[2,3,6]. These studies have shown that in the sodium dodecylsulfate (SDS)–polyethyleneoxide (POE) system, there is no interaction between SDS and POE for SDS concentrations C below a value which is the critical micellization concentration (CMC) of SDS in the presence of POE. For $CMC < C < C_2$, all added SDS forms micelles bound to the POE, whereas at $C > C_2$, bound and free SDS micelles coexist in the system. C_2 has the signification of a concentration of saturation of POE by SDS. To our knowledge there has been no detailed investigation of the effect of the polymer on the micelle aggregation number n (number of surfactant per micelle). Indeed, the dynamic nature of micelles prevents a measurement of this quantity by means of classical methods (light scattering, sedimentation, osmometry, etc. ...). The use of a recently developed fluorescence method has permitted us to circumvent these difficulties[7]. This paper is to report the first measurements of n for the SDS–POE and the SDS–poly-N–vinylpyrrolidone (PVP) systems. Data on the CMC and micelle ionization degree at the CMC for these systems and for the tetradecyltrimethylammonium bromide (TTAB)–POE and TTAB–PVP systems, are also given.

MATERIALS AND METHODS

The samples of POE and PVP had molecular weights of 20,000 and 57,000 respectively.

The CMC of the surfactants in the presence of polymer was obtained as the break in the equivalent conductivity vs $C^{1/2}$ curve (C=surfactant concentration). The micelle ionization degree at the CMC was taken as the ratio of the slopes of the specific conductivity vs C curves above and below CMC[8]. The CMC and micelle ionization degree were also determined for SDS in the presence of N-methylpyrrolidone (NMP), as a function of the NMP concentration. Indeed NMP can be considered as the repeat unit of PVP. The comparison of the results for NMP and PVP then allows one to assess the part of the change of the micelles properties due to the polymeric character of the additive.

The micelle aggregation number n was obtained from the analysis of the fluorescence decay curve of a micelle-solubilized fluorescence probe, pyrene, in conditions where the [pyrene]/[micelle] concentration ratio is around 1, that is, when the fractions of micelles containing no probe, one probe, and 2 or more than 2 probes are not very different[7,9,10]. This analysis also provided the value of the first order rate constant k_E for excimer formation within the micelle. k_E contains information on the microviscosity of the environment of the fluorescent probe, and on the distance over which the probes must diffuse to form an excimer within the micelles[10]. It is to be noted that the n and k_E values obtained as indicated above correspond to values which are averaged over all the sizes of the micelles present in the surfactant-polymer system. For the present systems, the time characterizing micelle size fluctuations is much shorter than the lifetime of pyrene in the micelles owing to the small size of the micelles[11]. In this situation, it has been shown[11] that the measured n is equal to the number average aggregation number. Nevertheless, at C values above C_2, where both bound and free micelles are present in the system, some difficulties arise concerning the precise meaning of n. Finally, the ratio I_1/I_3 of the intensities of the first and third vibronic peaks of the fluorescence spectrum of micelle solubilized monomeric pyrene was used to estimate the polarity sensed by pyrene in its micellar solubilization site[9,10], that is the micelle palisade layer[12].

RESULTS AND DISCUSSION

Figures 1 and 2 show the changes of CMC and micelle ionization degree β of the various surfactant-polymer systems, and of the SDS–NMP system, as a function of the concentration C_p of polymer or NMP (in g. per 100 ml of solution).

It can be seen that the additions of NMP and PVP bring about qualitatively differing changes of CMC and β for SDS. They illustrate well the difference between the effect on the CMC and β of molecular and macromolecular additives : progressive changes of CMC and β with the former, and stepwise changes with the latter, at the lowest concentration investigated. The fact that the effect of polymer additions becomes independent of the polymer concentration C_p, even at the lowest C_p investigated (0.025%), suggest that this effect is determined by the local concentration of the polymer repeat units in the polymer coil, rather than by the stoichiometric concentration C_p. Indeed the local concentration is independent of C_p, as long as C_p is not too large, i.e., when the coils are not interpenetrated. It is the local concentration which effect the surfactant association behavior.

Figures 1 and 2 show that the effect of the polymer is strongly dependent on the nature of the surfactant. Thus the CMC and β values for TTAB are hardly affected by the presence of POE or PVP, indicating no interaction between this cationic surfactant and the two investigated polymers. This result confirms those of other workers[3,13]. On the contrary both POE and PVP strongly decrease the CMC of SDS but increase the ionization of SDS micelles at the CMC, the changes induced by PVP being larger than those due to POE for a given C_p. Several conclusions can be inferred from these results.

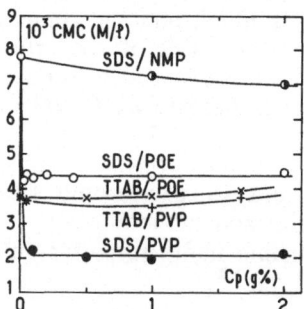

Figure 1: Variation of the CMC with the polymer concentration for the systems SDS–NMP (◐); SDS–POE (O); SDS–PVP (●); TTAB–POE (X); and TTAB–PVP (+).

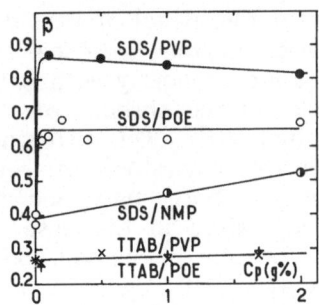

Figure 2: Variation of the micelle ionization degree at the CMC with the polymer concentration for the systems SDS-NMP (◐); SDS-POE (O); SDS-PVP (●); TTAB-POE (X); and TTAB-PVP (+).

1)   The striking difference of behavior between SDS and TTAB as far as changes of CMC and β are concerned suggest that the polymer-surfactant interaction occurs at the level of the surfactant ionic head group and does not involve the surfactant alkyl chain. This is turn means that the penetration of the polymer in the micelle will be restricted to the head group region, with little if any at all, penetration of the polymer in the micelle hydrophobic interior.   Thus, the polymer surfactant interaction can be looked at as an adsorption of the polymer chain on the micelle surface[2].

2)   The larger effect of PVP relatively to POE suggests a better penetration of PVP than POE in the SDS micelle palisade layer.

3)   The increase of the ionization degree of SDS micelles upon addition of POE explains the increase of relaxation rate of the Na^+ counterions observed in a previous NMR study[2]. Moreover, the very large values of β at the CMC indicate that the first SDS micelles formed on, and bound to, the polymer at low C must be of much smaller size, or aggregation number than the micelles formed in water.

This last conclusion is confirmed by the results of Figure 3 and Table I for the SDS-POE system which is the one for which the most comprehensive measurements have been performed. Figure 3 shows that the addition of POE indeed results in a decrease of the aggregation number of SDS micelles at a given C: the larger the C_p the smaller the value of n.   However the changes of n with C_p are progressive, contrarily to the variations of CMC and β. At a given C_p, n increases with C.   Qualitatively, the changes of n with C and C_p are similar to those found for SDS[10] and TTAB[4] in the presence of penthanol, at least at low C. At higher C values, n becomes nearly equal to the value found in the absence of polymer, independently of the POE concentration of the system. In this respect the effect of POE differs considerably from that of pentanol. Indeed, in the presence of pentanol at a given concentration, n increases monotonously with C from a value lower than the value 65 usually found in the absence of additives at low C to a value much above 65 at high C^{10}. The difference of behavior between pentanol and POE is easily understood by recalling that above the saturation concentration C_2 both free and bound micelles are in equilibrium in the system[2] while in the case of pentanol only mixed pentanol-SDS micelles are present in the solution. In fact the change of C_2 with C_p is such that at C = 0.5 $\underline{M}$ most of the micelles present in the system are free, even at C_p = 2%.[2] The fact that the fluorescence method measures a number average value of n then readily explains why at C = 0.5 $\underline{M}$ the value of n is nearly equal to that in the absence of POE.

It must be emphasized that the results of Figure 3 are the first ones reported on such systems. In some studies[6,14,15] the authors were led to conclude that for CMC < C < C_2 n was lower in the presence than in the absence of POE but a quantitative study of the effect could not be done owing to the lack of an appropriate method of determination of n.

The changes of the values of I_1/I_3 with C and C_p (see Table I) are in qualitative agreement with the changes of n. At low C, I_1/I_3 is increased in the presence of POE, indicated that the micelle-solubilized pyrene senses a more polar, water-penetrated palisade layer, than in the absence of POE, as would be expected for the small SDS micelles then present in the system. As C is increased at constant C_p, I_1/I_3 decreases because the micelles grow in size and reaches at high C a value very close to that in SDS micelles in the absence of POE, as to be expected from the fact that at such concentration most SDS micelles are free. It is to be noted, however, that the changes of I_1/I_3 with the POE concentration, are opposite to those found upon addition of medium chain length alcohols to micelles at a given C^{10}. Pentanol can be taken as a representative of these alcohols. There I_1/I_3 decreased upon increasing pentanol concentration[10].

The behaviors of POE and pentanol as far as changes of I_1/I_3 and n are concerned can be explained as follows. A pentanol molecule is solubilized in a micelle with its hydroxyl group anchored at the micelle surface, and its alkyl chain going through the palisade layer and reaching the micelle hydrophobic core. The solubilized pentanol thus affects the palisade layer, where pyrene is solubilized, over its whole thickness. Upon pentanol solubilization some of the water present in the palisade is released, resulting in a decrease of its overall polarity, which is sensed by pyrene[4]. On the contrary, the POE repeat units interacting with the micelles remain located near the micelle outer surface. Thus, if the only effect of the POE-micelle interaction were a release of some of the

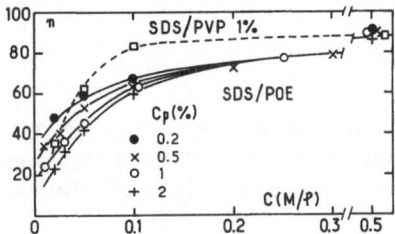

Figure 3: Variation of the micelle aggregation number n with the polymer concentration for the system SDS-PVP ($\square$, C_p = 1%) and the systems SDS-POE, with C_p = 0.2% ($\bullet$); (X) 0.5%; (O) 1%; and (+) 2%.

water present in the palisade layer[2], the polarity sensed by pyrene would be decreased, as in the case of addition of pentanol. However, as discussed elsewhere[4], the solubilized pentanol and to a similar extent, the adsorbed POE repeat units, increase the micelle ionization and thus the repulsions between head groups, which brings about a micelle breakdown, at least at low C. The resulting smaller SDS micelles would have a more open, water penetrated structure than in the absence of POE or pentanol. However, the micelle-solubilized pentanol fills the voids in the packing of the SDS chains and prevents the pentration of water. Thus the polarity felt by pyrene remains lower than in pentanol-free systems. This process cannot occur when pentanol is replaced by POE as the repeat units remain located very near the micelle surface. Thus the polarity felt by pyrene should be higher in the presence than in the absence of POE. As C is increased, n increases and I_1/I_3 tends towards its value in polymer-free systems, as is indeed observed.

Having thus examined the effect of POE on the values of I_1/I_3 and n for SDS micelles, one can now attempt to understand the more complex effect of POE on the rate constant k_E for the intramicellar excimer formation of pyrene. It must be recalled that this process involves the diffusion of an excited state pyrene and a ground state pyrene towards each other in a restricted space (the micelle volume). Thus k_E should decrease upon increasing size and viscosity of the micelle interior (microviscosity $\bar{\eta}$). It can be seen that as expected k_E decreases upon increasing C, at constant C_p, that is upon increasing n (and thus micelle size). On the contrary, at constant C, k_E decreases upon increasing C_p that is when n is decreased. To explain this last result it must be assumed that at high C_p/C the micelle adsorbed POE repeat units bring about an increase of the $\bar{\eta}$ sensed by pyrene in its motion within the palisade layer. The effect of the increase of $\bar{\eta}$ must overcome that due to the decrease of micelle size, thus resulting in a net decrease of k_E. Recall that additions of pentanol always resulted in a decrease of $\bar{\eta}$ but also of k_E[10]. The increase of $\bar{\eta}$ in the presence of POE may be due to a hindrance brought to the motion of the pyrene molecules in

Table I: Effect of the SDS and Polymer Concentrations on the Aggregation Number of SDS in the presence of Polymer and on the values of k_E and I_1/I_3 of Micelle-solubilized Pyrene at 20°C

C_p (g/100ml)	C (M)	n	I_1/I_3	$10^{-7}k_E$ (s^{-1})
POEG 0.2%	0.01	31 ± 4		
	0.02	48 ± 4	1.37	1.74
	0.05	59 ± 5	1.30	1.61
	0.1	67 ± 5	1.30	1.55
	0.5	91 ± 7	1.24	1.56
POEG 0.5%	0.01	34 ± 4	1.46	1.74
	0.025	40 ± 4	1.40	1.51
	0.05	53 ± 5	1.36	1.26
	0.1	63 ± 5	1.31	1.30
	0.2	72 ± 6	1.31	1.31
	0.3	79 ± 7	1.26	1.33
POEG 1.0%	0.01	24 ± 4	1.48	1.40
	0.03	36 ± 4	1.40	1.31
	0.05	45 ± 4	1.37	1.26
	0.1	63 ± 5	1.34	1.23
	0.25	77 ± 6	1.30	1.24
	0.5	90 ± 7	1.24	1.24
POEG 2.0%	0.02	23 ± 4	1.48	1.35
	0.03	31 ± 4	1.41	1.30
	0.05	42 ± 4	1.38	1.13
	0.1	60 ± 5	1.36	1.10
	0.5	86 ± 7	1.24	1.01
PVP 1.0%	0.01		1.90	
	0.02	31 ± 4	1.77	
	0.05	59 ± 5	1.60	0.49
	0.1	81 ± 7	1.50	0.57
	0.5	88 ± 8	1.26	0.65

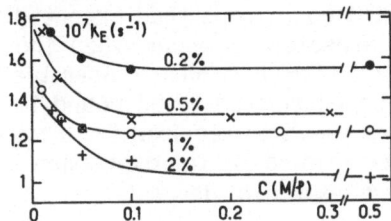

Figure 4: Variation of the rate constant k_E for pyrene excimer formation for the SDS-POE systems with C_p = 0.2% (●); (X) 0.5%; (O) 1%; and (+) 2%.

the palisade layer owing to the presence of the POE repeat units which are chemically linked, contrarily to the water or pentanol. In line with this interpretation, all the k_E values determined for the SDS-POE system were found to be smaller than for pure SDS micelles (1.8-2 x 10^7 s^{-1} whichever the SDS concentration.

One result, however, cannot be easily understood. It concerns the limiting values of k_E at high C (see Figure 4). At high C values, say 0.5 M, most of the SDS micelles are free[2], even at C_p = 2%. One would have thus expected the k_E vs C plots to tend towards the value of k_E in SDS micelles in the absence of POE, that is about 1.8 x 10^7 s^{-1}. In fact the results show that at high C, the k_E values become nearly constant or increase very slightly with C, but remain smaller than in the absence of POE. A possible explanation to this observation is as follows. At C > C_2, when polymer bound micelles and free micelles are present in the solution, the measured k_E value is an average over the values $k_{E,b}$ and $k_{E,f}$ for bound and free micelles, respectively. In the concentration range CMC < C < C_2, where the system contains only bound micelles the results show that $k_{E,b}$ is smaller than $k_{E,f}$. Thus any averaging over $k_{E,b}$ and $k_{E,f}$ should lead to average k_E values smaller than in the polymer free solution, as is observed. Also, as the fraction of bound micelles increases with C_p, the measured k_E is expected to decrease for increasing C_p, as is also observed. This interpretation, however, requires that $\bar{k}_E$ (average k_E) be not given by a simple expression such as : $\bar{k}_E$ = $k_{E,b}$ x_b + $k_{E,f}$ (1 - x_b) where x_b is the fraction of bound micelles. Indeed, at C_p = 2%, x_b = 0.25 and even with $k_{E,b}$ << $k_{E,f}$, an unlikely possibility, $\bar{k}_E$ would be significantly larger than the experimental value (about 1.35 x 10^7 against 10^7 s^{-1}). The same remark applies to all limiting values of k_E. Thus another effect appears to be involved in determining the value of k_E at high C, and modifies the equation. Most likely this effect has to do with the dynamics of the system. Free and bound micelle are in equilibrium, and a given micelle, during its lifetime (up to 10^{-2}s) is likely to exchange between free and bound states. Indeed, the POE-SDS interaction is at the level of the surfactant head group and the POE-SDS complex is thus expected to be very labile. A complex situation is expected to occur when the time t_E required for excimer formation ($t_E = 1/k_E$) becomes comparable to the time t_B during which a micelle can be considered to be bound to the polymer. In this situation, t_B as well as the time required for a micelle to associate with a polymer chain will be involved in the expression of k_E, and may explain the low value of k_E measured at large C.

Finally, we can briefly consider the effect of PVP on SDS micelles. We recall that the CMC and β changes upon addition of PVP suggested that this polymer interacts more strongly than POE with SDS. The results of Table I show that the value of I_1/I_3 at low C in the PVP-SDS system, is close to that in water. Also, the values of k_E are small and they increase with C, contrarily to the POE-SDS system. These two facts suggest a specific interaction between PVP and pyrene which may keep pyrene exposed to water (large I_1/I_3), and slows down its diffusion in the micelle (small k_E). More investigations on the PVP-SDS systems appear necessary before a definite conclusion can be drawn concerning this system.

CONCLUSIONS

The results presented in this paper clearly show that for the investigated polymer-surfactant systems the interaction between the micelles and the polymer repeat units occurs at the micelle surface. The SDS micelles bound to POE at C below the polymer-surfactant stoichiometry are more water-penetrated and smaller than the micelles formed in the absence of polymer. From the aggregation number of the bound micelle determined in this work one can give a fairly complete description of the system. Thus for instance, at C = 0.05 $\underline{M}$ and C_p = 1%, it can be inferred from the value n = 45 that each POE chain of molecular weight 20,000 has an average of 2 micelles bound to it. Thus about 220 repeat units are available for each bound micelle, and the molecular weight of the POE-SDS complex is of about 46,000. If the polymer chain were completely stretched the intermicellar distance on the chain would be about 800 Å. This value is much larger than the distance 120 Å between two micelles, in the absence of polymer, if C = 0.05 $\underline{M}$ and n = 45. Thus the polymer chain must present considerable foldings both on itself and around the bound micelles.

Note Added in Proof

Additional measurements of CMC and n have been performed on the SDS-POE systems at fixed SDS and POE concentrations : 0.05 $\underline{M}$ and 1%, respectively (that is, systems where all SDS micelles are bound to the polymer), with POE samples of various molecular weights. The

Table II.   Effect of the POE Molecular Weight on the CMC and Micelle Aggregation Number

M_W[a]	10^3CMC(M/1)	n
1,000	5.8 + 0.2	43 + 4
2,000	5.4 + 0.2	41 + 4
6,000	4.7 + 0.2	39 + 4
20,000	4.3 + 0.2	40 + 4[b]

[a] Molecular weight in Dalton

[b] This value of n has been determined with a different setting of the fluorescence lifetime apparatus, and more than one year after the one reported in Table I for the same system.   The difference between the two values is within the experimental error.

results are listed in Table II.   They show an increase of the CMC as the POE molecular weight is decreased in agreement with other investigations.[2,6]   They also show that n remains nearly constant, or increases only very slightly, as the polymer molecular weight is decreased from 20,000 to 1,000.   Notice that for the lower molecular weight samples, there is less than one micelle per POE chain.   Nevertheless, the possibility that several chains interact with one micelle cannot be discarded.

REFERENCES

1.   T. Isemura and A. Imanashi, J.Polym. Sci., 1958, 33, 337. T. Takagi, K. Tsujii and K. Shirahama, J. Biochem., 1975, 77, 339. I. Satake and J. Yang, Biopolymers, 1976, 15, 2263.
2.   B. Cabane, J. Phys. Chem., 1977, 81, 1639 and references therein.
3.   G. Kresheck and W. Hargraves, J. Colloid Interface Sci., 1981, 83, 1 and references therein.
4.   R. Zana, S. Yiv, C. Strazielle and P. Lianos, J. Colloid Interface Sci., 1981, 80, 208.
5.   See "Surface Phenomena in Enhanced Oil Recovery", D. O. Shah Ed., Plenum Press, New York 1981.
6.   T. Sasaki, K. Kushima, M. Matsuda and H. Suzuki, Bull. Chem. Soc. Jpn, 1980, 53, 1964.
7.   S. Atik, M. Nam and L. Singer, Chem. Phys. Lett., 1979, 67, 75.
8.   R. Zana, J. Colloid Interface Sci., 1980, 78, 330.
9.   P. Lianos and R. Zana, Chem. Phys. Lett., 1980, 72, 171 and 1980, 76, 62; J. Phys. Chem., 1980, 84, 3339; J. Colloid Interface Sci., 1981, 84, 100.
10.   P. Lianos, C. Strazielle, J. Lang and R. Zana, J. Phys. Chem., 1982, 86, 1019.

11.   M. Almgren and J. E. Lofroth, J. Chem. Phys., 1982, 76, 2734.
12.   J. K. Thomas, Acc. Chem. Res., 1977, 10, 133; Chem. Revs., 1980, 80, 283.
13.   S. Gravsholt, Proc. 5th Int. Cong. Surface Active Subst., Barcelona, 1968, Vol. 2, 993.
14.   T. Gilanyi and E. Wolfram, Colloids Surf., 1981, 3, 181.
15.   K. Shirahama, M. Tohdo and M. Murahashi, J. Colloid Interface Sci., 1982, 86, 282.

VISCOMETRIC INVESTIGATION OF COMPLEXES BETWEEN POLY-ETHYLENEOXIDE AND SURFACTANT MICELLES

R. Nagarajan and B. Kalpakci*
Department of Chemical Engineering
The Pennsylvania State University
University Park, PA 16802

*Sohio Research and Development
Cleveland, OH 44115

ABSTRACT

Nonionic polymers in aqueous solutions may associate with surfactant micelles to form polymer-micelle complexes. In this work, the polymer-micelle association is examined by viscometrically monitoring the conformational changes induced in the polymer. Polyethylene oxide with a flexible backbone and dextran with a rigid backbone have been examined in solutions of anionic, cationic, and nonionic surfactants. It is found that no polymer-micelle association takes place in case of dextran. For polyethylene oxide, the extent of polymer-micelle association is large for anionic micelles and small for cationic micelles. It decreases with increasing bulkiness of the surfactant head group. We conclude that the physical factors governing the polyethylene oxide-micelle association are (i) the extent of augmented shielding from water provided by the polymer segments to the hydrocarbon core of the micelles and (ii) the steric and electrostatic interactions between the polymer segments and the surfactant head group at the micellar surface. For these reasons, polymer-micelle complex formation is not favored if the polymers have rigid backbones, if the surfactants have bulky head groups, or if the polarity of the surfactant head group is similar to that of the polymer.

INTRODUCTION

The interactions between surfactant molecules and synthetic polymers in aqueous solutions are of interest for many chemical, pharmaceutical, mineral processing and petroleum engineering applications. The mutual presence of polymer and surfactant molecules alter the rheological properties of solutions, adsorption characteristics at the solid-liquid interfaces, stability of colloidal dispersions, the solubilization capacities in water for sparingly soluble molecules, and liquid-liquid interfacial tensions.[1,2] The ability of the surfactant and the polymer molecules to influence the above solution and interfacial characteristics is controlled by the state of their occurrence in aqueous solutions. Generally, the aqueous solution is composed not only of singly dispersed surfactant and polymer molecules but also of surfactant aggregates such as micelles or vesicles as well as intermolecular complexes between the polymer and the surfactant.

Various topologies of polymer-surfactant complexes can be visualized,[3] depending on the nature of the interaction forces operative between the solvent, the surfactant and the polymer and also based on their stereochemical features. One may consider the following structures (Figure 1):

Type 1 - No surfactant is bound to the polymer and the latter remains as free polymer in solution.

Type 2 - Single surfactant molecules are bound linearly along the length of the polymer molecules.

Type 3 - A single surfactant molecule binds at more than one binding site of a single polymer molecule, giving rise to intramolecular bridging. Alternatively, a single surfactant molecule binds to more than one polymer molecule giving rise to intermolecular bridging.

Type 4 - The polymer molecule along with a layer of surfactant molecules bound on it is solubilized in the interior of a surfactant micelle.

Type 5 - Clusters of polymer segments and surfactant molecules associate to form pseudomicelles such that the hydrocarbonaceous regions of the polymer segments and of the surfactant are shielded from having contact with water.

Type 6 - Pure surfactant micelles associate with the polymer molecule in such a way that the polymer segments partially penetrate and wrap around the polar head group region of the surfactant micelles. A single polymer molecule can associate in this manner with one or more surfactant micelles depending on the polymer and micellar properties.

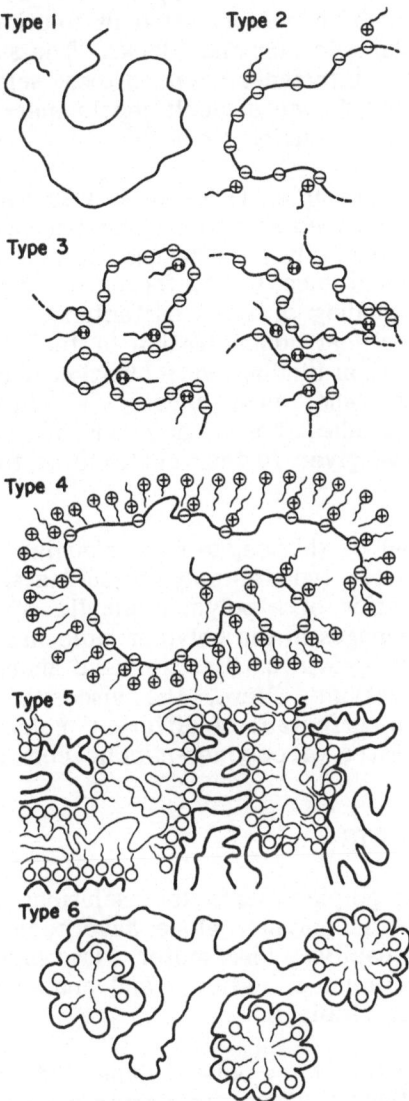

Figure 1.   Schematic representations of polymer-surfactant complexes described in the text.   In the block copolymer-surfactant complex (Type 5), the dark lines represent the polar blocks of the polymer molecule whereas the lighter lines represent the non-polar blocks.

Of the above, two types of complexes are considered to form in solutions of nonionic polymers.   One type (Type 5) occurs in block copolymers consisting of polar and nonpolar blocks.   In this case, the polymer molecule assumes a conformation in solution characterized by self-segregation of polar and nonpolar blocks.   The surfactant molecules locate themselves at the interfaces between these segregated regions.   A quantitative description of such a block copoloymer–surfactant complex has been developed in our earlier paper.[4,5]

The other type of complex (Type 6) is visualized as consisting of the polymer molecule wrapped around spherical surfactant micelles with the polymer segments partially penetrating the polar head group region of the micelles.   Indirect clue to its formation is provided by studies which show that the binding of some surfactants to nonionic polymers occurs only above a critical concentration of the surfactant.[6-11]   The existence of such a critical binding concentration implies that binding is a cooperative process and must involve a cluster of surfactant molecules.   Recent studies based on nmr and neutron scattering measurements[12,13] have given further evidence to the topology of such a complex.

The principal goal of this paper is to examine the physical forces responsible for this latter type of polymer–surfactant micelle association. Since the formation of a polymer–micelle complex gives rise to gross conformational changes in the polymer molecule, a measurement of the solution viscosity provides the simplest means for monitoring polymer–micelle association.   Here, the viscosity data on solutions containing polymer and surfactant of different molecular structures are used to explore the nature of polymer–micelle complex formation.

EXPERIMENTAL RESULTS

The viscosities of polymer–surfactant solutions as well as those of the surfactant and of the polymer alone have been measured at 25°C using a Cannon–Fenske capillary viscometer.   Two high molecular weight nonionic polymers and anionic, cationic and nonionic surfactants have been used in this study (Table I).

The relative viscosity of the polyethylene oxide–surfactant solution, $\eta_{rel} = \eta(\text{PEO} + \text{surfactant})/\eta(\text{PEO})$, is plotted in Figure 2 as a function of the surfactant concentration for different surfactants.   For anionic SDS, η_{rel} remains invariant at 1 up to a concentration of 4×10^{-3} M, indicating absence of any association.   Beyond 4×10^{-3} M SDS, the relative viscosity shows a large increase up to about 2.6×10^{-2} M SDS. This can be attributed to the association of SDS micelles with the PEO segments and the resulting expansion of the PEO molecules.   Beyond 2.6×10^{-2} M SDS, a reduction in the relative viscosity is observed.   This is because, when saturation binding of SDS to PEO is reached (say at 2.6

Table I

POLYMERS AND SURFACTANTS EVALUATED

Polyethylene Oxide, PEO (Union Carbide)

Nonionic Polymer with a Flexible Backbone.

M. Wt. 5×10^6.

Dextran, DEX (Pharmacia Fine Chemicals)

Nonionic Polymer with Rigid Backbone. M. Wt. 2×10^6.

Sodium Dodecyl Sulfate, SDS (Pfaltz and Bauer)

Anionic Surfactant, Eq. Wt. 288, $a_p = 17\text{Å}^2$, CMC=8×10^{-3}M.

Sodium Dodecyl Benzene Sulfonate, SDBS (Alcolac)

Anionic Surfactant, Eq. Wt. 348, $a_p = 17\text{Å}^2$, CMC=1.72×10^{-3}M.

8-Phenyl Hexadecane Benzene Sulfonate, UT-1 (University of Texas

at Austin)

Anionic Surfactant, Eq. Wt. 404, $a_p = 17\text{Å}^2$, CMC=9.6×10^{-4}M.

Has two alkyl chains per surfactant.

Dodecyl Amine Hydrochloride, DAC (Eastman Kodak)

Cationic Surfactant, Eq. Wt. 222, $a_p = 12\text{Å}^2$, CMC=1.4×10^{-2}M.

Didodecyl Dimethyl Ammonium Bromide, DDAB (Eastman Kodak)

Cationic Surfactant, Eq. Wt. 463, $a_p = 17\text{Å}^2$, CMC=6×10^{-5}M.

Has two alkyl chains per surfactant.

Ethyl Hexadecyl Dimethyl Ammonium Bromide, EHD (Eastman Kodak)

Cationic Surfactant, Eq. Wt. 378, $a_p = 35\text{Å}^2$, CMC=8×10^{-4}M.

Isooctyl Phenoxy Polyoxyethanol, Triton X-100 (Rohm and Haas)

Nonionic Surfactant, Eq. Wt. 648, $a_p = 42\text{Å}^2$, CMC=1.6×10^{-4}M.

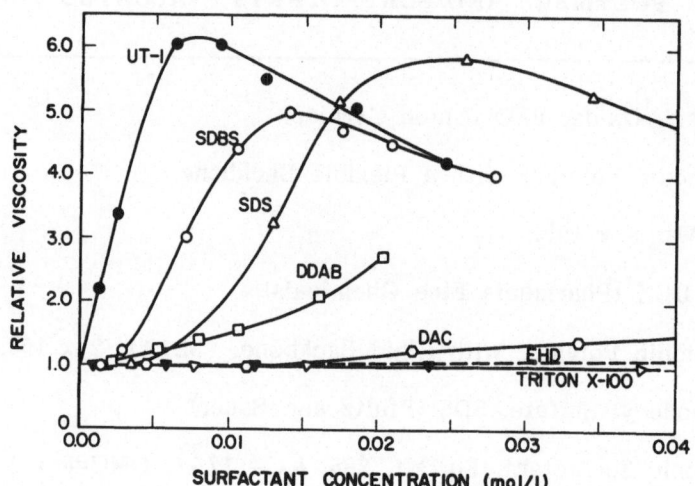

Figure 2.   Influence of various types of surfactants on the relative
             viscosity of the solutions containing 1000 ppm polyethylene
             oxide-surfactant micelle complexes.

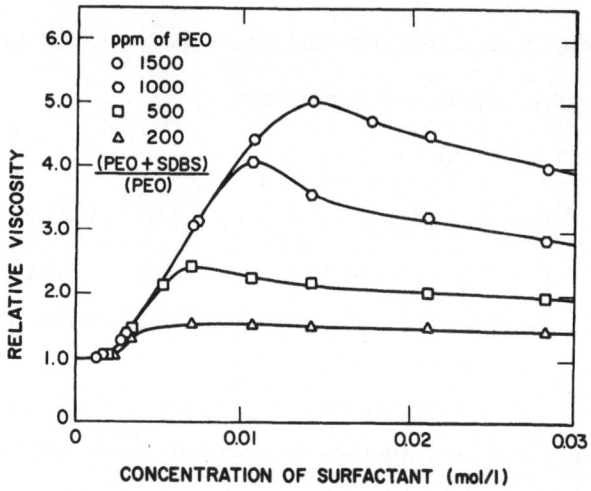

Figure 3.   Influence of polymer concentrations on the relative vis-
             cosity of solutions containing polyethylene oxide and
             anionic sodium dodecyl benzene sulfonate micelles.

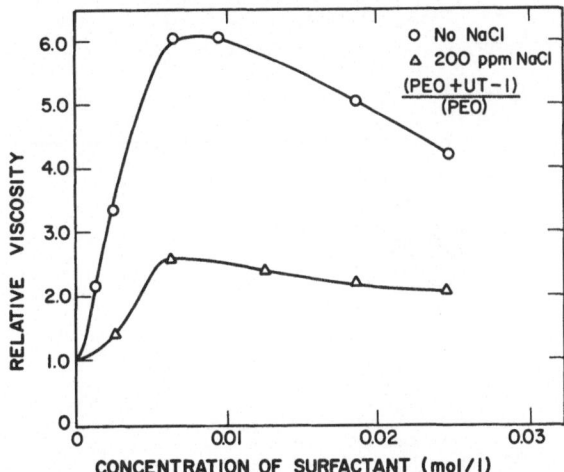

Figure 4.   Influence of electrolyte NaCl concentration on the relative viscosity of solutions containing 1000 ppm polyethylene oxide and 8-phenyl hexadecane benzene sulfonate micelles. The polyelectrolyte type behavior of the polymer-micelle complex can be noted.

$\times~10^{-2}$ M SDS), further addition of SDS results in an increase in the concentration of singly dispersed SDS molecules and of free SDS micelles unattached to the polymer. This gives rise to an increase in the ionic strength of the solution and a consequent reduction in η_{rel}, as is expected for polyelectrolyte solutions.[14] The behavior of the other anionic surfactants SDBS (Figure 3) and UT-1 (Figure 4) in solutions containing PEO is very similar to that of the PEO + SDS solution. As is to be expected, the critical binding concentrations and the concentrations corresponding to saturation binding of the surfactant are however different from those of SDS. In contrast to the behavior of these anionic surfactants, all the cationic surfactants display relative viscosities close to one implying low degrees of binding to the PEO, with EHD demonstrating virtually no binding. Also remarkable is the relative viscosity of unity and hence the complete absence of binding shown by the nonionic surfactant Triton X-100.

Figure 5 shows the viscosity ratio of the Dextran-surfactant solution, $\eta_{ratio} = \eta_{(dextran + surfactant)}/\eta_{(dextran)} \times \eta_{(surfactant)}$, for solutions containing SDBS and EHD. The data show that no binding occurs in case of both anionic SDBS and cationic EHD with nonionic dextran polymer. The absence of any surfactant binding is also confirmed by surface tension measurements (Figure 6) which show that the surface tensions of SDBS and SDS solutions in the absence of and in the presence of dextran are identical.

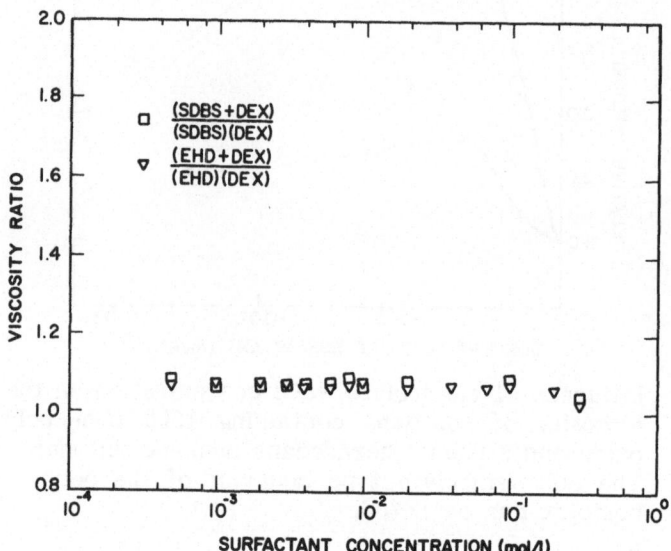

Figure 5.   Viscosity ratios of solutions containing (a) non-ionic dextran
             polymer and anionic sodium dodecyl benzene sulfonate
             micelles and (b) dextran polymer with cationic ethyl
             hexadecyl dimethyl ammonium bromide micelles.   The
             polymer concentration is 1000 ppm in both cases.

In Figure 7, the relative viscosity of SDS + PEO solutions are plotted as a function of the concentration of n-butanol and Triton X-100 which are used as additives. Both the additives form mixed micelles with SDS and the composition of the mixed micelles changes as the concentration of the additives is altered. When no additives are present the PEO + SDS solution is at a SDS concentration of 2.6×10^{-2} M corresponding to saturation binding of SDS to PEO. As the concentration of the additives is increased, η_{rel} decreases indicating a decrease in polymer-micelle association. The viscosity data show that the nonionic surfactant is very effective in causing a decrease in polymer-micelle association. These viscometric results are interpreted in the following section in terms of the binding and competitive micellization processes.

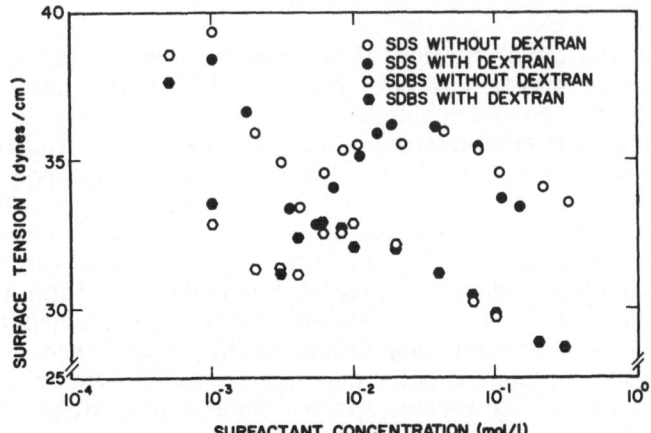

Figure 6.   Surface tension of (a) solutions containing 1000 ppm polyethylene oxide and SDS on SDBS micelles and (b) solutions
containing SDS on SDBS micelles but in the absence of any
polymer.

THERMODYNAMICS OF POLYMER-MICELLE ASSOCIATION

The physical picture of the polymer-surfactant micelle complex
described earlier suggests that the following factors are relevant to the
association process: (i) the penetration of polymer segments in the polar
head group region of the micelles augments the shielding from water of
the hydrocarbon core of the micelle, (ii) the crowding of the polymer
segments along with the surfactant head groups at the micelle surface
increases mutual steric repulsions, and (iii) the dipoles of the polymer
segments electrostatically interact with the surfactant head groups at
the micellar surface. Further, since the micelles associating with the
polymer are quite similar to the free micelles, the formation of the
polymer-micelle complex is also influenced by factors determining
surfactant micellization.

For an aqueous solution containing free micelles of size M and
polymer-micelle complexes in which n micelles of size λ are associated
with each polymer molecule, the total mole fraction of the surfactant
(S_T) is given[3-5] by

$$[S_T] = [S_f] + \frac{n\lambda K_b \; [S_f]^{\lambda} P_T}{1 + K_b \; [S_f]^{\lambda}} + M \; K_m[S_f]^M \qquad (1)$$

where S_f is the mole fraction of the singly dispersed surfactant, P_T is the total mole fraction of the polymer, K_b is the intrinsic binding constant for the polymer-micelle complexation, and K_m is the equilibrium constant for micellization. In eq. (1), the three terms on the right hand side represent the singly dispersed surfactant, polymer bound surfactant and surfactant in free micelles, respectively. The critical binding concentration is approximately equal to $(K_b)^{-1/\lambda}$ and the critical micelle concentration (CMC) is close to $(K_m)^{-1/M}$. However, if polymer-micelle association occurs in a given solution then the formation of free micelles takes place at a surfactant concentration much beyond the usual CMC. The relative magnitudes of K_b, K_m, λ and M determine whether or not polymer-micelle complex formation occurs and what is the composition of the various species present in solution.

Taking into account the physical factors relevant to polymer-micelle association, the binding and micellization equilibrium constants can be related:

$$(-RT \; \ln K_b^{1/\lambda}) - (-RT \; \ln K_m^{1/M}) = \Delta G^o_{\substack{\text{inter-}\\ \text{facial}}} + \Delta G^o_{\text{steric}} + \Delta G^o_{\substack{\text{electro-}\\ \text{static}}} \qquad (2)$$

where $\Delta G^o_{\text{interfacial}}$ is the free energy contribution due to the augmented shielding of the hydrocarbon core of the micelles by polymer segments, $\Delta G^o_{\text{steric}}$ is the contribution representing steric repulsions between polymer segments and surfactant head groups and $\Delta G^o_{\text{electrostatic}}$ accounts for the electrostatic interactions between the polymer dipoles and surfactant head groups.

In the case of dextran polymer, the rigidity of its backbone does not permit the polymer segments to penetrate the polar head group region of the micelles. As a result, the various free energy contributions in eq. (2) are all practically zero and $K_b^{1/\lambda} = K_m^{1/M}$. Assuming $\lambda = M$, from eq. (1) we note that as long as $nP_T << 1$, for all values of S_f, the formation of free micelles is favored compared to that of the polymer-micelle complex. For 1000 ppm dextran of a molecular weight of 2 million, $P_T \approx 10^{-8}$ and $n \approx 10$ to 10^3. Therefore, no association of surfactant micelle to the polymer can occur as is indicated by the viscometric and surface tension data. This behavior is expected for all polymers which because of their relative rigidity cannot penetrate into and thus modify the nature of the head group region of the micelles.

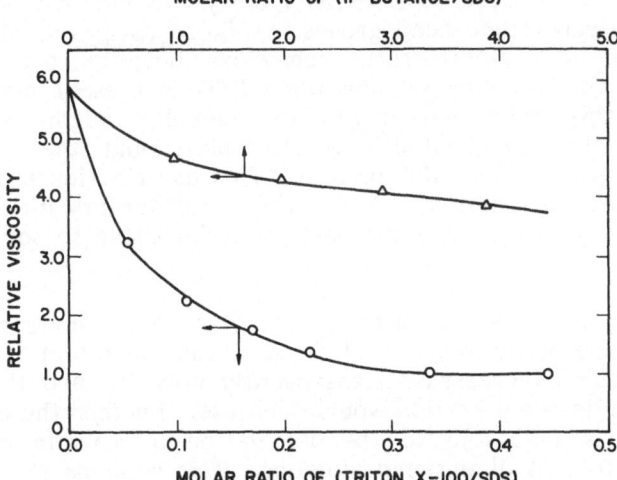

Figure 7. Relative viscosity of solutions containing 1000 ppm poly-ethylene oxide and 2.6×10^{-2} M sodium dodecyl sulfate as a function of the amounts of n-butanol and Triton X-100 added to the solution.

In contrast, because of the flexible nature of polyethylene oxide, the polymer segments can penetrate the polar surface of the micelles and modify the nature of that interface. Assuming that polymer segments shield about 10 $\mathring{A}^2$ (per surfactant molecule) of the hydrocarbon core area of the micelle from water, $\Delta G^o_{interfacial}$ is approximately -1.2 RT, taking the hydrocarbon water interfacial tension to be 50 dynes/cm. ΔG^o_{steric} can be estimated in a manner similar to that used in our micellization model.[15]

$$\Delta G^o_{steric} = -RT \ln[(a - a_p - a_{pol})/(a - a_p)], \qquad (3)$$

where a is the surface area per molecule of the hydrocarbon core of the micelle, a_p is the cross-sectional area of the polar head group of the surfactant, and a_{pol} is the projected area of the polymer segments at the micellar interface per surfactant molecule. The area $(a - a_p)$ is a measure of the freedom of movement of head groups at the micelle surface in the absence of polymer segments and $(a - a_p - a_{pol})$ is the corresponding quantity when the polymer segments are present. For a = 66 $\mathring{A}^2$ corresponding to the largest spherical micelle and assuming a_{pol} = 10 $\mathring{A}^2$, ΔG^o_{steric} is about 0.20 RT for a_p = 10 $\mathring{A}^2$, 0.35 RT for a_p = 30 $\mathring{A}^2$ and 1 RT for a_p = 50 $\mathring{A}^2$. The electrostatic contribution can be estimated by treating the electron deficient oxygen atom of the ether linkage in PEO as free ions which modify the electrostatic repulsion between the surfactant head groups. $\Delta G^{o}_{electrostatic}$ is therefore the difference in the electrostatic repulsions between surfactant head groups[15] at the micellar surface when PEO is present and when it is absent. By this approximation, PEO of molecular weight 5×10^{6} at 1000 ppm is equivalent to about 0.05 M electrolyte, and the corresponding $\Delta G^{o}_{electrostatic}$ is -0.12 RT for SDS. This contribution is favorable to polymer-micelle association. For cationic surfactants interacting with PEO, $\Delta G^{o}_{electrostatic}$ is +ve and hence is unfavorable to polymer-micelle association.

The approximate estimates for the free energy contributions show that as the size of the polar head group of the surfactant increases, the steric repulsion increases. Consequently both K_{b} and the extent of polymer-micelle complexation would decrease. Further the electrostatic contribution is favorable to the association of anionic micelles and unfavorable to that of cationic micelles. This explains the viscometric data of PEO in the presence of various surfactants shown in Figure 2. When mixed micelles of SDS + Triton X-100 form, the number of Triton molecules per micelle increases as the ratio of Triton X-100 to SDS is increased. As a result, because of the bulky head group of Triton X-100, the extent of PEO-mixed micelle association sharply decreases as shown by the viscosity data of Figure 7. All the above estimates are obtained, for the illustrative purposes of this paper, based on $a_{pol} = 10 \text{ Å}^{2}$. However, a better estimate of this projected area may be possible from a consideration of the atomic dimenisons and bond movements of the PEO segments and the surfactant head groups. Nevertheless, the qualitative conclusions of this paper remain valid for modified estimates for a_{pol}, as well.

CONCLUSION

The formation of polymer-micelle complexes is governed by (i) the extent of augmented shielding provided by polymer segments to the hydrocarbon core of the micelles, and (ii) the steric and electrostatic interactions between the polymer segments and the surfactant head groups at the micellar surface. For these reasons, polymer-micelle complex formation is not favored if the polymers have rigid backbones, if the surfactants have bulky head groups or if the polarity of the surfactant head group is similar to that of the polymer.

REFERENCES

1. M. M. Breuer and I. D. Robb, Chemistry and Industry, 13, 531 (1972).
2. I. D. Robb in "Anionic Surfactants - Physical Chemistry of Surfactant Action," (Ed., E. H. Lucassen Reynders), Marcel Dekker, New York (1981).
3. R. Nagarajan, Polymer Preprints, 22, No. 2, 33 (1981).
4. R. Nagarajan, Chemical Physics Letters, 76, 282 (1980).
5. R. Nagarajan and M. P. Harold, in "Solution Behavior of Surfactants," (Eds., K. L. Mittal and E. J. Fendler), Plenum Press, New York (1982). V. 2.
6. S. Saito, J. Colloid Interface Sci., 24, 227 (1967).
7. M. L. Fishman and F. R. Eirich, J. Phys. Chem., 75, 3135 (1971).
8. M. J. Schwuger, J. Colloid Interface Sci., 43, 491 (1973).
9. M. L. Smith and N. Muller, J. Colloid Interface Sci., 52, 507 (1975).
10. T. Gilanyi and E. Wolfram, Proc. Int. Conf. Colloid Surf. Sci., V. 1 (1975), p. 633.
11. M. N. Jones, J. Colloid Interface Sci., 23, 36 (1967).
12. B. Cabane, J. Phys. Chem., 81, 1639 (1977).
13. B. Cabane, in "Solution Behavior of Surfactants," (Eds., K. L. Mittal and E. J. Fendler), Plenum Press, New York (1982). V. 1.
14. F. Oosawa, Polyelectrolytes, Marcel Dekker, New York (1971).
15. R. Nagarajan and E. Ruckenstein, J. Colloid Interface Sci., 71, 580 (1979).

COMPLEX FORMATION BETWEEN IONIC SURFACTANTS AND POLYMERS IN AQUEOUS SOLUTION

T. Gilányi and E. Wolfram
Department of Colloid Science
Loránd Eötvös University
Puskin u. 11-13, H-1088 Budapest, Hungary

ABSTRACT

The polymer–surfactant interaction has been treated as formation of micelle-like aggregates involving also a fragment of the polymer chain. The model adequately explains the shape of binding isotherms of ionic surfactants by polymers, the critical surfactant concentration of interaction and its dependence on polymer concentration as well as the appearance of two break-points in the colligative properties vs. surfactant concentration curves. It has been concluded from potentiometric dodecyl sulfate activity measurements that the amount of surfactant in complex increases with the equilibrium surfactant concentration until it reaches the critical micelle formation concentration. The aggregation number of surfactant in poly(vinyl-alcohol)–NaDS, poly-(vinyl-pirroli-done)–NaDS and poly-(ethylene-oxide)–NaDS complex in 0.1 M NaNO$_3$ is between 40 and 50, which is lower than that of the free micelles, and for poly-(vinyl-alcohol)–NaDS system it increases with the ionic strength of the solution. Unlike to neutral polymers, the binding sites of the gelatin are occupied by two DS–ions.

The intrinsic viscosities were determined by dilution at constant polymer–NaDS complex composition. The hydrodynamic volume of the polymer is increasing with the NaDS-content of the complex. In 0.1 M NaNO$_3$ solution a minimum appears in the intrinsic viscosity vs. complex composition curves for the poly-(vinyl-pirrolidone)–NaDS and poly-(vinyl-alcohol vinyl-acetate) copolymer–NaDS system.

The light scattering of poly-(vinyl-pirrolidone)–NaDS (0.1 M NaNO$_3$) solutions is interpreted in terms of mixed solvent theory and also by regarding the complex molecules as copolymer ones. The complex was found to be homogeneous as to composition.

INTRODUCTION

The interaction between macromolecules and surfactants in aqueous solution has been extensively investigated for a few decades. First, the protein-surfactant systems were stressed because of their biological importance. With the appearance of well-defined synthetic polymer models, the research was extended to the neutral polymer-surfactant systems, too, with the promise of a deeper insight into the more complex protein-surfactant interaction.

Above a definite (critical) concentration, ionic surfactants interact with neutral polymers. Equilibrium dialysis measurements give evidence of direct interaction, i.e., binding of surfactant by the polymer. According to this the idea has prevailed that the polymer-surfactant interaction is an adsorption process, i.e. the polymer chain as a linear adsorbent (without geometric interface) binds the surfactant molecules from solution. From another point of view, the polymer-surfactant interaction can be considered as complex formation. With increasing surfactant concentration, the number of surfactant molecules per polymer chain increases, from which it can be concluded that the complex does not form with definite stoichiometry. In a third view, somewhat similar to adsorption, the polymer-surfactant interaction is regarded as a preferential solvation of the polymer from a binary water-surfactant mixed solvent. The underlying supposition in this picture is that statistically all the polymer segments are solvated by surfactant molecules to the same degree.

A fundamental question as to the nature of polymer-surfactant interaction arises from the large difference between the mass of polymer and that of surfactant molecules. The adsorption concept emphasizes that the polymer is a large unit and it has many sites available for surfactant binding. In the complex formation model, the stress is on the elementary process between one polymer site and the surfactant molecule, disregarding the fact that the polymer sites are linked with each other in the polymer chain.

In order to describe the polymer-surfactant interaction, the important points to be dealt with are as follows:

-    a quantitative description of the experimental "binding isotherms", and the forces operating in the interaction;

-    the micro-structure of the polymer-site - surfactant contact and the characterization of the polymer - surfactant complex molecules as a whole.

A stepwise sequence of several chemical equilibria (multiple equilibria) has been extensively used for the macromolecule-ligand interaction.[1] In a simple approach, if the binding sites are identical and indistinguishable, the binding isotherm reduces to an equation which is equivalent to the Langmuir adsorption isotherm or to the mass action equation applied for a simple one-step equilibrium (binding site, S - surfactant molecule, A):

$$S + A \rightleftharpoons SA$$

The existence of a critical surfactant concentration for the interaction between neutral polymers and ionic surfactants is contradictory to the concept that the surfactant interacts in its individaul, i.e. monomeric form.   The critical interaction concentration indicates the cooperative nature of the process for the surfactant.

Cabane's idea[2] is that the polymer is adsorbed in the surface layer of the ionic surfactant micelles and the process is governed by the

$$P + mic. \rightleftharpoons (P-mic.)$$

equilibrium. It is neglected, however, that the aggregation number and the structure of the micelles may be different in the free and polymer-complex state.   The aggregation number of the surfactants being clustered with the polymer was estimated by Shirahama for the poly(ethylene oxide)-sodium dodecyl sulfate system and it was found to be smaller than that of the surfactant micelles.[3]

A number of observations revealed similarities between micelle formation and polymer-surfactant complex formation. In both cases the interaction is enhanced by increasing the hydrophobic chain length of the surfactant.[4] As far as polymer is concerned, all studies confirm that the more hydrophobic the polymer, the stronger is the complex formation.[5]

The role of the hydrophobic character of the polymer and surfactant as well as the cooperativity of the process suggest that the main driving force of the polymer-surfactant complex formation is hydrophobic interaction. On the other hand, anionic surfactants, as a rule, interact more strongly with macromolecules as compared with cationic ones of the same hydrophobic moiety.[6] The nature of the head group and the type of counter-ion also have an influence on the interaction.   Non-ionic surfactants either do not interact with macromolecules at all, or their interaction is very weak,[7] pointing to the role of electrostatic interactions, or possibly the steric effect of the surfactant's polar group.

The micro-structure of the surfactant aggregates in the polymer complex can be considered as that of micelles with a hydrophobic core inside, with some segments of the polymer chain being also involved in the aggregate.   Evidence for this structure has been given by NMR measurements.[2]   The combination of the polymer segments with the surfactant aggregate is likely to be similar to mixed micelles, i.e. the hydrophobic fragments are oriented in the hydrophobic core and the hydrophilic groups remain in the surface layer of the surfactant aggregate. The present ideas about the local structure of the polymer-surfactant complex are more or less speculative.

In this work, a quantitative model of the polymer-surfactant interaction is summarized and experimental studies by potentiometric, viscosity and light scattering measurements are presented for the interaction of sodium dodecyl sulfate with polymers in aqueous solution.

MODEL FOR POLYMER-SURFACTANT COMPLEX FORMATION

If an aggregate forms from m surfactant molecules, A, on an active center, S, of the polymer (which may contain many segments of the polymer chain), then the equilibrium is given as

$$S + mA \rightleftharpoons SA_m . \tag{1}$$

The product of this elementary step of the surfactant-polymer complex formation, SA_m, is called a "subaggregate" as distinguished from the entire polymer-surfactant complex molecule which may contain many subaggregates.

The first question in this treatment is the meaning of the polymer-active center. The polymer chain contains many centers available for interaction with surfactant. Insofar as the centers are identical and indistinguishable, the polymer-surfactant complex formation can be treated as a multiple chemical equilibrium:

$$P + mA \rightleftharpoons PA_m$$

$$PA_m + mA \rightleftharpoons PA_{2m}$$

$$PA_{im} + mA \rightleftharpoons PA_{(i+1)m}$$

.

.

.

where P denotes the polymer molecule. A rigorous treatment of these series of equations leads to the result that the process can be simply described by the single equilibrium given by Equation 1. In other words, assuming that the binding sites are identical, the equilibrium is governed by the number of binding sites and not that of the polymer molecules. Although the number of active sites in the solution is proportional to the polymer concentration, the active centers in the same polymer chain cannot be "diluted" by varying the polymer concentration. For a consequent definition of the active center activity, especially in order to compare the thermodynamic parameters of complex formation by various polymers the dependence of complex subaggregate formation on the polymer molecular mass should be analyzed.

In the case of ionic surfactants, the aggregation of surfactant ions in a small volume results in a large charge density and the development of electric double layer as well as ion-pair formation (counter-ion binding to the surface of colloidal aggregates). The free-energy change of complex formation can be formulated by separating electric and non-electric contributions and that of ion-pair and double layer formation within the electric part, as it has been done in the pseudo-phase separation model of ionic micelle formation.[8]   According to this treatment, the degree of counter-ion binding (the "true" degree of dissociation) should be known, which in our opinion cannot be determined. Experimentally only the mean ionic activity of the surfactant solution or the electric properties of micelles can be studied, and in both cases the counter-ion binding and the electric double layer formation manifest themselves together. For this reason, instead of the separation of electric contributions, the apparent number of bound counter-ions is used in which both the counter-ion binding and double layer formation are involved.

In aqueous systems containing both polymer and surfactant, two independent equilibria are to be taken into account. First, the polymer-surfactant complex formation:

$$S + mA^{\ominus} + nK^{\oplus} \rightleftharpoons (SA_m K_n)^{\ominus(m-n)} \equiv \text{compl.},$$

second, the free micelle formation:

$$m'A^{\ominus} + n'K^{\oplus} \rightleftharpoons (A_{m'}K_{n'})^{\ominus(m'-n')} \equiv \text{mic.}$$

Applying the mass action law,

$$[\text{compl.}] = K_{\text{compl.}} [S] [A^{\ominus}]^m [K^{\oplus}]^n \tag{3}$$

and

$$[\text{mic.}] = K_{\text{mic.}} [A^{\ominus}]^{m'} [K^{\oplus}]^{n'} \tag{4}$$

where compl. means the polymer-surfactant subaggregate containing m surfactant anions ($A^{\ominus}$) and n apparently bound counter-ions ($K^{\oplus}$) and m' and n' have the same meaning for the micelle formation.

Introducing the apparent degree of dissociation ($\alpha = \frac{m-n}{n}$ and $\alpha' = \frac{m'-n'}{m'}$) and using the total number of active polymer sites $[S]_0 = [S] + [\text{compl.}]$ Equations 3 and 4 can be written as

$$\frac{[\text{compl.}]}{[S]_o - [\text{compl.}]} = K_{\text{compl.}} \left[A^{\ominus}\right]^m \left[K^{\oplus}\right]^{m(1-\alpha)} \tag{5}$$

and $\quad \left[\text{mic.}\right] = K_{\text{mic.}} \left[A^{\ominus}\right]^{m'} \left[K^{\oplus}\right]^{m'(1-\alpha')} . \tag{6}$

In order to solve these equations, the parameter $Y = [A^{\ominus}] [K^{\oplus}]^{(1-\alpha)}$ was introduced and the activity coefficients for dilute solutions were taken to be 1. Hence

$$\left[K^{\oplus}\right] = \left[A^{\ominus}\right] + m\alpha \left[\text{compl.}\right] + m'\alpha' \left[\text{mic.}\right] . \tag{7}$$

Equations 5-7 were solved for a series of appropriate Y values and the result expressed in molal concentration of surfactant monomer units as a function of the total surfactant concentration (c). In Figure 1, the mean ionic activity of surfactant $([A^{\ominus}] [K^{\oplus}])^{1/2} = a_{\pm}$, the concentration of the surfactant monomer anions, the counter-ions, the polymer-surfactant complex and the free micelle vs. the total surfactant concentration are plotted. The parameters of Equations 5 and 6 were chosen such as to be characteristic for the sodium dodecyl sulfate and poly-(vinyl-alcohol) system ($K_{\text{compl.}}^{1/m} = K_{\text{mic.}}^{1/m'} = 2.7 \times 10^6$, $\alpha = \alpha' = 0.3$, $m = 10$, $m' = 70$ and $[S]_0 = 2.10^{-4}$ in mole fraction).

The results given in Figure 1 can be summarized as follows. With increasing surfactant concentration first individual surfactant ions and counter-ions are only present in the system. Above a critical concentration (c_{cr}) practically all additional surfactant goes into the form of polymer-surfactant complex. When the concentration becomes greater than a second critical value (c_M'), the surfactant ions are distributed between monomeric, complex and micellar forms.

It is important to note that at c_M', which frequently appears as a (second) breakpoint in some physico-chemical properties vs. surfactant concentration plots, and is regarded as the saturation value of polymer with surfactant, the polymer is not necessarily saturated. In the present example the bound amount of surfactant at c_M' is only 11% of the saturation value. As a result of the micelle formation the surfactant activity increases very little above c_M', consequently [compl.] cannot further increase due to the constancy of the mean activity of surfactant. The equilibrium constants for the complex formation can be chosen so that at the beginning of the micelle formation the amount of surfactant in complex is close to that for saturation. In this case the mean activity of surfactant as a function of the total surfactant concentration shows an inflexion below c_M'.

From Figure 1 the existence of a narrow concentration range can be established, below which no complex formation can be detected, and

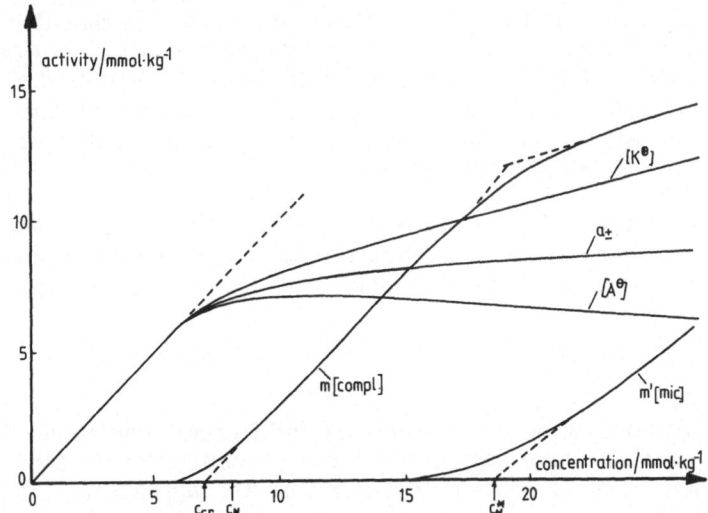

Figure 1.   The activity of different reaction species (surfactant-ion, counter-ion, complex and micelles) as a function of the surfactant concentration. Calculated by Equations 5-7 with $K^{1/m} = K_{mic.}^{1/m'} = 2.7 \times 10^6$; $\alpha = \alpha' = 0.3$; $m = 10$; $m' = 70$ and $[S]_0 = 2 \cdot 10^{-4}$ in mole fraction.

above which practically all additional surfactant goes into the form of polymer-surfactant complex. The model calculations as given above suggest the possibility of a rough estimation of a critical concentration for complex formation even in the case of complex subaggregates containing only 10 surfactant ions. Neglecting the small change in surfactant activity during complex formation, the process can be treated as a phase-separation. For the critical concentration

$$c_p c_{cr}^{m(2-\alpha)} = K_{compl.}' \qquad (8)$$

where c_p means the polymer concentration.

Up to this point the interaction of the subaggregates with each other in the same polymer coil and the contribution of the conformational change of the polymer have been neglected. This approximation is justified only if the number of subaggregates tends to zero and, consequently, the interaction between them vanishes.

The long-range interaction between the subaggregates and the polymer conformational change can be formally taken into account by allowing $K_{compl.}$ to depend on the subaggregate concentration.

The main feature of the proposed mechanism is that it is expressed in terms of complex subunit (subaggregate) formation in excess form. Equations 5 and 8 permit one to determine the standard free energy change of the process and the aggregation number of the surfactant clusters in polymer complex, as well as the total number of the active centers in the polymer molecules.

The number of surfactant ions in the subaggregates can be determined from the dependence of c_{cr} on the polymer concentration with Equation 8 or if $[S]_0 >> [compl.]$ by Equation 5 in the form

$$\log \frac{c_{compl.}}{c_p} = const. + mlgY \ . \tag{9}$$

In the presence of a large excess of indifferent electrolyte Equation 9 is very useful because the counter-ion concentration is constant and it is not necessary to know α to determine the aggregation number:

$$\log \frac{c_{compl.}}{c_p} = const. + mlg[A^\Theta] \ . \tag{10}$$

The description of the polymer-surfactant interaction outlined above is based on the conception that the complex formation is a cooperative process of surfactant ions and reflects the general experimental finding, viz., the existence of a critical surfactant concentration for interaction. It also suggests a decrease of c_{cr} with increasing polymer concentration, which was experimentally found by many authors, as well as the appearance of two breakpoints in the different colligative properties vs. surfactant concentration curves.

EXPERIMENTAL

Materials

Sodium dodecyl sulfate (NaDS, Merck) was recrystallized twice from hot benzene-ethanol (1:1) mixture. Double distilled water, $NaNO_3$ and NH_4NO_3 of reagent grade were used. Poly-(vinyl-alcohol) with 12% acetate content (Rhodoviol MS 135) was used as copolymer model (PVA-Ac) and after removal of its acetate residue by hydrolysis (the final acetate content is 0.65%) and purifying by dialysis against distilled water, it served as poly-(vinyl-alcohol) (PVA) model. M_w was found to be 1.3×10^5 by light scattering. Poly-(vinyl-pyrrolidone) (PVP, Fluka product, $M_w = 1.03 \times 10^6$), poly-(ethylene-oxide) (PEO, Carbowax-6000, Union Carbide, $M_n = 4.2 \times 10^3$), dextran (T-500, Pharmacia Fine Chemicals, $M_w = 5 \times 10^5$), gelatin (Rousselot Kuhlmann), amylose and amylopectine (Koch-Light Laboratories Ltd.) were used without further purification.

Methods

Potentiometric measurements have been performed using sodium-ion responsible glass electrode (Radelkis OP-Na-7111) and Au/Hg/Hg$_I$DS dodecyl sulfate electrode. The e.m.f. measurements were carried out without transference as well as with Ag/AgCl reference electrode joined by sat. NH$_4$NO$_3$ bridge. For details of the method see ref. 9.

Viscosity measurements were carried out by Ubbelohde-type capillary viscosimeter. The surface tension and kinetic energy corrections were neglected.

The intensity of scattered light was measured by Brice-Phoenix OM-2000 photometer at 436 nm as a function of scattering angle (45° - 135°). The solutions were filtrated by Millipore membrane filter. A stock solution of 0.4% PVP made in NaDS solution was diluted at constant chemical potential of the solvent. The composition of the solvent for dilution was determined from potentiometric and equilibrium dialysis measurements as well. Refractive index increments were determined by Brice-Phoenix differential refractometer.

All measurements were performed at 25.0 $\pm$ 0.1^{o}C.

RESULTS AND DISCUSSION

Potentiometric measurements

In Figure 2 the sodium-ion, dodecyl sulfate ion activities and the mean activity of the NaDS solution are plotted against the total surfactant concentration at a constant (0.224%) PVA concentration. The activity coefficients are relative to that of a 5.0×10^{-3} mol.kg^{-1} NaDS solution (a^*). The shape of the experimental activity curves is similar to that obtained by the model calculations. At about 6 mmol.kg^{-1} concentration, which is still less than the critical micelle formation concentration (c_M = 8.1 mmol.kg^{-1}), complex formation occurs between PVA and NaDS, and above this point the mean activity increases only slightly.

In order to apply Equation 5, α was determined by Botre's method[10] and its value was found to be 0.3, which is higher than for micelle formation (α_{mic}=0.18). In Figure 3, the log ($c_{compl.}/c_p$) vs. logY function is plotted (where $c_{compl.}$ is the concentration of the surfactant in complex and c_p is the polymer concentration). According to Equation 5, at small values of $c_{compl.}/c_p$ (i.e., if the polymer molecules are far from being saturated with surfactant), the plot gives a straight line. From the slope the aggregation number of PVA-NaDS complex sub-aggregates is found to be 13 $\pm$ 2. At high complex concentration log

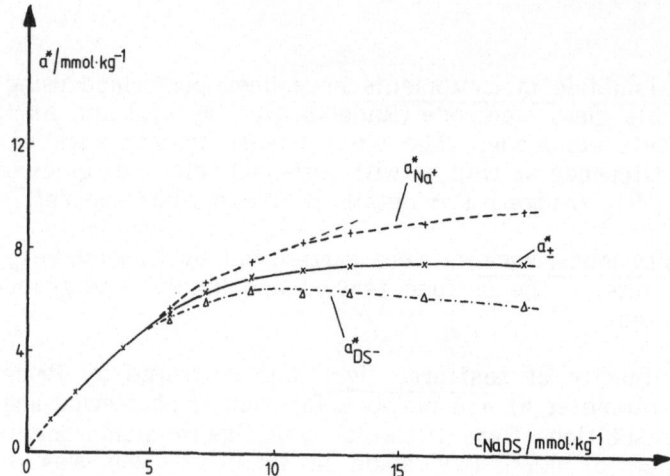

Figure 2.   The relative activity of DS-ions, sodium-ions and the mean
activity vs. c_{NaDS} function (c_{PVA} = 2.24 g.dm^{-3}).

($c_{compl.}/c_p$) is expected to reach a limiting (saturation) value and also
to decline from linearity owing to the enhanced electrostatic repulsion
between the surfactant ions being in the same polymer coil.   Contrary
to this, the values of log ($c_{compl.}/c_p$) tend to infinity.   In this range of
surfactant activity, the micelle formation becomes favorable and the
micelle concentration is also involved in the experimental $c_{compl.}$
values, because $c_{compl.}$ is calculated from the difference between the
total and equilibrium surfactant concentration.   The linearity of the log
($c_{compl.}/c_p$) vs. logY function up to the beginning of micelle formation
suggests that this representation of data is not sensitive to the energy
contribution arising from the electrostatic interactions between the
sugaggregates, or it is compensated by an increase in aggregation
number with the surfactant activity.

     In Figure 4 the equilibrium concentration of dodecyl sulfate ions is
plotted against the total surfactant concentration for the PEO-NaDS
system in the presence of 0.1 M NaNO$_3$.   While the PEO-NaDS complex
formation starts around 1 mmol.kg^{-1}, the micelles appear in a detectable
amount only if the free surfactant ion concentration exceeds the value
of 1.3 mmol.kg^{-1} (see the O PEO curve).   The upper limit of the free DS$^{\ominus}$
concentration is the value of c_M = 1.44 mmol.kg^{-1}.   In the presence of
the polymer, the equilibrium DS$^{\ominus}$ concentration shows two inflexion
points, the position of which is shifted to higher surfactant concentration
with increasing polymer content.   From the analysis of Equation 5 at
constant counter-ion activity, it follows that

$$\left[\frac{dc_{DS}\,\theta^{\ast}}{d\left(c_{compl.}/c_p\right)}\right]_{min.} = \left(c_{compl.}/c_p\right)^{-1}_{infl.} \quad \text{and} \quad (11)$$

$$\left(c_{compl.}/c_p\right)_{infl.} = \frac{1}{2}\left(c_{compl.}/c_{p\ sat.}\right)(1-1/m) \quad (12)$$

where the subscript sat. refers to the amount of bound surfactant at saturation. According to Equations 11 and 12, the inflexion of the c_{DS} vs. $c_{compl.}/c_p$ function gives half of the saturation value of bound surfactant if $m \gg 1$.

In Figure 5 the $\dfrac{\Delta c_{DS}\theta}{\Delta\left(\dfrac{c_{compl.}}{c_p}\right)}$ increments calculated from the data of Figure 4 are plotted as a function of $c_{compl.}/c_p$. Within the experimental error, all the points determined at different polymer concentrations fall to the same curve. From the minimum the half-saturation of the PEO with NaDS is found to be 5.5 ± 0.5 mmol.kg^{-1}. The second inflexion in the plot of the equilibrium $\overline{DS}\theta$ vs. total NaDS concentration function results in a maximum in the differential function. In the absence of micelle formation the increment should increase monotonously after its minimum (dotted line). However, as a consequence of the micelle formation the monomer activity may not increase beyond c_M, therefore, the increment decreases and tends to zero (as it was mentioned $c_{compl.}/c_p$ may exceed the saturation value, because the experimental c_{compl} involves $c_{mic.}$)

In Figure 6 log ($c_{compl.}/c_p$) is plotted against log $c_{DS}\theta$ for the different polymers and NaDS systems in presence of 0.1 M NaNO$_3$. The slope of the linear region of the curves gives the aggregation number of the surfactant subaggregates. Correcting for activity the aggregation numbers are beween 40 and 50, for all the neutral polymer concerned which is about half of that for ordinary micelles under similar conditions, and they seem to be characteristic for the surfactant rather than for the polymer in question. On the other hand, for the PVA-NaDS system the aggregation number of the surfactant subaggregates in 0.1 M NaNO$_3$ solution is significantly higher than without excess salt. This trend of the increasing aggregation number with the counter-ion concentration is the same as in the case of the free micelle formation.

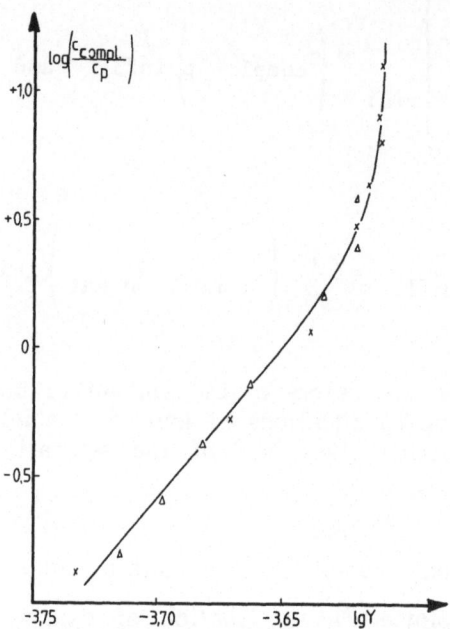

Figure 3.   The plot of $\log(c_{compl.}/c_p)$ vs. $\log Y$ function for the PVA-NaDS complex (c_{PVA}: x 2.24 g.dm^{-3}; $\Delta = 4.48$ g.dm^{-3})

Contrary to the PVA-NaDS system, the binding of NaDS to PVP and PEO in 0.1 M NaNO$_3$ solution shows a saturation region preceding the micelle formation (Figure 6).   At around $c_{DS}\Theta = 1.44$ mmol.kg^{-1} (which is the c_M) the equilibrium dodecyl sulfate ion concentration does not increase further and $c_{compl.}/c_p$ tends to infinity.

The standard free energy of the transfer of surfactant ions into the subaggregates at constant 0.1 M NaNO$_3$ concentration was found to be -10.6 kT, -11.0 kT, -11.1 kT and -11.3 kT for the micelle formation, PVA-NaDS, PEO-NaDS and PVP-NaDS systems, respectively.   These values are practically the same, which suggests that the polymer-surfactant complex formation is a process similar to the micelle formation.   In the case of the PVP and PEO complex the number of polymer segments per one surfactant subaggregate at the saturation is 130 and 120, respectively.   The small difference between the standard free energies of the complex and micelle formation suggests that only a small fraction of the polymer segments is in contact with the surfactant molecules.

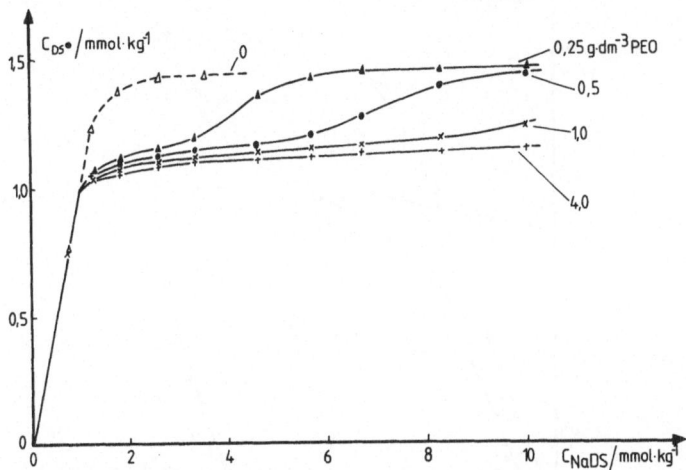

Figure 4.   The equilibrium DS-ion concentration as a function of the total NaDS concentration for the PEO-NaDS complex in 0.1 M $NaNO_3$ solution. c_{PEO}: ▲ 0.25 $g.dm^{-3}$; ● 0.5 $g.dm^{-3}$; x 1.0 $g.dm^{-3}$; + 4.0 $g.dm^{-3}$.

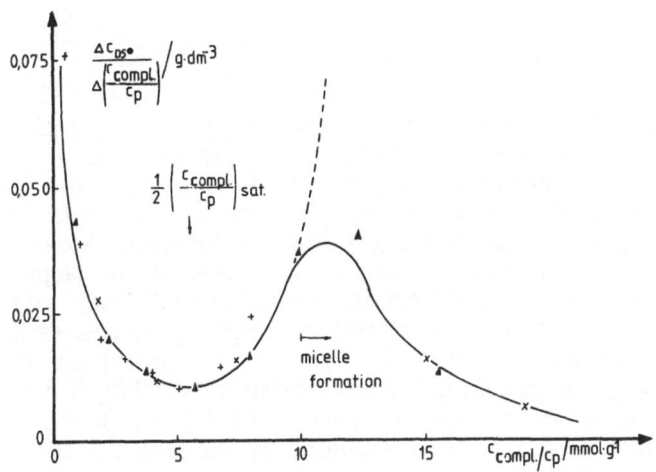

Figure 5.   The $\dfrac{\Delta c_{DS}{}^{\ominus}}{\Delta\,(c_{compl.}/c_p)}$ increments vs. $c_{compl.}/c_p$ for the PEO-NaDS complex in 0.1 M $NaNO_3$ solution. c_{PEO}: x 0.25 $g.dm^{-3}$, ▲ 0.5 $g.dm^{-3}$, + 1 $g.dm^{-3}$.

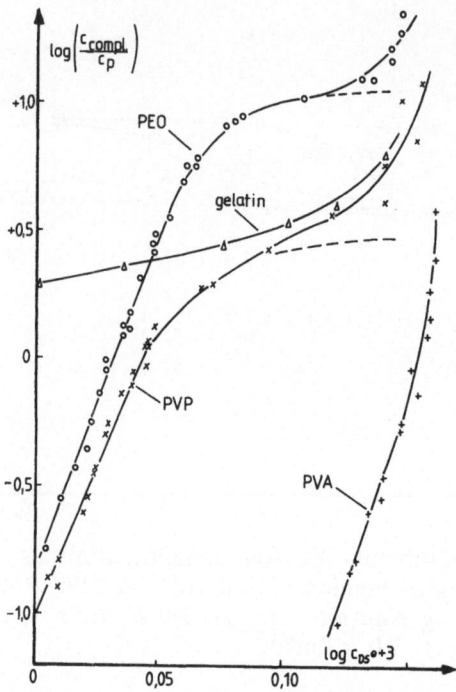

Figure 6.   The log $(c_{compl.}/c_p)$ vs. log $c_{DS}^{\ominus}$ function for different polymers.

In Figure 6 the data for the interaction between gelatin and NaDS in 0.1 M $NaNO_3$ at neutral (uncontrolled) pH are also plotted. The amount of bound surfactant is detectably high at low equilibrium concentration range (not shown in the Figure), i.e. it is 0.3 mmol/g gelatin at $c_{DS}^{\ominus}$ = 0.3 mmol.kg^{-1}, and it increases slowly with the equilibrium surfactant concentration. The value of the slope corrected for activity is 2.07 $\pm$ 0.09, which means that DS-ions interact with gelatin in form of dimers. Formation of large aggregates as in the case of the neutral polymers cannot be detected, the upward curvature of the log $(c_{compl.}/c_p)$ vs. log $c_{DS}^{\ominus}$ curve above 1 mmol.kg^{-1} concentration can be attributed to micelle formation and it also suggests that the binding of NaDS on gelatin is interrupted by the free micelle formation. The dimer-interaction can be interpreted by a significant energy contribution from charge–charge interaction between the positive charges of gelatin and DS-ions. It is to be expected, for sterical reasons, that the electrostatic interaction between the polyelectrolyte and ionic surfactant is favourable to the formation of small surfactant associates, otherwise when the interaction is driven mainly by hydrophobic inter- action the larger micelle-like surfactant aggregates are stabilized.

In the case of dextran (0.1-1%) and amylopectin (0.05%) the potentiometric measurements do not indicate any interaction with NaDS, the activity of NaDS solutions is the same with and without these polymers. An amylose solution of 0.025% was turbid and it did not clear up even in the presence of 20 mmol.kg^{-1} NaDS. However, the potentiometric measurements showed binding of NaDS according to a Langmuir isotherm with a saturation value of 0.27 mmol/g amylose. The binding of NaDS on amylose is regarded as an adsorption on solid amylose particles and not as an interaction between them in solution. It is concluded that the investigated polysaccharides do not interact with NaDS in aqueous solution.

Viscosity measurements

Viscosity is one of the most frequently applied method to study the polymer-surfactant interaction. The hydrodynamic data are expressed in various ways: viscosities relative to the solvent (water) or to the surfactant solution; furthermore, specific viscosity measured as a function of polymer concentration at constant surfactant concentration and as a function of surfactant concentration at constant polymer content as well as measurements at a constant polymer/surfactant concentration ratio can be found in the literature. In some cases, efforts were made in order to determine the intrinsic viscosity of the polymer-surfactant complex by extrapolation from the linear region of the η_{sp}/c_p vs. c_p function in spite of the fact that at low polymer concentration it shows anomaly.[11]

Most of the viscosity measurements are aimed at indicating the interaction or at comparing the strength of the interaction of the same polymer with different surfactants. Saito and coworkers[12] have come to the conclusion that the counter-ions are of primary importance in the strength of the polymer-surfactant complex both because of their influence on bulk water structure and their different degree of binding. It is noted here that the strength of interaction deduced from the extent of the viscosity increase in polymer solution is unreliable, because it reflects the conformational change of the polymer rather than the strength of the complex.

In the absence of interaction between polymer and surfactant there are three contributions in the relative viscosity of the solutions: those from the polymer, the surfactant micelles and the surfactant monomers, respectively. The contribution of surfactant monomers is negligible, but that of micelles may be relatively significant, especially in case of low molecular mass polymers. In order to interpret the conformational change of the polymer during complex formation, a consequent choice of the reference medium is of primary importance. Referring to polymer-free surfactant solution is unsuitable, because the surfactant solution has a different composition as a solvent with and without polymer. In a 25 mmol.kg^{-1} aqueous NaDS solution the micelles result in a 4% increase in viscosity. However, in the presence of a sufficient amount of polymer there are no micelles in the system (the surfactant is distributed between monomer and polymer-complex form), so the reference to the surfactant solution becomes unrealistic. An ill-defined reference medium can even lead to negative specific viscosity.[13] Therefore, in what follows, data relative to water (or 0.1 M $NaNO_3$ solution) are used, which express the true specific viscosity of the polymer-surfactant complex as far as surfactant concentrations are less than c_M^*.

In Figure 7 η_{sp}/c_p at constant PVA concentration is plotted as a function of the NaDS concentration. The change in the viscosity of solution reflects the three characteristic concentration regions of surfactant. Up to about 6-8 $mmol.kg^{-1}$ there is no interaction between PVA and NaDS, but exceeding this value the viscosity increases. This increase is significantly higher than that calculated by assuming additivity of the specific viscosities of the components (dotted line in Figure 7 for the case of 0.187% PVA).

An inflection appears in the η_{sp}/c_p vs. c_p function, the position of which shifts to higher surfactant concentration with increasing polymer content. The potentiometric measurements have showed that, if the NaDS content of the PVA-NaDS complex reached the value of 3.5 mmol NaDS/g PVA (in this case $c_{NaDS}=c_M^*$), any further increase in NaDS concentration resulted in the appearance of NaDS micelles in the system. The NaDS concentrations with respect to c_M^* are shown by arrows in Figure 7. It can be concluded that the inflections coincide with the beginning of micelle formation.

The increase of the hydrodynamic volume of PVA molecules owing to the interaction with NaDS can be interpreted by the formation of a polyelectrolyte-type complex, i.e. the polymer coil expands, as a result of the electrostatic repulsion between the charged surfactant sub-aggregates. The increase in viscosity also suggests that the complex molecules are not compact, but the polymer keeps its random coil conformation.

The intrinsic viscosity of the complex can be determined by dilution of the system neither at constant polymer/surfactant concentration ratio nor at constant surfactant concentration, because the composition of the complex (the number of surfactant ions in one polymer coil) and the ionic strength of the solution are to be kept constant under dilution. Therefore, the dilution of the polymer-surfactant solution was performed by using as a diluent the surfactant solution being in dialysis equilibrium with the PVA-NaDS complex. The equilibrium surfactant concentration was found to be independent of the amount of the complex in the system, so the dilution was carried out at constant ionic strength as well.

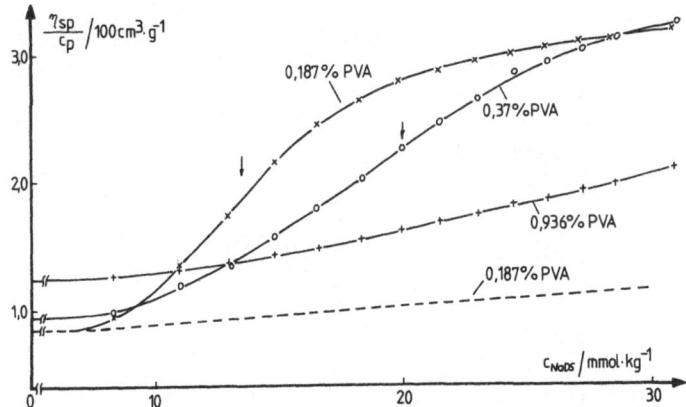

Figure 7.   The η_{sp}/c_p values as a function of the NaDS concentration for the PVA-NaDS system (the dotted line is calculated by assuming additivity of specific viscosities of the components at 0.187% PVA).

In Figure 8 and η_{sp}/c_p vs. c_p function at different constant PVA-NaDS complex composition is plotted. A linear relationship was found up to 2.4 mmol NaDS/g PVA. The intrinsic viscosity of the PVA determined from Figure 8 increases with the NaDS content of the complex.

In Figure 9 the relative intrinsic viscosities, $[\eta]/[\eta]_0$, where $[\eta]_0$ refers to the polymer without NaDS, are plotted against the polymer-NaDS complex composition. The relative change in the hydrodynamic volume of the polymer depends on the ionic strength of the solution, and on the number of charges due to one polymer molecule as well as on the distribution of the surfactant over the polymer coil, i.e., on the aggregation number of surfactant subaggregates. The order of PVP > PVA > PVA-Ac > PEO for the increase of $[\eta]/[\eta]_0$ is in accordance with the molecular mass of the polymer, and seems to suggest that the expansion of polymer coil is roughly ruled by the number of surfactant ions in it.

In Figure 10 the value of $[\eta]/[\eta]_0$ are given in the presence of 0.1 M NaNO$_3$ as a function of the polymer-NaDS complex composition. At high ionic strength of the solutions, the electrostatic interaction between the surfactant subaggregates is diminished. Within the experimental error, the hydrodynamic volume of PEO does not change, in the case of the PVA there is a slight increase, while the curve for the PVP and PVA-Ac shows a shallow minimum. The appearance of the minimum suggests that besides the expansion of the polymer due to the electrostatic repulsion, an opposite effect is also to be taken into

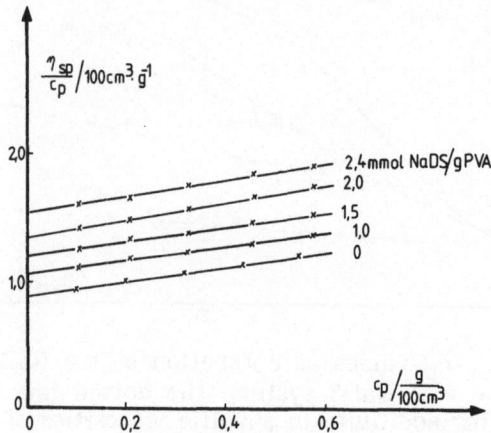

Figure 8.   The values of η_{sp}/c_p as a function of c_p at different PVA-
NaDS complex composition.

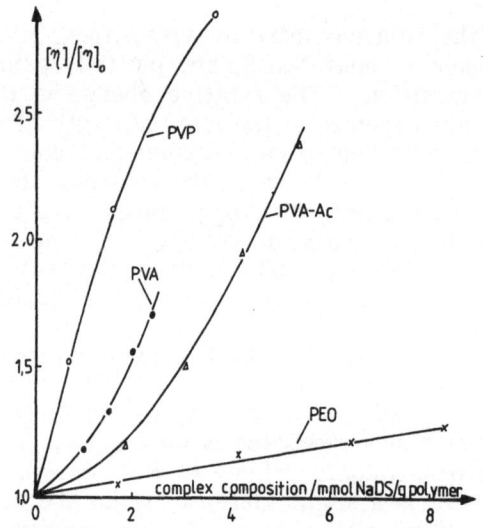

Figure 9.   The relative intrinsic viscosity of the polymer–NaDS complex
as a function of the complex composition.

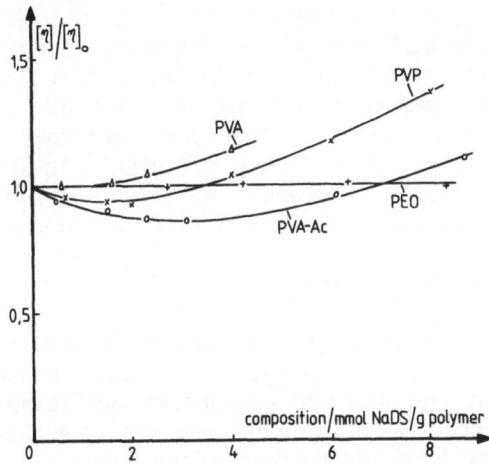

Figure 10.   The relative intrinsic viscosity of the polymer-NaDS complex
as a function of the complex composition in the presence of
0.1 M NaNO$_3$.

consideration.   It is clear that at high amount of excess NaNO$_3$ the
minimum cannot be interpreted by the effect of counter-ions on bulk
water structure or by counter-ion binding.   The reason for the decrease
in the hydrodynamic volume of the polymer during complex formation
must be in the structure of the polymer-surfactant complex.

One possible explanation is an apparent shortening of the polymer
chain if a "twisted-up" section of the polymer is involved in the
subaggregates, or if loops are formed by interconnection of polymer
segments being far away from each other in the chain.   On the other
hand, the polymer-surfactant complex can be considered as a copolymer
molecule in which the interchain interactions are governed by polymer
segment-polymer segment, subaggregate-subaggregate and polymer seg-
ment-surfactant subaggregate interactions.   The conformation of the
polymer coil is mainly determined by the electrostatic interaction
between the surfactant subaggregate which is always repulsive.   At small
number of subaggregates in the coil and at large ionic strength, a long-
range interaction between the polymer segments and surfactant subaggregates may succeed.   The surfactant subaggregates represent a high
local electric field within the coil and also a high counter-ion density in
their electric double layer.   The thickness of the electric double layer
at 10^{-3}-10^{-1} M ionic strength extends over 10-1 nm, which means that
many polymer segments can be in permanent interaction with a single
surfactant subaggregate.

Hence there is no possibility to get even a qualitative conclusion about the role of the different parameters which determine the behavior of the polymer-surfactant complex as a whole. It is concluded, however, that the viscosity measurements are not suitable to compare the strength of complexes, or, in some cases, not even to indicate the interaction itself. The polymer-ionic surfactant complex seems to be a unique model to study the behavior of polyelectrolytes in which the charges are not individually distributed over the chain.

Light scattering measurements

The aqueous solution of polymer and surfactant can be treated as ternary system (polymer and mixed solvent), or as binary system (polymer-surfactant complex and equilibrium surfactant solution). According to the theory of multicomponent systems the total scattered light is a composite contribution from three terms: τ_1 from the density fluctuations, τ_2 from the concentration fluctuations of polymer including the surfactant preferentially adsorbed on the polymer, and τ_3 from the concentration fluctuations of the diffusible small molecules (surfactant monomers and water). It was shown to a good approximation that τ_2 can be experimentally obtained as the difference between the total scattered light of the solution and that of the mixed solvent without polymer. The simplest method of obtaining the weight average relative molecular mass of the polymer (M_w, in the following called molecular mass) is to determine the refractive index increment at constant chemical potential of the solvent, $(dn/dc_p)_\mu$, which is experimentally achieved by using the solution and the solvent being in dialysis equilibrium[4]:

$$K \left(\frac{dn}{dc_p} \right)_\mu^2 \cdot \frac{c_p}{R_o} = \frac{1}{M_w} , \qquad (13)$$

where c_p is the polymer concentration, R_o is the Rayleigh ratio due to the concentration fluctuation of polymer, extrapolated to zero concen-

tration and zero scattering angle, and $K^* = \dfrac{2\Pi n_o^2}{\lambda_o^4 N}$ in which n_o

is the refractive index of the solvent, λ_o is the wavelength of the incident light in vacuo and N is the Avogadro number.

Neglecting all other restrictions for the system under study, the mixed solvent theory of light scattering can be applied only if the solution composition is the same in the vicinity of every polymer molecules, viz., the polymer-surfactant complex is homogeneous with respect to composition. In order to study the heterogeneity of composition of polymer-surfactant complex, the latter was considered as a copolymer solved in pure solvent. The polymer-surfactant molecules are built up of $M_{w,p}$ polymeric and $M_{w,NaDS}$ surfactant component, the average composition by weight is defined as

$$W = \frac{M_{w,p}}{M_{w,p} + M_{w,NaDS}} \cdot$$

The composition of the copolymer was kept to be constant under dilution and the refractive index increment in Equation 13 was given as $\nu = (dn/dc_{compl.})_{\mu}$. Applying Equation 13 for the copolymer solution defined above it yields an apparent molecular mass of the polymer-surfactant complex $(M_{w,compl.}^*)$ which differs from the true molecular mass $(M_{w,compl.})$ according to the equation[15]:

$$M_{w,compl.}^* = M_{w,compl.} + 2P \frac{\nu_p - \nu_{NaDS}}{\nu} + Q \frac{\nu_p - \nu_{NaDS}}{\nu}, \qquad (14)$$

where ν_p and ν_{NaDS} are the refractive index increment of the polymer and NaDS, respectively, P and Q are heterogeneity parameters defined as:

$$P = \Sigma \nu_i M_i \delta w_i \qquad \text{and} \qquad Q = \nu_i M_i \delta w_i^2 \quad,$$

in which γ_i is the relative concentration of the complex molecules of composition w_i, and δw_i is the deviation from the average composition w.

The usual procedure in applying Equation 14 is to change the refractive index increments by varying the solvent. In our case the solvent cannot be exchanged, therefore, in order to estimate the heterogeneity of the complex at a given set of refractive index increments, $M_{w,compl.}$ was determined independently from

$$M_{w,compl.} = M_{w,p}(1+\alpha),$$

where α is the bound amount of NaDS per unit mass of polymer, calculated from dialysis equilibrium.

One more restriction should be mentioned while interpretating light scattering data of the polymer-surfactant solutions. The complex is a polyelectrolyte and to ensure the applicability of the fluctuation theory the non-electroneutral fluctuation must be neglected as compared with neutral fluctuation. For this reason, and in order to diminish the range of electrostatic intermolecular interaction, 0.1 M excess $NaNO_3$ has been used throughout. Under such circumstances, light scattering data yield regular Zimm diagrams.

In Table 1 a compilation of light scattering results is given for the PVP-NaDS system. The weight average molecular mass of the PVP, measured in NaDS (0.1 M $NaNO_3$) solutions has been found to be independent of the amount of NaDS or, what is the same statement, the

Table I

The parameters of the PVP-NaDS complex (in 0.1 M NaNO$_3$) from light scattering measurements

composition $\dfrac{\text{mmol NaDS}}{\text{g PVP}}$	w —	$\left[\dfrac{dn}{dc}\right]_{comp.}$ μ cm$^3\cdot$g^{-1}	$M_{w,PVP}$ x10^{-6}	$M^*_{w,compl.}$ x10^{-6}	$M_{w,compl.}$ x10^{-6} from dialysis	$\overline{(\rho^2)}^*_z$ x10^{11} cm^2
0 (PVP)	1.0	0.185	1.03	1.03±0.06	(1.03)	3.45±0.15
0.96	0.783	0.172	1.00	1.28±0.10	1.31	2.83±0.17
1.86	0.651	0.165	1.04	1.59±0.09	1.58	1.72±0.08
2.86	0.548	0.159	1.05	1.92±0.13	1.88	2.75±0.09
3.70	0.484	0.155	1.03	2.12±0.08	2.13	4.74±0.13

apparent molecular mass of the "copolymer" complex coincides with the true one (from dialysis), what means that $P = Q = O$ (since $\nu_{PVP} \neq \nu_{NaDS}$). The PVP-NaDS complex molecules are homogeneous with respect to composition within the sensitivity of light scattering to composition heterogeneity. In the last column, the z-average mean square of radius of gyration is shown. For a copolymer having components of different refractive index increment the radius of gyration determined from light scattering is also an apparent one $\left(\overline{\rho^2}\right)^*_z$. As a function of the NaDS-content of the complex, $\left(\overline{\rho^2}\right)^*_z$ shows a minimum at about 2 mmol NaDS/g PVP, which is in qualitative accordance with the change in the hydrodynamic volume of the complex.

REFERENCES

1. Steinhardt, J.; Reynolds, J. A.:   Multiple Equilibria in Proteins, Chap. II., Acad. Press, New York-London, 1969.
2. Cabane, B.:   J. Phys. Chem., 1977, 81, 1639.
3. Shirahama, K.:   Colloid Polym. Sci., 1974, 252, 978.
4. Arai, H.; Murata, M.; Shinoda, K.:   J. Colloid Interface Sci., 1971, 37, 223.
5. Breuer, M. M.; Robb, I. D.: Chem. Ind. 1970, 13, 530.
6. Gravsholt, S.:   Proc. 5th Int. Congr. Surface Active Subst., Barcelona, 1968, Vol. 2. p. 993.
7. Saito, S.:   J. Colloid Sci., 1960, 15, 283.
8. Shinoda, K.; Hutchinson, E.:   J. Phys. Chem., 1962, 66, 577.
9. Gilanyi, T.; Wolfram, E.:   Colloids and Surfaces, 1981, 3, 181.
10. Botre, C.; Crescenzi, V. L.:   J. Colloid Sci., 1953, 8, 593.
11. Isemura, T.; Imanishi, A.:   J. Polymer Sci., 1958, 33, 337.
12. Saito, S.; Taniguchi, T.; Kitemura, K.:   J. Colloid Interface Sci., 1971, 37, 154.
13. Saito, S.; Yukawa, M.:   J. Colloid Interface Sci., 1969, 30, 211.
14. Strazielle, C.:   in "Light Scattering from Polymer Solutions", Chap. 15. Ed. by Huglin, M. B. Acad. Press, London-New York, 1972.
15. Bushuk, W.; Benoit, H.:   Canad. J. Chem., 1958, 36, 1616.

COMPLEXES OF CATIONIC POLYMERS AND ANIONIC SURFACTANTS

E. D. Goddard and
P. S. Leung
Union Carbide Corporation
Specialty Chemicals and Plastics Division
Tarrytown, New York 10591

ABSTRACT

Measurements of relative viscosity (Saybolt) and shear dependent viscosity (Haake) were made on solutions containing 1% of a cationic cellulose ether polymer with various ratios of added sodium dodecylsulfate (SDS). By contrast with results previously obtained with 0.1% polymer solutions, evidence is found for inter- rather than intramolecular interaction which is brought about by polymer bound surfactant ions. The intermolecular interaction manifests itself as high solution viscosity at low SDS to polymer ratios. At high concentrations of SDS, viz., beyond the region where precipitates of polymer and surfactant form, the results are consistent with a polymer chain associated with many surfactant micelles in a globular structure. The viscosity, in this range, is much lower. Shear dependent rheology was found useful in distinguishing between chain extension and the formation of intermolecular polymer/micelle structures. The latter are favored at low shear. However, the rheology observed shows that chain extension in fact is favored under certain conditions. Small angle neutron scattering results provided information which is consistent with these rheology interpretations.

INTRODUCTION

In previous publications on the interaction between cationic polyelectrolytes and anionic surfactants, we have described the solubility,[1] surface tension,[2] electrophoresis,[2] and dye solubilization[3] characteristics of their mixtures. We also reported briefly on their viscosity behavior.[3]

Ohbu et. al[4] studied the effect of the degree of quaternary ammonium substitution of cationic cellulose ethers on their binding of sodium dodecyl sulfate (SDS) using equilibrium dialysis, dye solubilization, solution density and NMR spin-lattice relaxation measurements. Structures were proposed for different SDS/polymer compositions. The concentration of polymer they studied was such that, at its nominal CMC, SDS was present in stoichiometric excess over the polymer.

The present paper reports a more extensive rheological study of a series of mixed aqueous solutions of a cationic cellulose ether, at a higher fixed concentration, and the anionic surfactant SDS, at varying concentration. To summarize the solubility characteristics of such systems, there are three interaction zones which occur with increasing SDS concentration:

1.   Where the polymer is in excess - a clear zone.
2.   Where the polymer and surfactant ratio is close or equal to the "stoichiometric interaction" (charge neutralization) value and precipitation occurs.
3.   A clear resolubilization zone, where the insoluble complex is resolubilized by excess surfactant.

At the level (1%) of polymer chosen for these studies, when the surfactant is at its nominal CMC, the system would still be in the precipitation zone. As we will show, the absolute concentration of polymers in these mixed systems can have a profound effect on the structure of the entities in solution.

EXPERIMENTAL

The cationic cellulose ether used is a Union Carbide Corporation product sold under the tradename UCARE® Polymer JR-400. Its molecular weight is approximately 400,000. The sodium dodecyl sulfate is a high purity specimen obtained from British Drug Houses. Viscosity measurements on the solutions were carried out with a Saybolt viscometer at 25°C. The normal density corrections were applied. Shear dependent viscosity was measured at 25°C using the Haake Rotovisco viscometer, R-V3, with the MK 500 measuring head. The sensor system used was a coaxial cylinder type fitted with the MVI bob, having a radius of 20.04 mm. The radius of the stator is 21 mm. With this attachment, the range of shear rate and of viscosity is from 0.5 to 3000 sec^{-1}, and from 8 to 10^6 centipoise, respectively.

RESULTS AND DISCUSSION

Previous measurements[3] on 0.1% Polymer JR-400 systems had shown that the relative viscosity decreased on initial addition of SDS

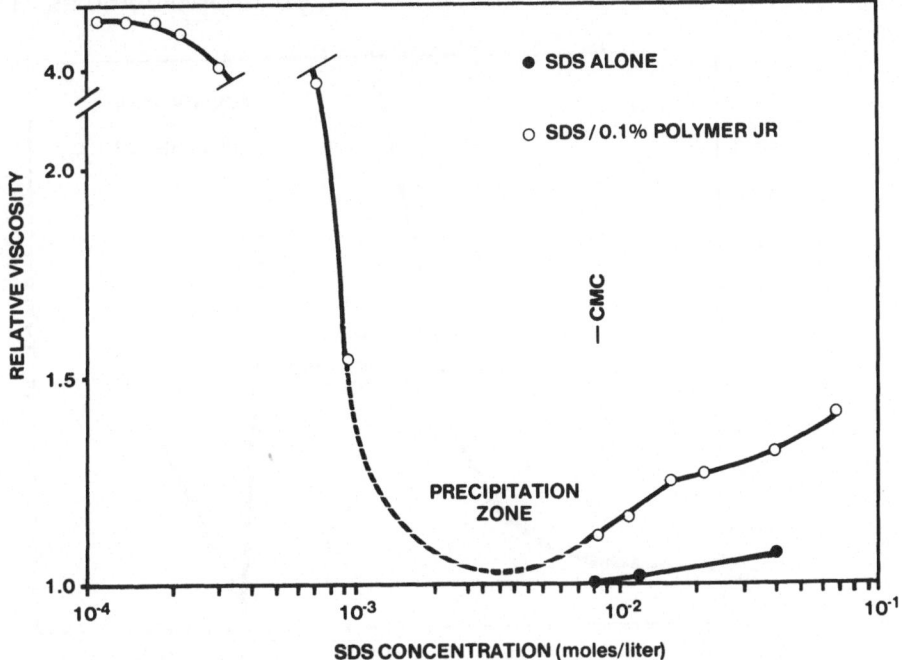

Figure 1. Relative viscosity of a system containing 0.1% Polymer JR and various concentrations of SDS (from Ref. 3).

(Figure 1). The viscosity continued to decrease as more SDS was added and the zone of neutralization and precipitation was entered. Further addition of SDS led to a small, gradual increase in viscosity. Goddard et. al[1] reported a pronounced increase in surface activity when SDS was bound to the polymer. The apparent increase in surface activity in the initial region is consistent with the reduced solubility of the polymer which is observed as complex formation is brought about by progessive charge neutralization and eventual precipitation. However, the concentration of polymer in these early experiments was not sufficiently high to allow association between bound surfactant molecules on different polymer chains. On the contrary, the decrease in viscosity together with the reported increase in dye solubilization reflects intra-polymer association of chains of the electrostatically bonded surfactant molecules. At SDS concentrations beyond the precipitation zone, i.e., where the surfactant complex starts to be resolubilized, the reported increase in dye absorption over the polymer-free system occurs at an SDS concentration which is below its formal CMC. This increase can be attributed to the formation of a "double" layer and some clustering of bound surfactant.

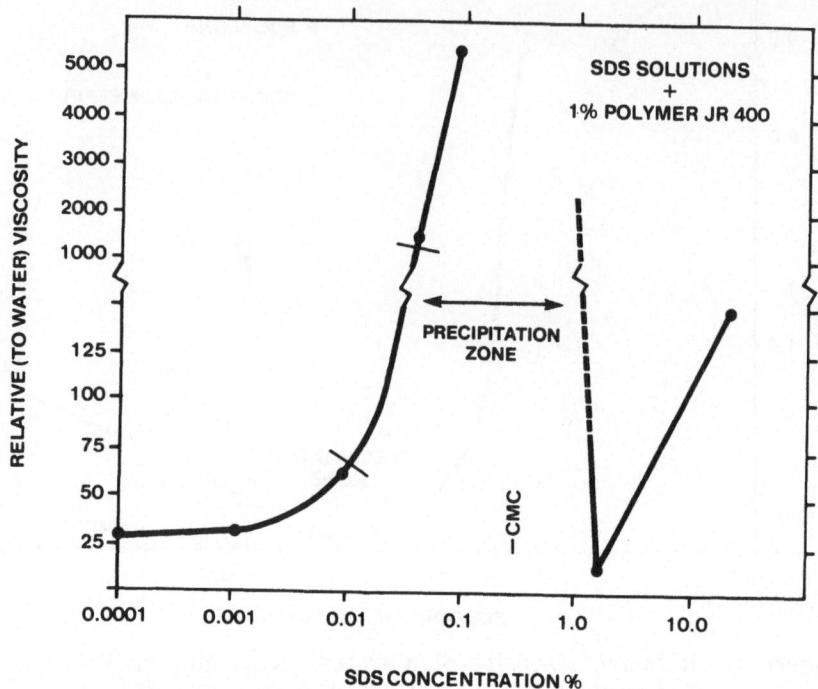

Figure 2. Relative viscosity (Saybolt) of a system containing 1% Polymer JR and various concentrations of SDS.

At the 1% Polymer JR-400 concentration employed in the current study, on adding the first small amount of SDS to the solution, the viscosity <u>increases</u> drastically. See Figure 2. At least two possible mechanisms involving a structural change could cause this increase of viscosity. One is chain expansion; the other is the setting up of intermolecular structures. As the viscosity does not increase when SDS is added to a solution of low polymer concentration, the chain expansion interpretation is not very plausible. In any event, it is seen from Figure 2 that the increase in viscosity is evident with as little as 0.001% SDS added.

It is thus reasonable to conclude that, at this polymer concentration (1%), <u>inter</u>-polymer/polymer interaction plays the dominant role. The interaction involves the "tails" of bound SDS molecules on <u>separate</u> polymer molecules. Such alkyl group association is expected to be relatively weak and should be sensitive to shear. In fact, variable shear rate rheology measurements provide definitive information for selecting

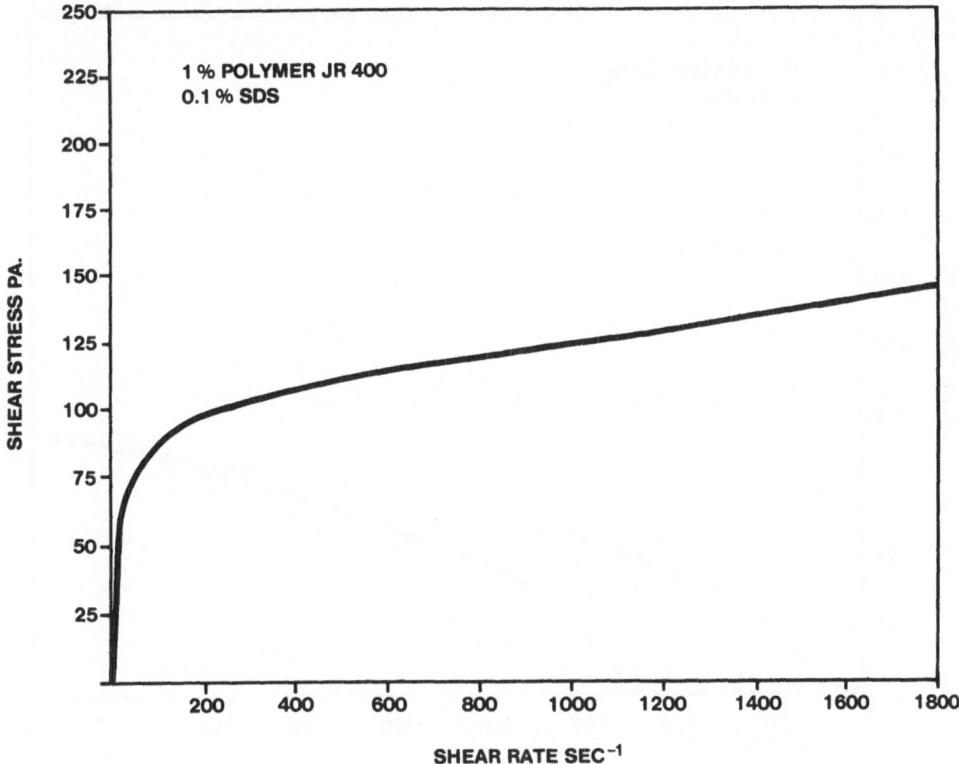

Figure 3. Shear stress versus shear rate plot (Haake) for a solution
containing 1% Polymer JR and 0.1% SDS.

between the chain expansion and the intermolecular structuring mechanisms. Figure 3 shows such data for a solution containing 1% Polymer
JR and 0.1% SDS. The shear stress increases sharply with shear rate
in the low share rate range. However, in a second region of shear (up
to about 200 sec^{-1}) the rate of stress increase drops off progressively.
At still higher shear, the rate of increase is sharply reduced, being
more or less linear with the rate of shear stress, i.e., exhibiting
Newtonian behavior in this region. The differential viscosity is
calculated to be about 22 centipoise, sharply reduced from the low
shear viscosity (in excess of 5000 cp). The rheological behavior was
measured at two other temperatures, viz., $35^{\circ}C$ and $50^{\circ}C$ for the same
system. It was found that the viscosity was almost independent of

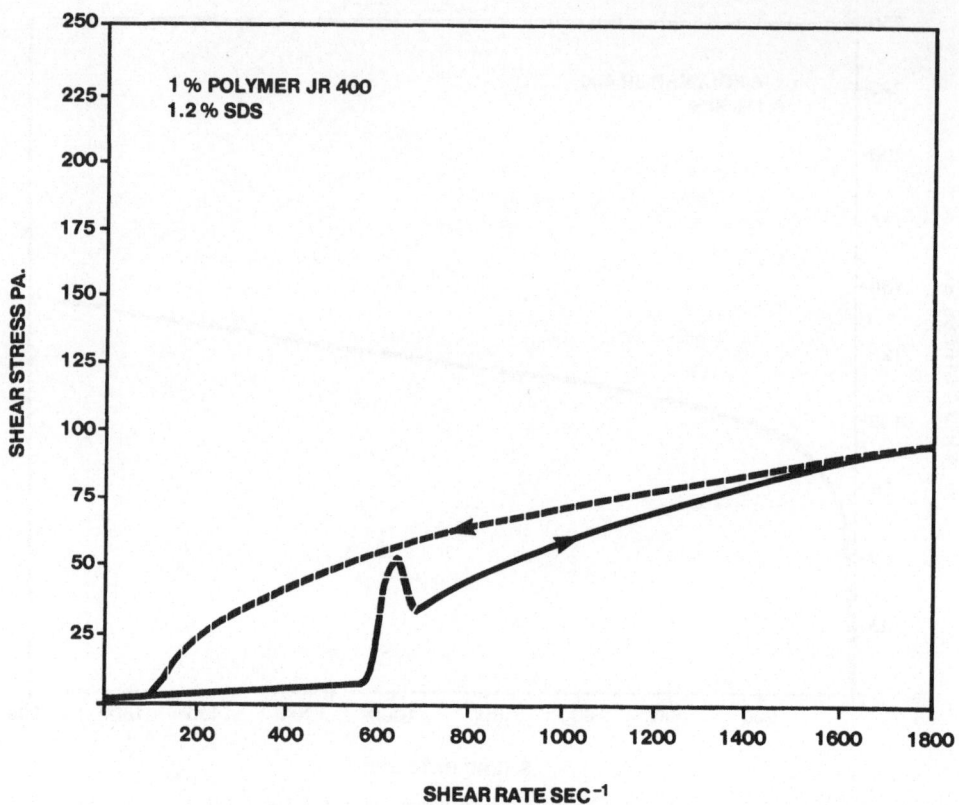

Figure 4.   Shear stress versus shear rate plot for a solution containing
            1% Polymer JR and 1.2% SDS.

temperature, implying a very low activation energy of flow.   This
confirms that the intermolecular association force for the system is
weak.   It should be noted that, in contrast to the situation prevailing
at low (0.1%) concentration of polymer, at the 1% polymer level, the
concentration of SDS needed to effect complete charge neutralization
and maximum precipitation exceeds its nominal CMC.   At some point,
beyond the charge neutralization, micelle formation is expected to
occur.   This is the case, for example, at 1.5% SDS where resolubilization of the precipitate has occurred.   At this point the viscosity
of the system is still quite low.   As seen in Figure 2, the viscosity is
actually lower than that of the free polymer.   Precipitation is caused
by charge neutralization with the formation of a "monolayer" of anionic
surfactant on the positively charged polymer backbone.   However, as
excess surfactant is added to bring the system into the resolubilization
zone, at least part of the strongly bonded surfactant monolayer will act

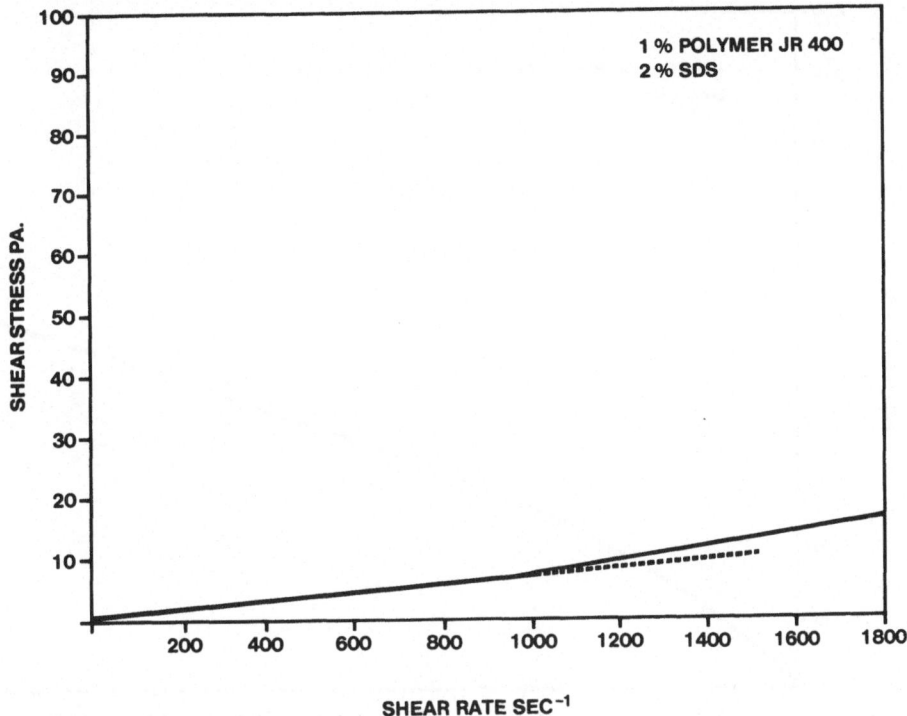

Figure 5.   Shear stress versus shear rate plot for a solution containing
             1% Polymer JR and 2% SDS.

as a nucleus to bind a second layer of surfactant, effectively forming
a bilayer and rendering the polymer negatively charged. A portion of
the first layer could also join with excess surfactant molecules in the
formation of "bonded" micelles.   The bilayer structure would be
expected to increase chain expansion and hence increase the viscosity.
On the other hand the chain-micelle association should result in a
relatively low viscosity and this is more in line with our observations.
As discussed earlier, such a structure may not be "feasible" for low
polymer levels at the SDS concentration just beyond the resolubilization
zone. It can be visualized that a structure with the polymer "wrapped
around" micelles would generate much less expanded entities in
solution. However, such structures would be expected to be very shear
sensitive: severing weak alkyl chain interactions in this situation would
expand the polymer chain and so contribute to a higher viscosity. Figure
4 shows that a sharp increase in shear stress and apparent viscosity
were indeed evident at a moderate shear rate.   Furthermore, pronounced hysteresis effects were found in the shear rate curves in this

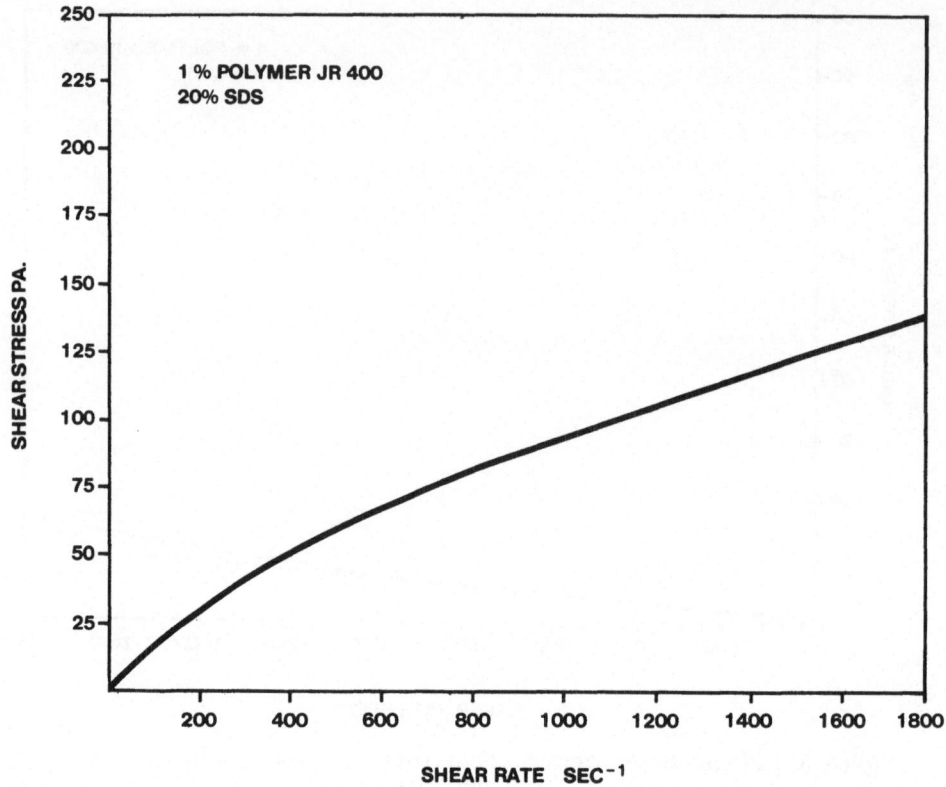

Figure 6.   Shear stress versus shear rate plot for a solution containing
            1% Polymer JR and 20% SDS.

region which indicates that the "starting" form of the globular
complexes is only recovered when the shear rate is reduced to a
particular, low level.

At higher concentrations of SDS (2%), when more micelles are
formed and bound to the polymer, the tendency for polymer unwrapping
under shear is less; this is seen in Figure 5 where a lower differential
viscosity and only a slight degree of rheopectic behavior is evident. We
should mention that rheopectic behavior is very unusual for homogeneous systems. It can only occur when shear sensitive structures are
present.

At much higher SDS concentration, as the micellar concentration
increases, the polymer/multimicelle aggregates will have grown larger
but be more close-packed. This is reflected in the viscosity increase
seen in Figure 2 above 2% SDS concentration.   In Figure 6,
pseudoplastic rheology is again observed.   The aggregates formed will

represent a balance between polymer binding together the micelles and the electrostatic self-repulsuion of these micelles.

A series of similar, mixed polymer, surfactant systems has been studied by small angle neutron scattering techniques in order to determine the size of the aggregates. The results obtained are consistent with the rheological interpretation presented here and will be reported separately.[5]

REFERENCES

1. E. D. Goddard and R. B. Hannan, J. Am. Oil Chem. Soc. 54, 561 (1977).
2. E. D. Goddard and R. B. Hannan, J. Colloid Interf. Sci., 55, 73 (1976).
3. E. D. Goddard and R. B. Hannan in "Micellization, Solubilization, and Microemulsion," K. L. Mittal (Ed.), Vol. 2, Plenum, NY (1977).
4. K. Ohbu, O. Hiraischi and Kashiwa, J. Am. Oil Chem. Soc., 59, 108 (1982).
5. E. D. Goddard, P. S. Leung, C. Han and C. J. Glinka, Colloids and Surfaces, 13, 47 (1985).

COOPERATIVE INTERACTION OF ANIONIC DYES WITH IMIDAZOLE-CONTAINING POLYMERS

J. S. Tan and T. M. Handel
Research Laboratories
Eastman Kodak Company
Rochester, New York 14650

ABSTRACT

The binding behavior of methyl orange to several homopolymers and copolymers containing imidazole and imidazolium salts was investigated as a function of polymer structure and counterion. The polymers were fractionated and characterized by light scattering, viscosity, and turbidimetry. Dye binding was measured by equilibrium dialysis, spectrophotometric titration, and microcalorimetry.

The extent of counterion binding to the polycations increases in the order Ac^- (acetate) $< Cl^- < Br^- < I^- < SCN^-$. The reverse order was observed, however, for the dye-binding strength of these polycations in the various salt media. These results demonstrate the competitive effect of the counterion on dye binding. The dye binding constant and the enthalpy of dye binding increase with decreasing [Polymer]/[Dye], suggesting a positive cooperative effect caused by dye stacking along the polymer chain. Incorporation of a nonbinding comonomer decreases binding strength, a result attributed to the lack of dye stacking. Our ΔG^O and ΔH^O data indicate that dye binding is driven both enthalpically and entropically. The constancy of ΔH^O in 0.01 M NaAc, NaCl, and NaSCN suggests that the electrostatic contribution to ΔH^O is not predominant. Van der Waals attraction between the bound dye molecules and the dye and the polymer side groups, however, may play an important role in dye binding.

INTRODUCTION

The interaction between anionic dye molecules and synthetic polycations has been investigated by several workers.[1-5] It is difficult, however, from existing data, to assess the relative importance of the various driving forces for binding. Several factors influence the binding of a planar aromatic anionic dye to a linear flexible polycation: van der Waals and coulombic effects, H-bonding, energy release upon desolvation, and the so-called hydrophobic bonding[6] or hydrophobic interaction.[7] The last effect is generally associated with an entropy gain frequently observed for the binding of organic molecules in aqueous media and is believed to be a result of the rearrangement of water molecules in the hydration shells surrounding the binding sites.

The importance of the hydrophobic interaction has been emphasized in many biological systems,[6,7] but the coulombic contribution has been shown to be small.[6,8,9] However, negative values for all the thermodynamic parameters, ΔG^o, ΔH^o, and ΔS^o, have been observed frequently for many small ligand/protein interactions[10] and the dimerization of dye molecules,[11] in which the reactions are entropically unfavorable.

The cooperative binding of small ligands to macromolecules has been extensively investigated for biopolymers,[12,17] but little is known about binding of anionic dyes to synthetic flexible vinyl polycations. Postive cooperativity has been suggested, however, for binding of cationic dyes to poly(styrenesulfonate),[18-20] based on analysis of spectrophotometric data by the treatment of Bradley and Lifson.[12,13]

To gain molecular insight into the dye-polymer binding mechanism, we chose to study methyl orange and several homopolymers and copolymers containing imidazole and imidazolium salts. The coulombic and van der Waals interactions were examined by varying the charge on the polymer, the quaternizing groups, and added salts. Dye binding was measured by spectrophotometric titration, equilibrium dialysis, and microcalorimetry. The importance of cooperativity caused by dye-dye interaction between the neighboring bound dye molecules, which is affected by the charge and the pendant group of the polymer, will be discussed.

EXPERIMENTAL SECTION

Poly(N-vinylimidazole) (PVI) and fully and partially quaternized benzyl chloride (BPVI) and methyl bromide (MPVI) salts were characterized by light scattering, intrinsic viscosity, and turbidimetry.[21] Methyl orange (MO) was obtained from Kodak Laboratory Chemicals.

Spectrophotometric Titration. A 3-mL portion of dye solution ([D$_t$] = 3.5 x 10^{-5} M) was titrated with microliter increments of polymer solution ([P$_t$] = 0.02 M in base mol/L), and the spectra were recorded on a Varian Superscan 3 spectrophotometer (the subscript t refers to the total concentration).

Equilibrium Dialysis. The experimental procedure was described earlier.[21a] The concentrations of the polymer (10^{-3} to 10^{-5} M) and the dye (5×10^{-4} to 5×10^{-6} M) and the extent of binding ($r = [D_b]/[P_t]$, where $[D_b]$ is the concentration of bound dye and $[P_t]$ is the total polymer concentration in base mol/L) cover a wider range than those of other systems.[4,5] The apparent binding constant was calculated according to $K_{app} = [D_b]/[P][D]$, where $[D]$ is the concentration of the free dye species and $[P]$ is the base molar concentration of the unbound homopolymer (for partially quaternized copolymers, $[P]$ is the molar concentration of the quaternary sites).

Microcalorimetry. Heats of binding were measured in an LKB 10700 batch microcalorimeter. A 4-mL portion of dye solution (5×10^{-4} M) was mixed with 2 mL of polymer solution at various $[P_t]/[D_t]$ values. Heats of dilution of the polymer were subtracted from the total heats of mixing. The enthalpy change ΔH^o was calculated in terms of kcal/mol of complex. The quantity of the complex formed at each mixing, or the value for r, was calculated from the dialysis experiment.

RESULTS AND DISCUSSION

Cooperative Binding by Dye-Dye Interaction. The binding between methyl orange and polycations containing imidazolium salts is cooperative because of the dye-dye interaction of the neighboring bound dye molecules promoted by the dye-polymer site interaction. Evidence for this dye-dye interaction comes from the following observations.

1. Spectral Changes. Figure 1 shows the spectra of MO + MPVI for various overall compositions $[P_t]/[D_t]$. Figure 1a demonstrates the increase of the 370 nm band and simultaneous decrease of the 465 nm band with increasing concentration of bound dye at low $[P_t]/[D_t]$. Similar spectral results were reported by other workers,[2,3] who attributed the 370 nm band to the dimeric form of the bound dye. Upon further increase of $[P_t]/[D_t]$, however, the monomer band at 465 nm is gradually recovered at the expense of the dimer, corresponding to a separation of the dye molecules along the polymer chain (Figure 1b). Similar titration behavior is shown by MO + BPVI, although the dimer band is less sharp and appears at 390 nm. For the complex of MO with both polymers, the stoichiometry was determined to be unity from a plot of the absorbance vs. $[P_t]/[D_t]$.

2. Binding Isotherms. Equilibrium dialysis results were reported by Takagishi et al.[4] for methyl orange and polycations containing a sulfone and a quaternized piperidyl group on the backbone chain. On the basis of a limited range of r values (most runs were made at r = 0-0.1 except for one polycation, which was run at r = 0-0.5), these workers assumed a constant binding constant and used the Klotz plots to obtain all the

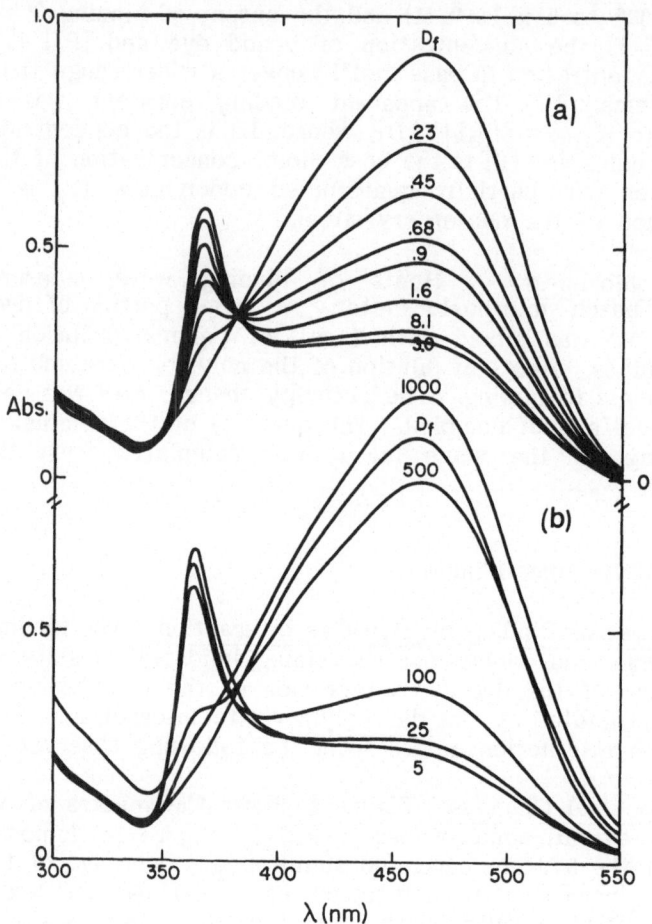

Figure 1. Spectral change for MO + MPVI at various $[P_t]/[D_t]$ in 0.01 M
NaAc. (a) $[P_t]/[D_t]$ = 0-8.1   (b) $[P_t]/[D_t]$ = 5-1000.

thermodynamic functions.   This approach is unsuitable for our present
systems, however.

The apparent binding constants K_{app} are plotted against r in Figure
2 for MPVI + MO and PVI + MO. The change of K_{app} with r as shown
here is similar to the change of dissociation constant for polyelectrolytes
upon neutralization and suggests neighboring bound dye–dye interaction.
Data similar to those of Figure 2 were also obtained for BPVI and the
partially quaternized copolymers.

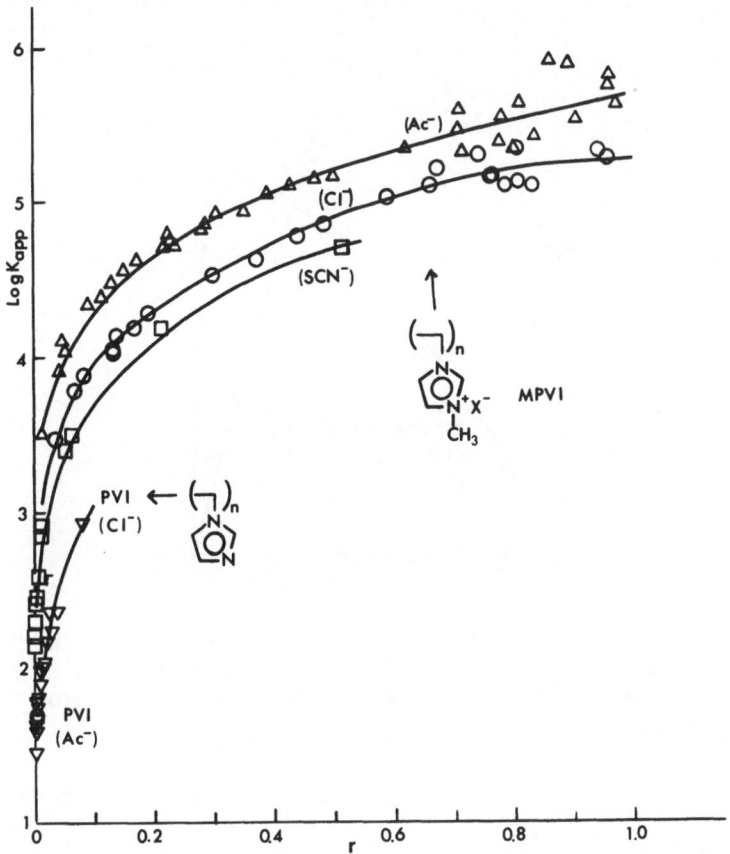

Figure 2. Log K_{app} vs. r (= $[D_b]/[P_t]$) for MO + MPVI in 0.01 M NaAc, NaCl, and NaSCN, and for MO + PVI in 0.01 M NaAc and NaCl.

Figure 3 shows the Klotz plots according to the expression

$$[P_t]/[D_b] = \frac{1}{r} = \frac{1}{nK[D]} + \frac{1}{n} \tag{1}$$

for the data of Figure 2. These isotherms show sharp rises in the region where excess polymer is present, i.e., high $[P_t]/[D_b]$ or low [D] region. This is attributed to the decreasing probability that there will be neighboring dye-dye interaction in the presence of excess empty sites on the polymer chain. The lateral displacement of the isotherms to higher [D] for MO + MPVI in 0.01 M NaCl and NaSCN compared with that in 0.01 NaAc reflects the decrease in the extents of dye binding as a result of counterion competition in the order $SCN^- > Cl^- > Ac^-$.[21b] Similar shifts in the isotherms were observed for the partially quaternized copolymers, where dye-dye interaction is suppressed by the weakly binding imidazole comonomers along the polymer chain.

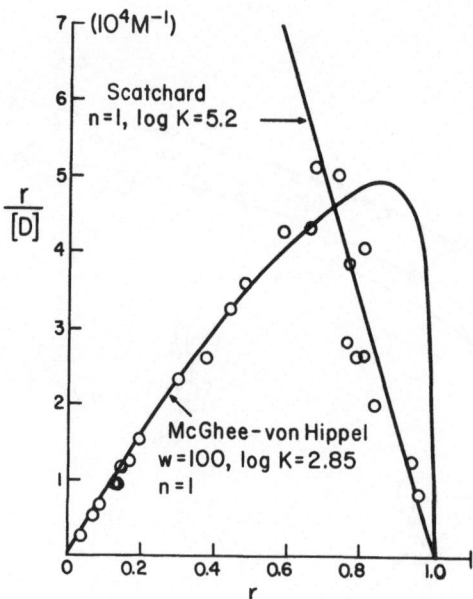

Figure 4. The Scatchard plot, r/[D] vs. r, for MO + MPVI in 0.01 M
NaCl. The Scatchard model fits in the region $0.7 \leqslant r \leqslant 1.0$
and the McGhee-von Hippel model fits in the region $0 \leqslant r$
$\leqslant 0.7$.

Further examination of the data points in Figure 4 showed that
Equation 2 was an appropriate representation for the data only in the
region of high saturation, $r = 0.7-1.0$, with $n = 1$ and $\log K = 5.2$.
On the other hand, the data points in the region of low saturation,
$r = 0-0.7$, are best described by the McGhee-von Hippel expression[16] for
cooperative binding (a modification of Equation 2), with $n = 1$, $\log K_O$
$= 2.85$, and $w = 100$. In this region, an incoming dye molecule
preferentially binds to a site with an occupied neighboring site. This
positive cooperativity does not persist beyond $r = 0.7$, however, possibly
because of steric hindrance. Hence $\log K_O$ remains relatively unchanged
at high r, and a linear Scatchard plot is more suitable.

Similar results were obtained for BPVI + MO in 0.05 M NaCl
($n = 1$, $\log K_O = 4.0$, $w = 53$ for $r = 0-0.6$, and $n = 1$, $\log K_O = 5.8$ for
$r = 0.6-1.0$) and the partially quaternized copolymers.

The extent of dye binding to the three homopolymers studied at a
constant ionic strength increases in the order PVI < MPVI < BPVI. The
stronger affinity of BPVI for the dye molecule is attributed to the
stronger intrinsic dye-site interaction between the aromatic groups of
the dye and the polymer. Partially quaternized copolymers show less

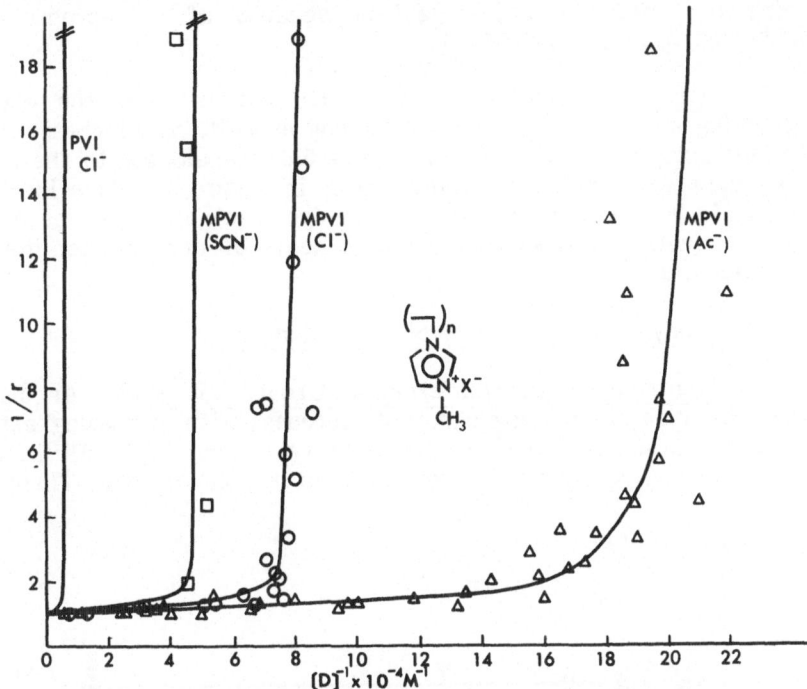

Figure 3. The Klotz plots, $1/r$ vs. $[D]^{-1}$, for MO + MPVI in 0.01 M NaAc, NaCl, and NaSCN, and for MO + PVI in 0.01 M NaCl.

Because the binding process for our systems cannot be described by the independent single-site model as discussed above, we further examined our data by the expressions derived on the basis of the nearest-neighboring interaction model.

Rearrangement of Equation 1 gives the Scatchard equation:

$$r/[D] = K(n - r). \qquad (2)$$

The data points in the plot of $r/[D]$ vs. r for MPVI + MO in 0.01 M NaCl (Figure 4) shows a downward curvature, which is expected from the nonlinear nature of the Klotz plots shown in Figure 3. By taking into account the neighboring bound dye–dye interaction, Equation 2 can be modified[22,23] as

$$r/[D] = K_0(n - r)e^{-2Er/kT} \qquad (3)$$

where K_0 is the intrinsic binding constant, k is the Boltzmann constant, and E is the nearest-neighbor interaction energy. This expression was inappropriate for our data, however, because the plot of $\log r/[D](n - r)$ vs. r is not linear over the whole range of r.

binding per mole of quaternization because of a decrease in the neighboring dye-dye interaction.

3. Heats of Binding. Our ΔH^O results show the effect of neighboring dye-dye interaction on binding as well. The increase in $-\Delta H^O$ with increasing r for MO + MPVI (Figure 5a) is analogous to the increase in the apparent binding constant shown in Figure 2. This increase in exothermic heat with r, caused by dye-dye interaction on the chain, is consistent with the observed negative heats of dimerization for many dye molecules.[11]

Thermodynamic Functions of Dye Binding.

1. Contribution of Quaternizing Groups to ΔG^O. As discussed above, the apparent binding constant increases with increasing extent of binding (r) caused by positive cooperativity, hence ΔG^O, ΔH^O, and ΔS^O for the various polymers are compared at a given r value (Table I).

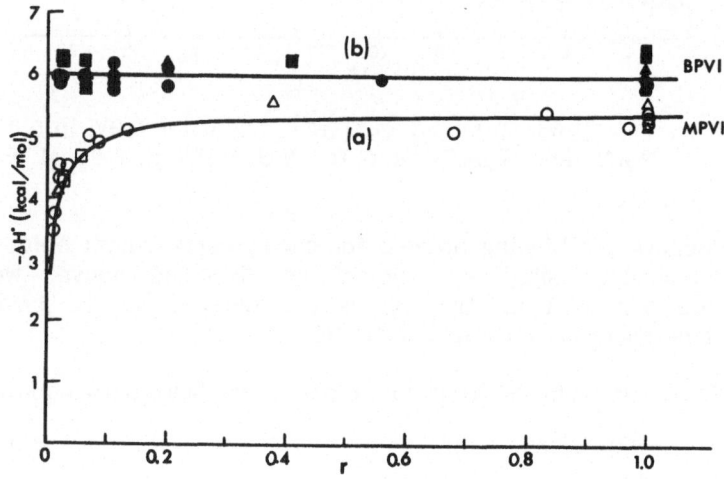

Figure 5.   Exothermic heats of binding $-\Delta H^O$ vs. r for

(a)   MO + MPVI   △ : 0.01 M NaAc;   ○ : 0.01 M NaCl;

                                □ : 0.01 M NaSCN

(b)   MO + BPVI   ▲ : 0.01 M NaAc;   ● : 0.01 M NaCl;

                                ■ : 0.05 M NaCl

Table I

Thermodynamic Functions of Dye Binding (25°C)

	Polymer	Medium	r	$-\Delta G^O$	$-\Delta H^O$ (kcal/mol)	$T\Delta S^O$
(a)	PVI + MO	0.01 M NaAc 0.01		2.3 ± 0.3	4.4 ± 0.3	-2.1
	MPVI + MO			4.6	3.1	1.5
	PVI + MO	0.01 M NaCl 0.01		2.6	4.4	-1.8
	MPVI + MO			4.2	3.1	1.1
(b)	MPVI + MO	0.01 M NaAc 1.0		7.8 ± 0.2	5.4 ± 0.2	2.4
	BPVI + MO			9.0	6.0	3.0
	MPVI + MO	0.01 M NaCl 1.0		7.1	5.4	1.7
	BPVI + MO			8.6	6.0	2.6

Because of the extremely weak binding capacity of the uncharged PVI polymer, data for this system are reported only for the low r region. Therefore, the uncharged PVI and charged MPVI polymers are compared at r = 0.01 (part (a) in Table I). Conversely, because of the strong binding capacity of BPVI, the binding data at low ionic strength (0.01 M NaAc and 0.01 M NaCl) are accessible only at high saturation, r = 0.8-1.0. Consequently, the two charged polymers are compared at r = 1.0 (part (b) in Table I).

The larger $-\Delta G^O$ values for MPVI than for PVI (part (a), Table I) are evidence for electrostatic contribution to the total driving force for dye binding. The fact that electrostatic effect is operative is also shown in Figure 2, where the stronger binding counterion SCN^- suppresses K_{app} more than the weaker binding counterion Ac^-.

The larger $-\Delta G^O$ values for BPVI than for MPVI (part (b), Table I) are attributed to the greater van der Waals dispersion forces between the dye and the benzyl group of the BPVI polymer. Comparison of $T\Delta S^O$ for the two charged polymers with the negative value for PVI shows that the overall effect of quaternizing groups is partly manifested in an entropy gain.

2. Contribution of Dye-Dye and Dye-Site Interactions to ΔH^O. As discussed earlier, the increase in $-\Delta H^O$ with r for MPVI + MO (Figure 5a) can be attributed to enchanced dye-dye interaction as the polymer chain is saturated with dye molecules. This interaction is independent of ionic strength and counterion, as shown by the data in Figure 5, suggesting that electrostatic force is not important for determining ΔH^O. The $-\Delta H^O$ value for the uncharged PVI + MO is larger than that for the charged MPVI + MO at r = 0.01 (Part (a), Table I), also suggesting that coulombic force is not a dominant factor for determining ΔH^O.

The constant value of $-\Delta H^O$ with r for BPVI + MO, on the other hand, suggests that the dispersion force between dye and polymer site may be more important than dye-dye interaction for this system. The larger $-\Delta H^O$ for BPVI than for MPVI over the whole range of r is also attributed to a greater dye-site interaction.

SUMMARY

The binding behavior of methyl orange to several homopolymers and copolymers containing imidazole and imidazolium salts was investigated as a function of polymer structure and counterion. Polymer-counterion interaction was characterized by light scattering, intrinsic viscosity, and turbidimetry. Dye binding was measured by equilibrium dialysis, spectrophotometric titration, and microcalorimetry.

The extent of counterion binding to these polycations increases in the order $Ac^- < Cl^- < Br^- < I^- < SCN^-$. The reverse order was observed, however, for dye binding to these polycations in the various salt media, demonstrating the effect of the counterion competition for the binding site on dye binding. The apparent binding constant increases with the extent of binding, $[D_b]/[P_t]$, suggesting positive cooperativity caused by bound neighboring dye-dye interaction. Incorporation of a nonbinding comonomer decreases binding strength, a result attributed to the lack of neighboring dye-dye interaction.

The thermodynamic functions of dye binding, ΔG^O calculated from binding constants and ΔH^O measured calorimetrically, show that dye binding is driven both enthalpically and entropically and is enhanced by the incorporation of quaternary sites containing an aromatic side group. The constancy of ΔH^O in 0.01 M NaAc, NaCl, and NaSCN suggests that electrostatic contribution to ΔH^O is not predominant but may promote dye-site and bound dye-dye van der Waals dispersion forces. A molecular modeling of the polymer-dye complexes was performed to corroborate the above findings.[24]

ACKNOWLEDGMENT

We are grateful to Dr. H. L. Cohen and Dr. I. S. Ponticello for the preparation of the polymers used in this study and to Dr. H. Yu and Dr. J. L. Lippert for discussions during this work.

REFERENCES

1. Klotz, I. M.; Royer, G. P.; Sloniewsky, A. R. Biochemistry 1969, 8, 4752.
2. Quadrifoglio, F.; Crescenzi, V. J. Coll. Interface Sci. 1971, 35, 447.
3. Reeves, R. L.; Harkaway, S. A. "Micellization, Solubilization, and Microemulsions"; Mittal, K. L., Ed.; Plenum Press: New York, 1977; p 819, Vol. II.
4. Takagishi, T.; Nakata, Y.; Kuroki, N. J. Polym. Sci. 1974, 12, 807.
5. Ando, Y.; Komiyama, J.; Iijima, T. J. Chem. Soc. Jpn. 1981, 3, 432.
6. Kauzmann, W. Adv. Protein Chem. 1959, 14, 1.
7. Franks, F. "Water, A Comprehensive Treatise"; F. Franks, Ed.; Plenum Press: New York, 1975, 4, 1.
8. Klotz, I. M.; Walker, F. M.; Pivan, R. B. J. Am. Chem. Soc. 1946, 68, 1486.
9. Lawton, J. B.; Phillips, G. O. J. Chem. Soc., Perkin Trans. 2 1977, 38.
10. Ross, P. D.; Subramanian, S. Biochemistry 1981, 20, 3096.
11. Coates, E. J. Soc. Dyers Colour. 1969, 85, 355.
12. Bradley, D. F.; Wolf, M. K. Proc. Natl. Acad. Sci. U.S.A. 1959, 45, 944.
13. Bradley, D. F.; Lifson, S. "Molecular Associations in Biology"; B. Pullman, Ed.; Academic Press: New York, 1968; p 261.
14. Schneider, F. W.; Cronan, C. L.; Podder, S. K. J. Phys. Chem. 1968, 72, 4563.
15. Crothers, D. M. Biopolymers 1968, 6, 575.
16. McGhee, J. D.; von Hippel, P. H. J. Mol. Biol. 1974, 86, 469.
17. Schwarz, G. Eur. J. Biochem. 1970, 12, 442.
18. Vitagliano, V.; Costantino, L. J. Phys. Chem. 1970, 74, 197.
19. Vitagliano, V.; Costantino, L.; Zagari, A. J. Phys. Chem. 1973, 77, 204.
20. Vitagliano, V. "Chemical and Biological Applications of Relaxation Spectroscopy"; Wyn-Jones, E., Ed.; D. Reidel Publ. Co.: Dordrecht-Holland, 1975; p 437.
21. (a) Tan, J. S.; Sochor, A. R. ACS Polymer Preprintss 1979, 20, 15.
    (b) Tan, J. S.; Sochor, A. R. Macromolcules 1981, 14, 1700.
22. Scatchard, G. Ann. N.Y. Acad. Sci. 1949, 51, 660.
23. Wagner, K. G. Eur. J. Biochem. 1969, 10, 261.
24. Orchard, B. J.; Tan, J. S.; Hopfinger, A. J. Macromolecules 1984, 17, 169.

ELECTRON TRANSFER PROCESS IN THE DOMAIN FORMED BY INTERMACROMOLECULAR COMPLEXES

Eishun Tsuchida and Hiroyuki Ohno
Department of Polymer Chemistry
Waseda University
Tokyo 160, Japan

ABSTRACT

Macromolecular chains interact with each other through secondary binding forces and form intermacromolecular complexes. The electrochemical behaviors of these complexes composed of electroactive polyviologen were studied and compared with those of component polymer to clarify the specific characteristics of polymer complexes.

The electron propagation through the polymer layers was interrupted strongly by the complex formation. As the migration of counter ions and certain flexibility of macromolecular chains to yield the close encounter of redox sites are necessary for the electron propagation in the matrix of electro-active polymers, the microdomain of polyion complexes was supposed to have a shrunk structure. The apparent total charge of polyion complexes could be changed widely in accordance with the redox reaction. It was expected that the selective uptake and release of multivalent ions could be easily realized by regulating the electrode potential. The catalytic reduction of molecular oxygen by the coated electrode was also investigated by means of rotating disk voltammetry. Within the pH range from 5 to 10, the coated electrode showed the excellent stability and high catalytic activity. Comparing with the bare electrode system, the catalytic reduction half wave potential shifted to the positive side with a maximum value of about 350 mV.

INTRODUCTION

Macromolecules aggregate with each other through the intra- and intermacromolecular interaction to form complexes in solution[1~5]. The importance of the microdomain formed by assembled molecules, macromolecules and intermacromolecular complexes was much interesting for its unique characteristics. The microdomain formed by the intermacromolecular complex has been analyzed by means of fluorescence technique and calorimetry[6~13]. It was revealed that segmental motion of every macromolecule decreased considerably by the complex formation[6], but their solvated state was still maintained[7]. To analyze the characteristics of microdomain from another point of view, the electron transfer and mass transfer processes through the microdomains should be studied by means of rotating disk voltammetry. This should clarify the size of cavity and folding of polymer chains in the polymer complex matrixes. Especially, intermacromolecular complexes containing electro-active sites such as viologen group are expected to provide interesting microdomain[14,15]. Although a few reports described the intermacromolecular complexes containing electro-active polymers[14~17], the electron transfer process in the microdomain of the complexes was scarcely studied[18,19]. In this paper, the application as well as the basic characteristics of such electro-active intermacromolecular complexes was studied in relation to the structure of their microdomains.

EXPERIMENTAL PART

Materials

Poly(xylylviologen) (PXV) was prepared from m-xylylene dibromide and 4,4'-bipyridine by the successive Menshutkin reaction in N,N-dimethylformamide (DMF) at 60°C for 72 h. PXV was obtained as precipitate in DMF solution after polymerization. The precipitate was filtered and washed with acetone to yield yellow powder after drying in vacuo. The intrinsic viscosity of PXV was 2.9×10^{-2} (dl/g) (in 0.5 M KBr aqueous solution at 25°C) corresponding to the average molecular weight ($\overline{M}w$) of about 8000[20].

Poly(alkyleneviologen) (PRV) was prepared from alkylene dibromide and 4,4'-bipyridine in DMF by the similar method as noted above. Poly(sodium styrene sulfonate) (PSS; Seimi Kagaku Co., Ltd.) was the same sample as used elsewhere[5]. The $\overline{M}w$ of PSS was 5×10^4.

Poly(methacrylic acid) (PMAA), poly(acrylic acid) (PAA) and poly-(L-glutamic acid) (PGA) were prepared and purified by the same method as reported previously[5,7].

"Nafion" membrane 125 (E. I. du Pont de Nemours & Co.) of perfluoropoly(sulfonic acid) was boiled in 0.20 M NaOH solution for 30 min, and washed with distilled water for several times. Swallen Nafion chips were dried under reduced pressure. Finely cut chips of the membrane were soaked in dimethyl sulfoxide (DMSO) and the resulting solution was boiled again for 3 h and filtered through a glass filter. The filtrate was diluted with given amount of ethanol to prepare the stock solution with the concentration of Nafion of 5.6 mg/ml.

Reagent grade $K_4Fe(CN)_6$ was recrystallized and used after complete drying. $K_3Fe(CN)_6$, $K_4Mo(CN)_8$ and $K_4W(CN)_8$ were prepared and purified by the conventional method.

Reagent grade sodium perchlorate and potassium chloride were used as supporting electrolytes without further purification.

Water was distilled twice.

Electrode Coating

Electrodes were prepared by attaching the cylindrical disk of basal plane pyrolytic graphite (BPG) (Union Carbide Co.) on the edge of stainless rod by conductive polymer paste, then they were sealed with heat-shrinkable olefin tube. All electrodes employed in this study had geometric area of 0.17 cm^2. Fresh electrode surfaces were produced by cleaving the disk of BPG with a scalpel. The polymers or polymer complexes were directly coated on the electrode by casting method as described previously[5].

Electrochemical Measurements

The cyclic voltammograms and rotating disk voltammograms were recorded with a Nikko-Keisoku potentiostat (DPG-S3), function generator (NFG-3), electrode rotator (RRDE-1) and a Watanabe Sokki X-Y recorder (WX-4403). The surface charge for the coated polymers was estimated by the manual integration of cyclic voltammogram. Sample solutions were bubbled sufficiently with argon to expel oxygen. All experiments were conducted at 25°C under prepurified argon atmosphere. Potentials were described with respect to a sodium chloride-saturated calomel reference electrode (SSCE).

RESULTS AND DISCUSSION

Electron transfer process in the microdomain of polyion complex

The electrochemical properties of poly(viologen)s have been ana-lyzed by means of cyclic voltammetry. The surface change obtained by the manual integration of redox wave was smaller than that expected from the coating aliquot due to the partial dissolution of polymer layers. Although detachment could be minimized by increasing the average molecular weight of applied polycations, it was difficult to prepare the poly(viologen)s with relatively high molecular weight because of phase separation during the polymerization (see experimental part). The use of perchlorate counter anions in the supporting electrolyte solution im-proved the stability of polymer film, as it formed the strong ion pair with the polycations. Further stabilization of poly(viologen) coatings was done by the intermacromolecular complex formation, which significantly decreased the solubility, with polyanions such as poly(sodium styrene sulfonate).

Physico-chemical properties of the microdomain of polymer com-plexes have been studied by means of ^{1}H-NMR[21], light scattering[21], or fluorescence polarization[6]. Here, the authors tried to evaluate the microdomain of polymer complexes by the electrochemical assay. The formation of polyion complex affected the redox behavior of poly-(viologen)s considerably. Fig. 1 shows the cyclic voltammograms for PXV-PSS complex coated electrode. The first redox peak shifted to positive side, and peak broadening was observed by the complex formation. It is clear that the redox behavior was restricted by the complexation. It is known that the electron transfer process must accompany the migration of counter ions to maintain electroneutrality. As in polymer complex microdomain, polyelectrolyte chains interacted with each other and decreased their free volume, they should thereby provide the domain with smaller porosity.

In spite of the high density of electroctive units, the electron transfer was interrupted in the complex domain due to the decrease in size of porosity which permitted fewer counter ions to pass.

The restriction of the electron transfer could also be notified from the scan rate dependence of the surface charge (Q) for the electrode coated with PRV or PRV-PSS complex. Fig. 2 shows the relation between scanning rate and the surface charges for the polymer coated electrode. Surface charge depended on the scanning rate and was reduced by the increase of scanning rate. PRV-PSS complex coated electrode showed this tendency strongly. With lower scan rate, the electron transfer between redox sites occurred sufficiently. On the other hand, redox reaction occurred only in a part of domain when potential was given with a higher scan rate because the electron transfer might be quite slow in the domain of the polyion complexes. The

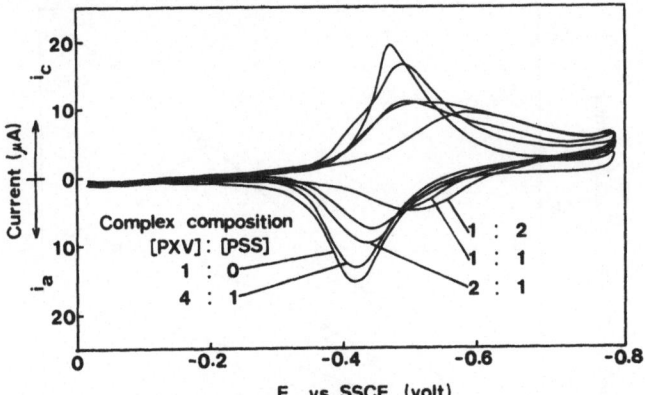

Figure 1. Effect of complex composition on the cyclic voltammograms of PXV-PSS complex-coated graphite electrode in 0.2 M NaClO$_4$ aqueous solution.

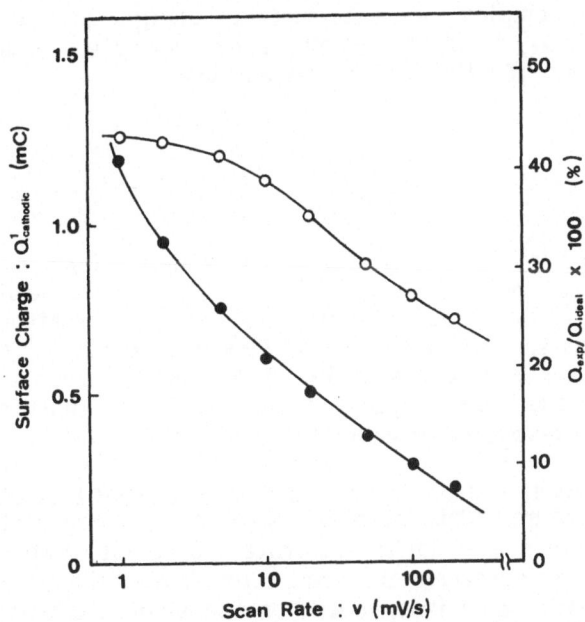

Figure 2. Relation between scanning rate and surface charge at the electrode coated with PRV (n=10) and PRV-PSS complex. o : PRV; ● : PRV-PSS complex, [NaClO$_4$] = 0.2 M, Γ_{PRV} = 1.5 x 10^{-7} unit mol/cm^2

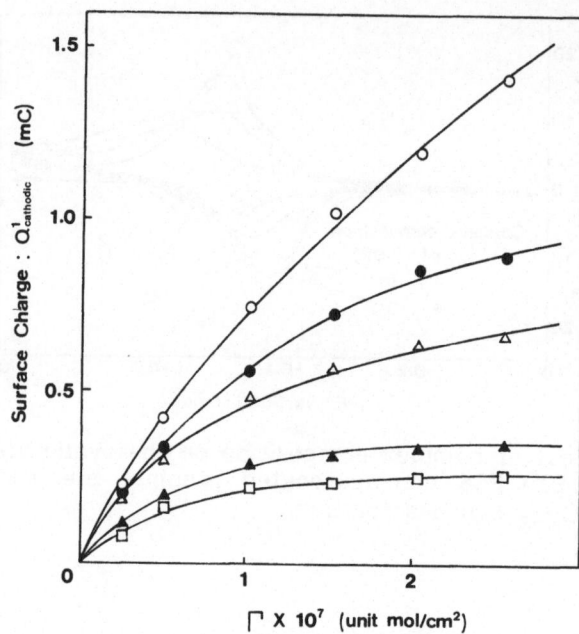

Figure 3.   Effects of scanning rate and coated amount of PRV (n=10)-
PSS complex on the surface charge of the electrode.
[NaClO$_4$] = 0.2 M; scanning rates were 1 (o), 5 (●), 10 (△), 50
(▲) and 100 (□) mV/sec, respectively.

polycation turned insoluble in water by the complex formation with the
oppositely charged polyelectrolyte[1~7,22], and it could easily be fixed
completely on the electrode.   In spite of its stability, surface charge of
the complex was smaller than that of PRV coated alone.   Only 7% of
viologen units was reduced in PRV-PSS complex at scan rate of 200
mV/sec.   This result also supported the prediction that the electron
transfer must be restricted considerably in the domain of the complexes.

Fig. 3 shows the effect of the amount of coated PRV-PSS complex
on its surface charge.   This increased with the increase of the amount
of PRV-PSS complex at a lower scan rate.   Gradient of these relations
was diminished by increasing the scan rate; finally the surface charge
reached a constant value independent on the amount of coated polyion
complexes.   Such a distinct scan rate dependence means that the
electron transfer in microdomain of the polyion complexes was slow and
it could not respond to the quick potential change of the given potential
at larger scan rate.

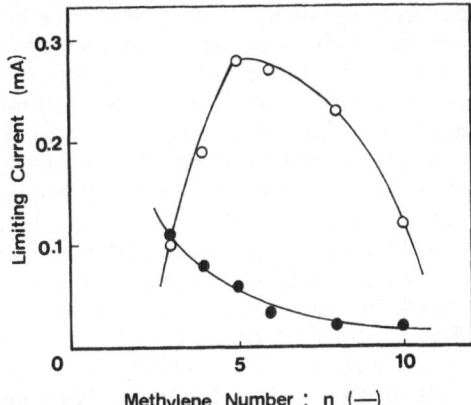

Figure 4. Dependence of limiting current on the methylene number (n) between two adjacent viologen units of PRV for the electrode coated with PRV-PSS complex. $\Gamma_{PRV} = 1.5 \times 10^{-7}$ unit mol/cm^2, rotating rate = 6400 rpm., [K$_3$Fe(CH)$_6$] = 2 mM, o : i_1, ● : i_2

The effect of the distance between two adjacent viologen units was studied by using a series of PRV with different methylene number (n) from 3 to 10. One could not find any distinct difference in the electrochemical behavior of a series of PRV[19]. However, the electron transfer and mass (ion) transfer processes in the domain of polymers were affected strongly by the methylene number between two adjacent viologen groups when PRV formed complexes with polyanions such as PSS. Fig. 4 shows the effect of methylene number on the limiting currents for the rotating disk electrode coated with PRV-PSS complexes. These electrodes were soaked into 0.2 M KCl aqueous solution containing 2 mM K$_3$Fe(CN)$_6$. Two limiting currents, i_1 and i_2, corresponded to the ion (Fe(CN)$_6{}^{3-}$) and electron transfer processes, were observed, respectively. Current i_1 was continuously observed when the electrode potential of about -0.1 V was given to the rotating disk electrode (rotating speed of 6400 rpm). At this potential, Fe(CN)$_6{}^{3-}$ migrated through the microdomain of PRV-PSS complex and was reduced directly at the graphite electrode. Thus the ion transfer process in the polymer complex could be followed by the change of current i_1. On the other hand, Fe(CN)$_6{}^{3-}$ was reduced by the viologen units when potential of -0.7 V was given. At this potential, the electrons were transferred from the electrode through the geared redox cycles of viologen units to the surface of the coated complex layers. This current, i_2, might therefore correspond to the electron transfer process through the microdomain of polyion complexes.

The limiting current, i_2, decreased monotonously with the increase of methylene number, n. This was explained by the repression of the electron transfer between the redox sites in polymer domain. Against this, i_1 varied through a maximum value. Spreading of the distance between two adjacent viologen units should form a loose complex domain, but on the contrary, longer hydrocarbon chain tended to assist the shrinkage of the domain through the hydrophobic as well as electrostatic interaction forces. So, relatively loose domain were provided by the PRV with suitable distance between two adjacent redox sites, i.e., n=5 or 6. The porosity of the microdomain of such a polyion complex was controlled by the subtle balance of hydrophobicity and charge density of PRV. A suitable porosity also depended on the size of substrate as well as the strength of interpolymer interaction, in other words, the primary structure of the applied polyanions should regulate the electrochemical properties of microdomain of interpolymer complexes.

Reversible concentration of multivalent ions in the microdomain of polyion complex

The redox reaction of PXV or PRV accompanied the changes of overall charge density of the polymer domain. This reversible redox reaction has been applied to the reversible uptake of multivalent microions from the very dilute solution. Recently, M. S. Wrighton et al[23] and F. C. Anson et al[24] have reported the selective binding or different diffusion behavior of complex metal ions through the polymer layer electrochemically. Selective concentration should also be performed because polyelectrolytes interact with several microions and the interaction force became larger with the increase of valency of the microions. $Fe(CN)_6{}^{4-}$, $Mo(CN)_8{}^{4-}$ and $W(CN)_8{}^{4-}$ were used as multivalent complex anions. PXV-coated graphite electrode was soaked into 0.2 M KCl aqueous solution containing 0.2 mM $K_4Fe(CN)_6$. The voltammogram for the ferrocyanide fixed in polymer domain through electrostatic interaction could be observed by potential sweep from -0.2 to +0.8 V. The peak current increased gradually with time and saturated after 8 min soaking. The ferro/ferri redox peak current obtained from this PXV-coated electrode was larger than that obtained from bare electrode. Ferrocyanide was accumulated and fixed selectively through electrostatic force in spite of a large excess of chloride ions (1000 times). This discrimination was explained by the increase of the equilibrium constant of the multivalent anions. Similar experimental data were obtained when $Mo(CN)_8{}^{4-}$ was used instead of $Fe(CN)_6{}^{4-}$. Results were noted in Table 1.

Selective adsorption was observed in the mixed solution of $Fe(CN)_6{}^{4-}$ and $Mo(CN)_8{}^{4-}$. Fig. 5 shows the voltammogram for PXV coated electrode soaked in a mixed solution of 0.2 mM $K_4Fe(CN)_6$ and 0.2 mM $K_4Mo(CN)_8$. $Mo(CN)_8{}^{4-}$ was predominantly adsorbed in PXV

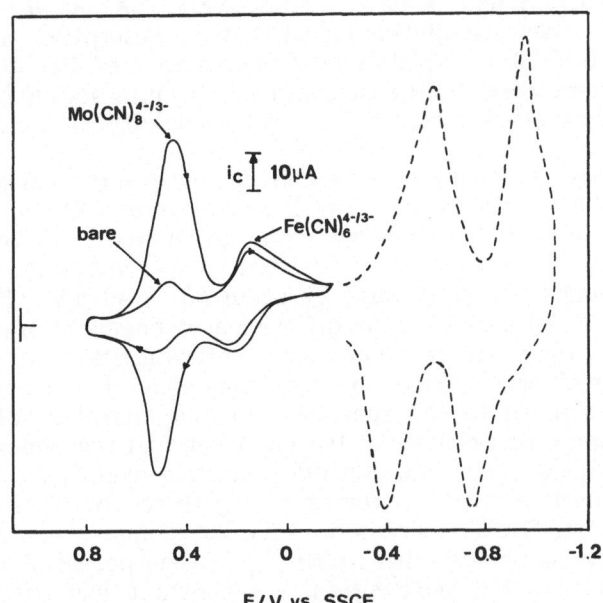

Figure 5. Cyclic voltammogram for the electrode coated with PXV soaked in the mixed solution of $Fe(CN)_6^{4-}$ and $Mo(CN)_8^{4-}$ at 25°C. Γ_{PXV} = 1.4 x 10^7 unit mol/cm^2, [$Fe(CN)_6^{4-}$] = 0.2 mM, [$Mo(CN)_8^{4-}$] = 0.2 mM, scanning rate = 190 mV/sec.

matrix, and the response due to $Fe(CN)_6^{4-}$ was as small as that obtained by the bare electrode. The quantities of adsorbed $Mo(CN)_8^{4-}$ were the same as those obtained by the PXV-coated electrode system soaked in only $Mo(CN)_8^{4-}$ solution. Thus, PXV shows apparent selectivity to the uptake of complex anion. This selectivity was considered to be based on the difference of the anion radius or hydrophobicity of these complex anions. Anson et al have found the same selectivity and related that to the difference of their diffusion coefficients[23]. The selectivity was confirmed by the same experiment by using $W(CN)_8^{4-}$ instead of $Mo(CN)_8^{4-}$. $W(CN)_8^{4-}$ was also concentrated predominantly than $Fe(CN)_6^{4-}$. Furthermore, there was no difference for the adsorption behavior between $Mo(CN)_8^{4-}$ and $W(CN)_8^{4-}$.

Incorporated multivalent complex anions in PXV matrix were desorbed by the decrease of cationic charge density of PXV. About half or all of cationic charges of PXV were reduced when potentials were stepped to -0.70 V and -1.20 V, respectively. At -0.70 V, $Fe(CN)_6^{4-}$ was desorbed theoretically (about 50%), but desorption efficiency of $Mo(CN)_8^{4-}$ at -0.70 V was half as much as that of $Fe(CN)_6^{4-}$ system. This might correspond to the strength of the interaction between PXV matrix and complex anions, or the hydrophobicity of $Mo(CN)_8^{4-}$.

The redox potentials of PXV-PSS complex were -0.50 V (the first wave), and -0.91 V (the second wave), and those of PXV-Nafion complex were -0.53 V and -0.93 V, respectively. Compared with those of PXV alone, that of the first wave was shifted to the cathodic side by about 60 ~ 90 mV, and the second wave by about 20 ~ 40 mV. The potential shift seemed to be based on the difference of chemical environment of PXV. Namely, the redox behavior was supressed a little by the shrinkage of domain which was induced by the complexation. Shrinkage of the domain corresponded to the supression of the migration of microions. PXV-PSS complex was coated on the electrode and was soaked in 50 μ M $Ru(NH_3)_6^{3+}$ aqueous solution. By the repetitive cycling from +0.80 V to -0.80 V, the peak current i_p, corresponding to reduction of $Ru(NH_3)_6^{3+}$ (E_p^c = -0.28 V vs. SSCE), increased. PXV-PSS complex coated electrode was completely saturated with $Ru(NH_3)_6^{3+}$ when potential was fixed at -0.80 V for 10 min. The peak current was constant over an hour, namely no desorption was observed during the experiment. It was recognized that theoretical amount of $Ru(NH_3)_6^{3+}$ was fixed in PXV-PSS complex domain.

The adsorbed multivalent complex cation could also be desorbed by regulating the overall charge density in the domain. The fixation of PXV in Nafion film was performed by the formation of polyion complex. Successive concentration change of the adsorbed $Ru(NH_3)_6^{3+}$ in the complex microdomain was shown in Fig. 6. As mentioned above, the domain of PXV-Nafion complex seemed to be shrunk loosely by comparing that of PXV-PSS complex. The peak current i_p, corresponding the oxidation of the fixed $Ru(NH_3)_6^{3+}$ (E_p^a = -0.30 V), increased when the potential was fixed for a certain period of time at -0.80 V (o) or -1.2 V (●), respectively. Especially, when viologen sites were reduced to nonionic, quantity of free sulfonic groups was doubled and the adsorption amount and efficiency increased (see Fig. 6, closed circles). Then the potential was canceled to 0.0 V, all of the adsorbed $Ru(NH_3)_6^{3+}$ was desorbed within a few sec. This rapid desorption seemed to be based on the cooperative interaction between Nafion and PXV so called "polymer effect"[1,2]. So the fixed $Ru(NH_3)_6^{3+}$ was put out immediately. From the result described above, it was deduced that complex ions could be adsorbed and desorbed reversibly on PXV-Nafion complex-coated electrode by only changing the applied potential. This knowledge makes almost all of other multivalent ions possible to concentrate from very

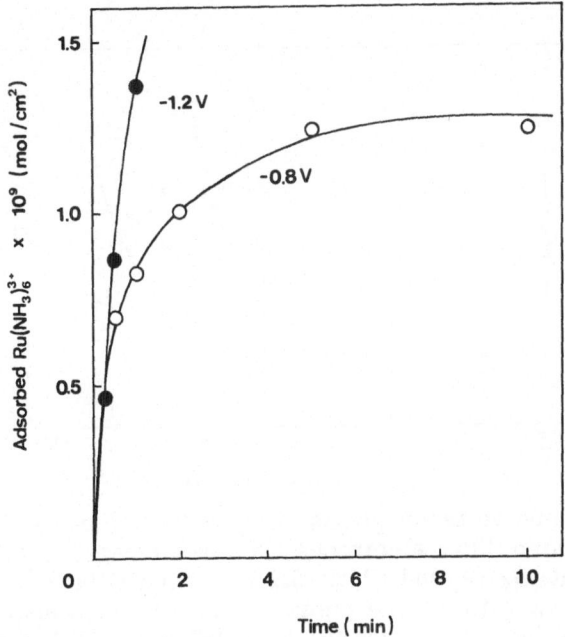

Figure 6. Adsorption of $Ru(NH_3)_6^{3+}$ in the domain of PXV–Nafion complex at 25°C. Γ_{PRV} = 1.5 x 10^{-9} unit mol/cm^2, $[Re(NH_3)_6^{3+}]$ = 50.0 μM. Polyion complex was directly formed by equimolar mixing of sulfonate groups of Nafion and cationic sites of PXV on the carbon electrode.

dilute solution. Though PXV–PSS complex film and PXV–Nafion complex film were much more stable than PXV alone, the polyion complexes were not so stable when PXV was reduced up to the nonionic state. Because PXV could no longer form complex with PSS or Nafion any more when PXV was reduced to the nonionic state, and every component polymer tended to dissolve in the bulk solution.

Application of microdomain for the interfacial redox catalyst

Electrocatalytic reduction of oxygen molecules was observed with graphite electrode coated with PXV–PSS complex. With respect to oxygen reduction, the catalytic activity of metal porphyrin complexes or metal phthalocyanine complexes adsorbed on the graphite surfaces has been reported by many workers and the mechanism was elucidated by using the rotating disk or rotating ring–disk voltammetry[25,26]. But there are a few reports concerning the activity of electrode coated with

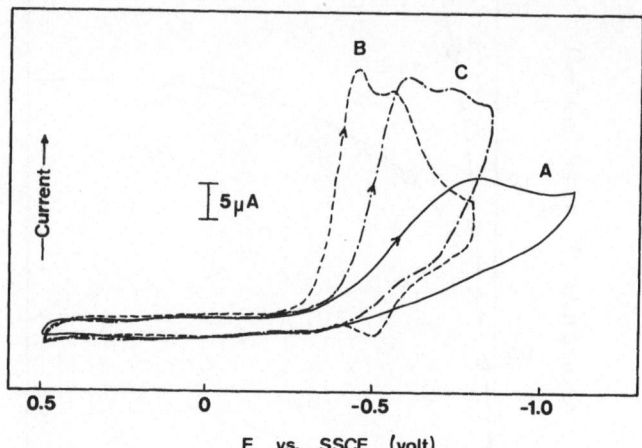

Figure 7. Cyclic voltammograms of O_2 reduction observed with the PXV coated BPG electrodes, in air-saturated solution of 0.2 M $NaClO_4$ (A and B), or 0.2 M CF_3COONa (C), adjusted to pH = 6.5 with 50 mM phosphate buffer, at scanning rate of 100 mV/sec : (A) bare; (B) PXV-PSS (C) PXV-Nafion.

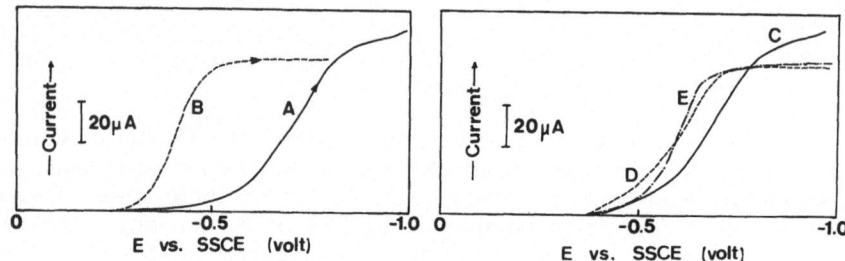

Figure 8. Current-potential curves for the reduction of O_2 at the viologen coated electrodes in air-saturated solution of 0.2 M $NaClO_4$ (A and B), or 0.2 M CF_3COONa (C, D and E), adjusted to pH = 6.5 with phosphate buffer : (A) bare; (B) PXV-PSS; (C) Nafion, (D) PXV-Nafion; (E) low molecular weight dimethyl viologen fixed in PSS matrix. Scanning rate = 2 mV/sec. Rotation speed of electrodes = 900 rpm.

metal-free compounds[14].   Fig. 7 demonstrated a comparison of the cyclic voltammograms of oxygen reduction by the PXV coated electrode with that obtained by a bare electrode. For Fig. 7 B observed in an air-saturated solution, the reduction current was enhanced greatly, the oxidation current decrease and cathodic peak potential was shifted to positive side by about 420 mV. Fig. 8 A shows a current-potential curve for the reduction of oxygen at a bare pyrolytic graphite rotating disk electrode, and Fig. 8 B shows their changes at the electrode coated with PXV-PSS film. The catalytic reduction of oxygen by radical cation unit of viologen bound to the microdomain of polymer complexes was confirmed from the positions of the half-wave potentials in Figs. 8 A and 8 B.   The PXV-PSS complex shifted the reduction wave to positive potential side but the limiting current remains proportional to the square root of the rotation rate. This result means that the microdomain of PXV-PSS complex was porous to allow oxygen molecules to migrate through it.

We have already evaluated the rate constant of the electron transfer reaction through the microdomain of PXV-PSS complex[14]. And the mechanism of oxygen reduction has already been speculated. As a result, oxygen molecule was reduced with two electrons to form hydrogen peroxide. When BPG electrode was coated with PXV-PSS film, the half-wave potential of O_2 reduction was -0.40 V and it shifted to positive side by 370 mV at maximum compared to that obtained with a bare electrode, and by about 90 mV compared with the average peak potential. This half-wave potential was independent of pH, indicating that the hydrogen ion does not participate in the rate-determining step of the oxygen reduction process[14].

The characteristics of the microdomain formed by interpolymer complex are quite charming for many fields and should provide us with possibilities for many more interesting application in the near future.

REFERENCES

1. A. S. Michaels, Ind. Eng. Chem., 57, 32 (1965)
2. H. J. Bixler and A. S. Michaels, Encycl. Polym. Sci. Technol., 10, 765 (1965)
3. E. A. Bekturov and L. A. Bimendina, Advances in Polym. Sci., 41, 99 (1981)
4. V. A. Kabanov and I. M. Papisov, Polym. Sci., USSR, 21, 261 (1979)
5. E. Tsuchida, Y. Osada and H. Ohno, J. Macromol. Sci., Phys., B 17, 683 (1981)
6. H. Ohno and E. Tsuchida, Makromol. Chem., Rapid Commun., 1, 585, 591 (1980)
7. H. Ohno, M. Shibayama and E. Tsuchida, Makromol. Chem., (1983) in press

8. K. Abe, M. Koide and E. Tsuchida, J. Polym. Sci., Polym. Chem. Ed., 15, 2469 (1977)

9. E. Anufrieva, Y. Ya Gotlib, M. G. Krakoviak and S. S. Skorokhodov, Vysokomol. Soedin, Ser. A 14, 1430 (1972)

10. E. Anufreiva, V. Pantov, V. Stepanov and S. Skorokhodov, Makromol. Chem., 180, 1843 (1979)

11. R. B. Cundall, J. B. Lawton, D. Murray and D. P. Rowlands, Polymer, 20, 389 (1979)

12. K. Abe, H. Ohno, A. Nii and E. Tsuchida, Makromol. Chem., 179, 2043 (1978)

13. H. Horiuchi and T. Ohshita, Kobunshi Ronbunshu, 38, 407 (1981)

14. N. Oyama, N. Oki, H. Ohno, Y. Ohnuki, H. Matsuda and E. Tsuchida J. Phys. Chem., 87, 3642 (1983)

15. H. Akahoshi, S. Toshima and K. Itaya, J. Phys. Chem., 85, 818 (1981)

16. M. Kato, N. Oki, H. Ohno, N. Oyama and E. Tsuchida, Polymer, 24, 846 (1983)

17. A. Factor and T. O. Rouse, J. Electrochem. Soc., 127, 1313 (1980)

18. H. Ohno, N. Hosoda and E. Tsuchida, Polym. Preprints, Japan, 31, 522 (1982)

19. H. Ohno, N. Hosoda and E. Tsuchida, Makromol. Chem., 164, 1067 (1983)

20. A. Factor and G. E. Heinsohn, Polym. Lett., 9, 289 (1971)

21. H. Ohno, K. Abe and E. Tsuchida, Makromol. Chem., 179, 755 (1978)

22. K. Abe, H. Ohno and E. Tuschida, Makromol. Chem., 178, 2285 (1977)

23. J. A. Bruce and M. S. Wrighton, J. Am. Chem., Soc., 104, 74 (1982)

24. R. J. Mortimer and F. C. Anson, J. Electroanal. Chem., 138, 325 (1982)

25. J. P. Collman, M. Marrocco, P. Denisevich, C. Koval and F. C. Anson, J. Electroanal. Chem., 101, 117 (1979); J. Am. Chem. Soc., 102, 6027 (1980)

26. J. Zagal, P. Bindra and E. Yeager, J. Electrochem. Soc., 127, 1506 (1980)

ABOUT THE CONTRIBUTORS

THEODOR W. P. ACKERMANN was born in 1925. He received the Diplomat in 1953 at the University of Gottingen and the Doctoral degree in 1956 at the University of Hamburg. His interests include biophysical chemistry, biochemical thermodynamics, and physical chemistry of nucleic acids. He is currently Professor and head of the Institute of Physical Chemistry of the University of Freiburg.

J. BEAUMAIS is at the Laboratory of Macromolecular Chemistry of the University of Rouen.

GUY C. BERRY is Professor of Chemistry and Polymer Science in the Department of Chemistry, Carnegie-Mellon University. He received his B.S. (Chemical Engineering), M.S. (Polymer Science), and Ph.D. degrees from the University of Michigan. He joined Mellon Institute in 1960 as a Fellow, becoming a Senior Fellow in 1964, and later, Professor at Carnegie-Mellon University. He was Visiting Professor at Colorado State University (1979), the University of Tokyo (1973) and the University of Kyoto (1983), and is associated with the University of Pittsburgh as an adjunct professor. His research interests include the physical chemistry of polymers in dilute solution and the rheology of polymers and their concentrated solutions.

JOHN BLACKWELL received his Ph.D. at the University of Leeds, England, and is a professor of macromolecular science at Case Western Reserve University, Cleveland, Ohio. His research interests are structure-property relationships of synthetic and biological polymers, and x-ray diffraction studies of polymeric materials.

G. BROZE obtained his Ph.D. (1980) at the State University of Liege (Belgium) under the guidance of Professor Teyssie and Dr. R. Jerome. He is now Senior Research Scientist at Colgate Palmolive Inc. (Herstal, Belgium).

ATTILIO CESÀRO was born in 1942 and received his doctorate in 1966 at the University of Naples. He did post-doctoral research at the University of California, Irvine. His research area is the physical chemistry of polysaccarides. He is currently Associate Professor of Polymer Chemistry at the Unviersity of Trieste.

SUMANA CHAKRABARTI received her Ph.D. from the University of Minnesota in 1982 and is currently employed by Dowell-Schlumberger, Tulsa, Oklahoma.

PATRICIA METZGER COTTS is currently a research staff member at the IBM Research Laboratory in San Jose, CA. She obtained her Ph.D. from Carnegie-Mellon University, under the direction of Professor Guy Berry. Her thesis research concerned the solution properties of rigid rodlike polymers in strong protic acids, the only known solvents for these polymers. Her research interests at IBM have continued in solution properties of synthetic polymers, specifically rigid and semi-rigid polymers, polyelectrolytes, and polymers which associate in solution. Techniques used to investigate these properties include static and dynamic light scattering, size exclusion chromatography and viscometry.

VITTORIO CRESCENZI is at the Institute of Physical Chemistry of the University of Rome.

FRANCO DELBEN graduated from the University of Trieste in 1967. His research interests comprise the physico-chemical properties of aqueous solutions and gels of polyelectrolytes. He is currently Associate Professor at the University of Trieste.

R. C. DOMSZY received his Ph.D. at the University of Manchester, Manchester, England. He was subsequently a Postdoctoral Associate at Florida State University. He is presently employed at Armstrong World Industries, Lancaster, PA.

C. O. EDWARDS received his Ph.D. at Florida State University. His present position is with General Tire and Rubber Co., Akron, Ohio.

TILBOR GILÁNYI was born in Hungary in 1944. He received his M.Sc. (1968) and Ph.D. (1969) in chemistry and was appointed to Assistant Professor at the Department of Colloid Science of Lorand Eotvos University, Budapest. Since that time, he has been working at the same department, recently as Associate Professor. In 1973 he worked at the Abo Academy, Finland, and in 1967 at the Lehigh University, Pennsylvania, as a co-worker of Professor A. C. Zettlemoyer. The main field of his research work is the polymer-surfactant interaction and the colloid stability of concentrated dispersions.

E. D. GODDARD received his B.S. (1945) and M.S. (1948) degrees from Rhodes University, South Africa, and his Ph.D. (1951) in physical chemistry from Cambridge University. From 1951-54 at Unilever, he worked on problems related to synthetic detergents. As Post-Doctorate Fellow at the National Research Council, Ottawa, (1954-56) he studied the energetics of micelle formation of surfactants. From 1956-59 he was a research chemist at Canadian Industries Limited, working in the field of synthetic fibers. From 1960-73 he was manager of the Physics and Physical Chemistry Research section at Lever Brothers Company,

Edgewater, New Jersey. He joined the Tarrytown Technical Center of Union Carbide in 1974, where, as Corporate Research Fellow, he heads a surface chemistry skill center.

TRACY M. HANDEL received her B.S. in chemistry (1980) from Bucknell University. She then worked as a research chemist from 1980-83 at the Research Laboratories of Eastman Kodak Company. She is presently a graduate student at the California Institute of Technology.

JOVANKA HUGUET is Charge de Recherche (Senior Scientist) at CNRS and is in charge of the bifunctional polybase program in Dr. Vert's group.

ALEX M. JAMIESON received his D. Phil. at the University of Oxford, England, and is currently professsor of macromolecular science at Case Western Reserve University, Cleveland, Ohio. His research interests are thermodynamic and mass transport of polymer fluids, quasielastic laser light scattering and structure-function relationships of biological polymers.

R. JEROME obtained his Ph.D. (1970) at the State University of Liege (Belgium), working with Professor V. Desreux. He is Assistant Professor at the University of Liege (Professor Teyssie's department) and teaches macromolecular chemistry at the University of Namur.

BAYRAM KALPAKCI received his B.S. degree in chemical engineering from Turkey and his M.S. and Ph.D. degrees from The Pennsylvania State University. His research interests include surfactant flow in porous media and surface chemical problems related to catalysis and lubrication. He is currently with Sohio Research and Development at Cleveland and is working on various aspects of enhanced oil recovery by chemical flooding and other technologies.

J. KRÁLÍČEK received his M.Sc., 1951, and Ph.D., 1959, at the Prague Institute of Chemical Technology, Czechoslovakia. He is Professor of Polymer Chemistry at the same Institute.

SHIGEO KUBOTA is currently assistant research biophysicist at the University of California, San Francisco. He received his B.S. in physics (1963), M.S. in biophysics (1965) and Ph.D. in biophysics (1968) from Hokkaido University, Japan. His research interests concern the syntheses and conformational studies of oligo- and polypeptides and their complexes with metallic ions.

J. LABSKÝ received his M.Sc., (1959) at the Prague Institute of Chemical Technology, Czechoslovakia and Ph.D., (1979) at the Institute of Macromolecular Chemistry, Czechoslovak Academy of Sciences, where he is senior scientist.

JACQUES LANG is Maitre de Recherche at the Centre de Recherches sur les Macromolecules, CNRS, Strasbourg, France. He received his D.Sc. degree in 1968 from the University of Strasbourg. His research has concerned fast kinetics of proton transfer reactions and micellar systems, and more recently he has been involved in the study of microemulsions.

PAK LEUNG is a research scientist at Union Carbide Corporation. He earned a B.S. in chemical engineering from National Taiwan University in 1957 and M.A. (1962) and Ph.D. (1967) from Columbia University. Since he joined Union Carbide, he has been working in the polymer physics area and in surface and colloidal applications.

PANAGIOTIS LIANOS is a professor at the University of Patras, Greece. He received his Ph.D. in 1978 from the University of Tennessee, Knoxville. He specializes in photophysics and is currently interested in using fluorescence probes in the study of organized assemblies.

LEO MANDELKERN received his undergraduate degree from Cornell University in 1942. After serving with the armed forces, he returned to Cornell and received his Ph.D. in 1949. He remained at Cornell in a postdoctoral capacity until 1952. Since 1962 he has been Professor of Chemistry and Biophysics at the Florida State University. In 1984 he was named R. O. Lawton Distinguished Professor of Chemistry. Other honors include the Arthur S. Fleming Award (1958), the American Chemical Society Award in Polymer Chemistry (1975), the Mettler Award in Thermal Analysis (1984) the Florida Award of the Florida Section of the American Chemical Society (1984).

TOMOYA MASAKI was a student of Saga University at the time of the research, and is now working in Product Development Division, Mochida Pharmaceutical Co., Ltd., Tokyo, Japan.

FRANTISER MIKEŠ received his M.Sc., 1964, and Ph.D., 1969, at the Prague Institute of Chemical Technology, Czechoslovakia. He was visiting scientist at the Polytechnic Institute of New York in 1980. He is Research Associate Professor at the Prague Institute.

WILMER G. MILLER is Professor of Chemistry, University of Minnesota, Minneapolis. His research interests are focused on physical chemical studies of polymers and surfactants.

R. NAGARAJAN is currently Assistant Professor of Chemical Engineering at the Pennsylvania State University. He received his Ph.D. degree in 1979 from the State Unviersity of New York at Buffalo. His research interests focus on surfactants, their mechanism of action and their applications. He has published more than 30 papers in the areas of thermodynamics of micelles, vesicles, solubilization, enhanced oil recovery and surfactant-polymer interactions.

KATSUTOSHI NITTA was born in 1942 and educated at Hokkaido University (BS 1964, MS 1966, PhD 1972). His research area is biopolymers. He is currently instructor of Polymer Science at Hokkaido University.

NOBUMICHI OHNO received his education at Hokkaido University (D.Sc., 1981). He was visiting research fellow at Dalhousie University. He is currently Associate Professor of Physical and Polymer Chemistry at Akita National College of Technology.

SERGIO PAOLETTI was born in 1948 and received his advanced degrees at the University of Trieste in 1972. He was visiting professor at Stevens Tech. in 1984. His research field encompasses both theory and experimental thermodynamics of polyelectrolytes and polysaccharides. He is currently Associate Professor of Polymer Chemistry at the University of Trieste.

E. PEFFERKORN is employed at the Centre de Recherches sur les Macromolecules. He received his B.Sci. and Ph.D. degrees in chemistry from the University Louis Pasteur of Strasbourg. The research of Dr. Varoqui and Dr. Pefferkorn encompasses studies of the properties of polyelectrolytes in solution and the physical-chemistry of interfacial properties of polymers.

GÜNTHER REHAGE was born in 1920. He received his education at the Technische Hochschule of Aachen. His interests are in physical and polymer chemistry, including gels, rubber elasticity and liquid crystals. He is currently Professor and Director of the Institute of Physical Chemistry at the Technical University.

R. RIZZO is at the Institute of Physical Chemistry of the University of Rome.

KATHLEEN SEIBEL is a medical student at the University of Minnesota, Minneapolis.

KEISHIRO SHIRAHAMA is Professor of physical chemistry of Saga University, Saga, Japan. He received a doctor of science from Kyushu University. His academic interest is directed to amphiphilic molecular assemblies such as surfactant-polymer complex, mixed micelle, and vesicle as well as random phenomena in physicochemical systems.

RANDY SHOGREN received his M.S. from Case Western Reserve University, Cleveland, Ohio. He is a research assistant in the Department of Macromolecular Science at Case Western. Research interests include structural studies of proteoglycans and mucin glycoproteins.

ULRICH P. STRAUSS, Professor of Chemistry, Rutgers University, the State University of New Jersey, graduated from Columbia University (A.B. 1941), then attended Cornell University to obtain a Ph.D. (1944). He was a Sterling Postdoctoral Fellow at Yale University from 1946-48. He has been on the Rutgers University faculty since 1948. His main areas of research are polyelectrolytes, polysoaps, DNA, specific counterion binding and hydrophobic polyacids and polyampholytes.

PETR ŠTROP received his M.Sc., 1969, and Ph.D., 1973, at the Prague Institute of Chemical Technology, Czechoslovakia. Since 1974 he has been working at the Institute of Organic Chemistry and Biochemistry, Czechoslovak Academy of Sciences as senior scientist.

SHINTARO SUGAI was born in 1926 and received his doctorate from Hokkaido University in 1960. His research interests include the conformation of globular proteins, ribosomal RNA and hydrophobic polyelectrolytes. He is Professor of Polymer Science at Hokkadio University.

KOJI TAKASHIMA is a graduate student of Saga University, and will shortly move to Kao Soap Co., Ltd., Tokyo, Japan.

JULIA S. TAN obtained her B.S. in chemistry (1961) from National Taiwan University, her M.A. in chemistry (1963) from Wesleyan University, and her Ph.D. from Yale University (1966). Dr. Tan was an assistant professor of chemistry from 1966-69 at Wesleyan University and then Research Fellow in biophysics at the University of Rochester (1969-70). She is research associate at the Research Laboratories, Eastman Kodak Company, where her research activity involves polyelectrolytes and reactive polymers.

LISBETH TER-MINASSIAN-SARAGA obtained her M.Sc. in physical chemistry from the Hebrew University (Jerusalem) and is Docteur des Sciences of Universite de Paris. As Directeur de Recherches (CNRS) she is in charge of a group involved in biophysical and biomedical applications of surface chemistry, such as monolayer and adsorption thermodynamics at air/liquid and solid/liquid interfaces; wetting of inorganic materials and plastics; and water contribution to monolayers and bilayers.

PH. TEYSSIE obtained his Ph.D. (1952) from the University of Leuven (Belgium) with Dr. G. Smets. Successively Research Fellow at Leuven, Professor at Lovanium University, Research Associate at Brooklyn Polytechnic Institute, and Senior Scientist at the French Petroleum Institute, he became Full Professor of organic and macromolecular chemistry at the State University of Liege (Belgium) in 1970, and also teaches at the University of Namur. His main interests include organic catalysis by coordiantion complexes, homo- and copolymerization catalysis, and molecular engineering of polymer materials properties.

DAVID A. TIRRELL is Associate Professor of Polymer Science and Engineering at the University of Massachusetts at Amherst. Dr. Tirrell earned his B.S. in chemistry at Massachusetts Institute of Technology in 1974 and his Ph.D. in polymer science and engineering at the University of Massachusetts in 1978. After a short stay at Kyoto University as a research associate, Dr. Tirrell joined the faculty of the Department of Chemistry at Carnegie-Mellon University. He accepted his present position in 1984. Dr. Tirrell's research interests are in polymer synthesis and bio-organic polymer chemistry. He served on the editorial boards of the Journal of Polymer Science and Industrial and Engineering Chemistry and as Secretary of Macromolecular Syntheses. He was a Sloan Fellow (1982-84) and holds a Presidential Young Investigator Award of the National Science Foundation (1984-89).

EISHUN TSUCHIDA received the doctoral degree at Waseda University in 1960 and has been a professor there since 1973. His research interests include: the electron transfer process of macro-molecular-metal complexes; oxygen carriers; polymeric catalysts for redox reaction; oxidative polymerization and chelating resins; interaction and complexation between macromolecules; and electro- and/or ionic-conductive macromolecules. He has published over 250 papers and has written or edited several books. He is a board member of the Society of Polymer Chemistry and the Chemical Society in Japan.

RAPHAEL VAROQUI holds a research position at the Centre de Recherches sur les Macromolecules at Strasbourg (France) where he has been employed since 1957 by the CNRS. He obtained his Ph.D. in chemistry in 1964 and did post-doctoral work at Rutgers University (New Jersey).

MICHEL VERT is Maitre de Recherche at CNRS. His main fields of interest are synthetic optically active polymers, bifunctional poly-electrolytes, biomedical polymers and chemical modifications of poly-mers. He has authored 70 papers and eight patents.

D. WAGNER is currently at the Institute of Physical Chemistry of the Technical University of Clausthal.

ERVIN WOLFRAM received his M.Sc. (1946) and Ph.D. (1949) in chemistry and physics from the Lorand Eotvos University, Budapest, the degree Dr. habil. (1959) of the Technical University of Dresden and the degree Dr. Sc. (1966) of the Hungarian Academy of Sciences of which he became a member in 1982. He is secretary of the IUPAC Commission on Colloid and Surface Chemistry. He is on the advisory board of J. Colloid Interface Sci., Colloids and Surfaces, Colloid and Polymer Science, and the book series Surface and Colloid Science. He organized of the first International IUPAC Conference on Colloid and Surface Science in Budapest in 1975. He was Unilever Visiting Professor of Colloid Science at the University of Bristol, England (1977-78). He is Pro-Vice-Chancellor of the Budapest University in charge of inter-national relations.

AKIHIKO YAMAGISHI research associate at the Department of Chemistry, Hokkaido University at Sapporo in Japan, was born in 1943. He received a D.Sc. at the University of Tokyo with Dr. Kenzi Tamaru. He joined the research team of Dr. Masatoshi Fujimoto at Hokkaido University in 1972. From 1977-79, he worked with Dr. Michael Szwarc of the State University of New York at Syracuse as a postdoctoral fellow. His main field of study is polymer-small molecule interation, stereochemistry of coordinated compounds, asymmetric induction and optical resolution of racemic mixtures.

JEN TSI YANG is professor of biochemistry at the University of California, San Francisco. He received his B.S. in chemistry from the National Central University, China (1944) and Ph.D. in biophysical chemistry from Iowa State University (1952). His research interests concern the conformation of biopolymers by means of circular dichroism and other biophysical methods.

RAOUL ZANA is Directeur de Recherche at the Centre de Recherches sur les Macromolecules, CNRS, Strasbourg, France. He received his D.Sc. degree in 1964 from the University of Strasbourg. His current research interests include equilibrium and dynamic aspects of micellar solutions, microemulsions and polyelectrolyte solutions.

INDEX

Absorption and emission spectra, 266
    polarity reporters, 268
    shift and half-band width, 272, 283
    pyridinium, spiropyran-merocyanine reporters, 268-276
    DNS reporter, 271

Aggregate, polymer-micelle, 414

Aggregation, 107, 112
    effect on light scattering, 107
    insensitivity of [ ], 112
    of halotelechelic polymers, 243
    of mucins, 217-220

Alginic acid, 184
    enthalpy of dissociation, 184

Alkali metal radioisotopes, 226

Amphiphilic polyacids, 225

Association, chain-micelle, 378, 413

Association, intersegmental, 89

Bilayer, phospholipid, 343

Binary solvents, 272, 276-286
    preferential sorption, 271, 277-281

Binding isotherm (see also, Dye binding, Conformation of poly(L-lysine) homologs), 303
    cooperativity, 306

Biomembranes, 344

Calorimetry, 95, 181
    of ionic polysaccharides, 181

Carrageenan, 166, 178

Cellulose ether, cationic, 407

Chain entanglement,
    effects on solution properties of mucins,
    viscosity of mucins, 211-220

Circular dichroism, (see also Conformation of poly (L-lysine) homologs), 312

Complex, polymer-surfactant, 384, 414

Conductivity, 359

Conformation of poly(L-lysine) homologs
    in aqueous solution, 315
    in alkyl sulfate solution, 316
    in lipid solution, 327
    binding isotherm, 323

Conformational transition
  amphiphilic polyacids, 225
  copolymers of maleic
    anhydride and alkyl
    vinyl ethers, 1-11
  pectic acid, 184
  polyelectrolytes, 174, 225
  polysaccarides, 160
  weak polybases

Copolymers of maleic anhydride,
  alkyl vinyl ethers, 1-11
  conformational transition,
    2
  effect of alkyl group
    size, 2-8
  fluorescence probe,
    dansyl, 2-5
  intrinsic viscosity, 5-7
  phase separation, 6
  free energy of
    conformational
    transition, 7-8
  potentiometric titrations,
    7-11
    representative sample
      subunit, 9
    species population
      distribution, 9-10
    cooperative unit size,
      11

Counterion binding, 161, 230,
  387, 417

Counterion condensation, 161
  description of Manning
    theory, 161

Cross-linked polymers, 286
  microenvironment polarity of

Crystallites
  fringed micelle, 125
  lamellar, 125

Crystallization, 121

Degree of dissociation (see also,
  ionization), 60, 233

Dextran, 352

Differential scanning
  calorimetry, 343, 344, 346
  dipalmitoyl phosphatidyl-
    glycerol, 346
  phosphatidylglycerols, 346

Diffusion coefficient, of mucins,
  211-220

Dodecylpyridinium bromide, 302
  surfactant electrode, 302

Dodecylpyridinium halide, 301
  DNA, 301
  binding isotherms, 301
  surfactant electrode, 301

Dye binding isotherms, 419, 421
  Klotz plots, 421, 422
  Scatchard plots, 422, 423
  McGhee-von Hippel model,
    cooperative model, 423
  exothermic heats, 424

Dye binding measured by
  spectrophotometric titration,
    418, 420
  equilibrium dialysis, 419
  microcalorimetry, 419

Dye-dye and dye-site interaction, 422, 426

Electric dichroism
  polymer-chelate complex,
    67-84
  polystyrenesulfonic acid,
    67-84
  Co(III) chelate;, 67-84

Electrical conductance of
  anionic polysoap solutions,
    226

Electron microscopy, 99, 126

Electrostatic enthalpy of mixing, 159

Electrostatic enthalpy, 170, 174
    conformational transitions, 174

Ethylene/butene, 129
    copolymers, 134

Ethylene/vinylacetate, 136

Fluorescence probing, 359

Frictional coefficient, 240

Fringed micelle, 121

Gaussian coil segment distribution, 237

Gelation, 121, 147

Gels:
    of rigid rods, 147
    of polybenzylglutamate–DMF, 148
    of hemoglobin S, 149
    of mucus, 150

Heats of dissociation, 170
    weak polyacids, 170

Heat of mixing of polyelectrolytes, 168

Hydrophobic domains, 13

Hydrophobic interactions, 2, 56, 67, 226, 301, 379, 384, 414

Hydrophobic polyelectrolyte, 13, 60, 226

Hydrophobicity
    polystyrenesulfonic acid, 67–84
    stereospecificity in, 78–79

Intensity correlation function, 192

Intermolecular aggregates, 191

Intermolecular complex
    rotating disk voltammetry, 429
    cyclic voltammetry, 432
    polymer coated electrode, 432
    electron transfer process, 432

Intermolecular interaction, 407

Inverted micelle, 226

Ion condensation (see also counterion)

Ion distribution, 236

Ionenes, 346
    dipalmitoyl phosphatidyl-glycerol

Ion–exchange isotherms, 233

Isotactic polymer, 352

i-polystyrene, 139

Light micrograph, 129

Light scattering, 191, 227
    copolymers, 402
    integrated intensity, 191
    of mucins, 215–222
    photon correlation, 191

Linear charge density, 167

Liposomes, bilayers, 345

Maleic acid - vinyl ether
    copolymer, 1, 226

Mark-Houwink relationship, 231

Methyl orange, 417,

Micelle - polymer interactions,
    357

Micelle size, 361

Microdomains, (see also
    conformations, see also
    solubilization)
        phase separation, 57
        phase diagram, 59
        phase aggregation, 57

Mixed bilayers
    cooperative association in,
        336
    hydration of, 336
    differential scanning
        calorimetry, 336
    ice molar enthalpy of
        melting, 337
    structural water, melting
        thermograms of, 340
    cooperativity, association
        of, 340

Mixed monolayers
    surface pressure, 334
    ionic composition of, 335
    ion pairs in, 336
    cooperative association in,
        336
    binding isotherm, 337
    cooperativity, 340
    association of, 340

Monomer-monomer, monomer-
    solvent interaction energies,
    238

Mono-molecular micelle, 226

Motional freedoms, 19

Mucin, pig submaxillary,
    211-222

Networks, 122, 147

Neutron scattering, 407

NMR measurements, 15
    spin-lattice relaxation time,
        23
    correlation time, 23
    chemical shift, 19

Pectic acid,
    heat of dilution, 184
    enthalpy of dissociation, 185

Phase separation, 348

Phase transition, 122

Phosphatidylcholines, 345
    dipalmitoyl

Phosphatidylglycerol, 345, 349
    diauroyl

Phospholipid, 348
    dilauroyl phosphatidyl-
        glycerol, 348
    dipalmitoyl phosphatidyl
        choline, 348

PMMA, 87

Polarity, 265, 266, 267
    polymer microenvironment,
        265, 266
    of PMA, PHEMA, PMMA,
        PBMA, PS, P-2VP, P-4VP,
        PVIm, 274, 275
    of poly[N-(2-hydroxypropyl
        methakrylamide)], 286

Poly(-ethylacrylic acid), 350

Poly(acrylamide), 352

Polyamines, 60

Polyanions, 1, 67, 87, 183, 346, 350

Polybenzylglutamate
    phase behavior, 143
    storage and loss modulus, 144
    aggregation, 145
    concentration dependence of elasticity, 146
    structure of gel, 151

Polycations, 60, 346, 407

Polyelectrolytes,
    anionic polysoaps, 226
    cationic, 60, 346, 418
    hydrophobic polyacids, 1-11
    counterion binding, 161, 230, 387, 417
    rodlike, 191
    pectic acid, 184
    amphiphilic, 1, 225
    enthalpy of, 174
    weak polyacids, 170

Polyethylene oxide, 129

Polyethylene, 122

Polyguluronic acid, 183
    heat of dilution, 183

Poly[N-(2-hydroxypropyl)-methacrylamide], 352

Polymannuronic acid, 183
    heat of dilution, 183

Polymer JR, 408

Polymer microenvironment
    polarity, 272
    soluble polymers, 272-286
    cross-linked polymers, 286, 288-291

Polymer microstructure, 352

Polymer-micelle complexes, 369-380; See also
    Thermodynamics of complexation
        topology, 370,371
        viscosity, 372, 374-377
        surface tension, 376
        polyethylene oxide, 374-375
        dextran, 376
        mixed micelles, 379

Polymers, halatotelechelic, 243
    saxs spectra, 257
    solutions, in non-polar solvents, 248
    ion pairs, 247
    aggregation, 255
    relaxation, 261

Poly(N,N-dimethylacrylamide), 352

Poly(N-vinylimidazole), 418
    fully, partially quaternized

Poly(N-vinylpyrrolidone), 352

Polysaccharides, 160
    ion-dependent conformational transitions, 160

Polysoaps, 225

Polystyrene, 197

Poly(tertiary amines), 52

Poly(vinyl alcohol), 352

Poly(vinyl butyral)
chain dimensions, 101
unperturbed dimensions
of, 116,
differential refractive index,
106
Mark Houwink relation in
THF, 114

Poly(1,4-phenylene-2,6-benzo-
bisthiazole), 191

Radius of gyration, 237

Raman spectroscopy, 349

Rational activity coefficients,
232

Rayleigh ratio, 193
polarized, 193
depolarized, 193

Rheology of
polybenzylglutamate, 148
hemoglobin S, 149
mucus, 150

Rheology, 88, 407

Rigid rod polymer-diluent
phase behavior, 143
gels (networks), 147

Rodlike macroions, 191

SEC/LAS, 108
sensitivity to aggregation,
108

Second factorial moment, 192

Sedimentation coefficient, 227

Self-association (see also
aggregation)

Semiempirical polarity
parameters, 266-268,
270-272

Sephadex, Sepharose, Dowex
1-X2, Spheron, 286, 288-291
hydrophobic chroma-
tography, 291

Site binding, 230

Sodium alginate, 182
heat of dilution, 182

Sodium dodecyl sulfate, 407

Solubilization
progesterone, 64

Spherulites, 134

Spinoidal decomposition, 123

Stereocomplex formation, 87

Stereo-irregular, 124

Stokes radius, 228

Supermolecular structure, 128

Surfactant, activity, 391

Syndiotactic polymer, 352

s-Polyvinylchloride, 139

Tacticity, 350

Temperature
dissolution, 125
melting, 125

Thermodynamic functions of
binding, 418, 424, 425

Tracer self-diffusion, 226

Transition, order–disorder, 345

Two-phase potentiometric
    acid–base titration, 232

Viscosity, 227, 397, 407
    polymer–surfactant, 397

Zwitterionic polymers, 346